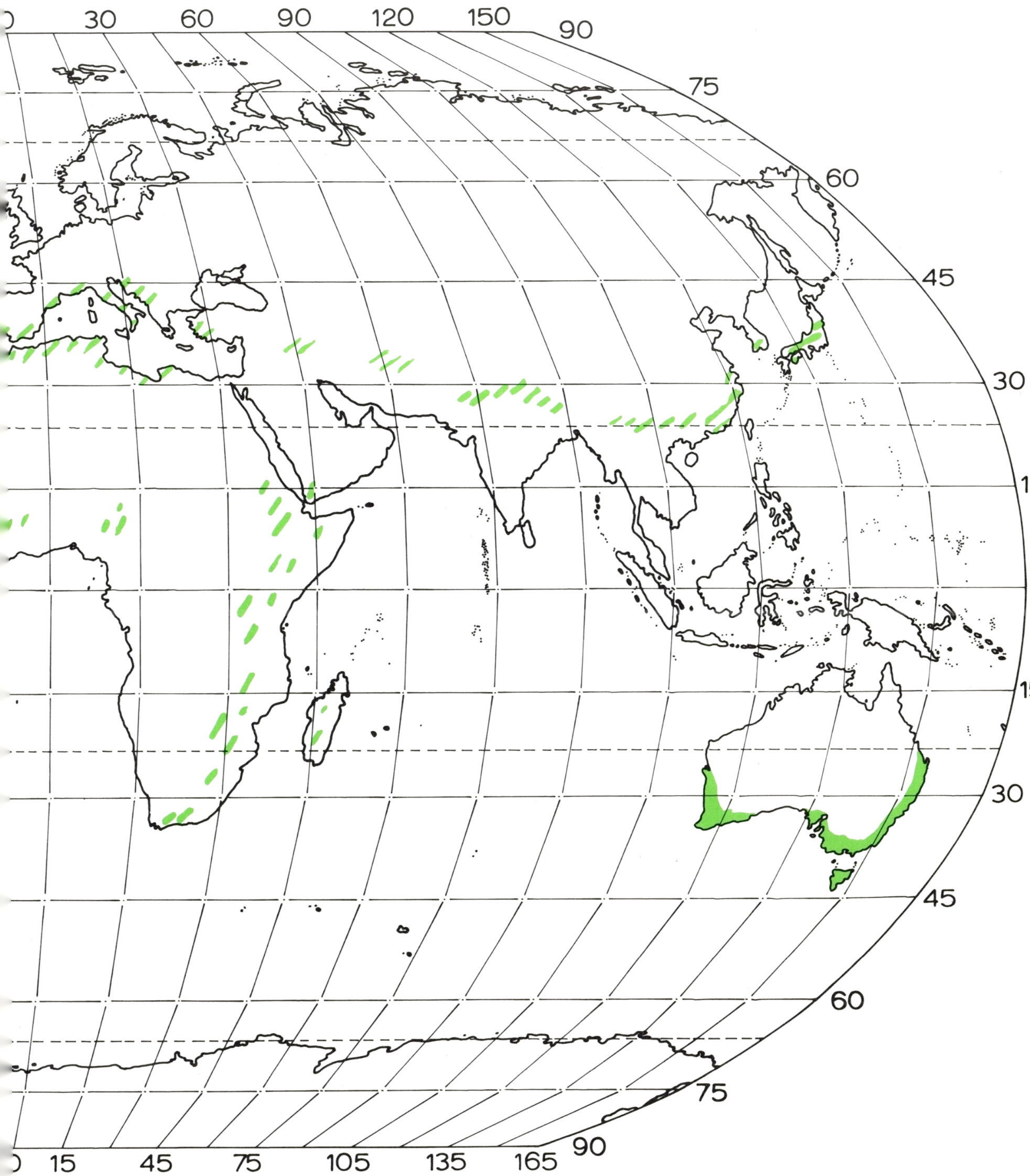

ed evergreen forest

ECOSYSTEMS OF THE WORLD 10

TEMPERATE BROAD-LEAVED EVERGREEN FORESTS

ECOSYSTEMS OF THE WORLD

Editor in Chief:

David W. Goodall

CSIRO Division of Land Resources Management, Wembley, W.A. (Australia)

ECOSYSTEMS OF THE WORLD 10

TEMPERATE BROAD-LEAVED EVERGREEN FORESTS

Edited by

J.D. Ovington
Australian National Parks and Wildlife Service
P.O. Box 636
Canberra City, A.C.T. 2601 (Australia)

ELSEVIER

Amsterdam — Oxford — New York 1983

ELSEVIER SCIENCE PUBLISHERS B.V.
1 Molenwerf
P.O. Box 211, 1000 AE Amsterdam, The Netherlands

Distributors for the United States and Canada:

ELSEVIER SCIENCE PUBLISHING COMPANY INC.
52 Vanderbilt Avenue
New York, N.Y. 10017, U.S.A.

Library of Congress Cataloging in Publication Data
Main entry under title:

Temperate broad-leaved evergreen forests.

(Ecosystems of the world ; 10)
Includes bibliographies and indexes.
1. Forest ecology. 2. Evergreens, Broad-leaved.
I. Ovington, J. D. II. Series.
QK938.F6T45 1983 581.5'2642 82-8891
ISBN 0-444-42091-6

ISBN 0-444-42091-6 (Vol. 10)
ISBN 0-444-41702-8 (Series)

Printed in The Netherlands

Ecosystems of the World

10. TEMPERATE BROAD-LEAVED EVERGREEN FORESTS

This volume has been published in November 1983.
Other volumes already published at that time are:

1. Wet Coastal Ecosystems (March 1977)
4A. Mires: Swamp, Bog, Fen and Moor. General Studies (January 1983)
4B. Mires: Swamp, Bog, Fen and Moor. Regional Studies (February 1983)
5. Temperate Deserts and Semi-Deserts (January 1983)
9A. Heathlands and Related Shrublands. Descriptive Studies (November 1979)
9B. Heathlands and Related Shrublands. Analytical Studies (February 1981)
11. Mediterranean-Type Shrublands (June 1981)
13. Tropical Savannas (January 1983)
14A. Tropical Rain Forest Ecosystems. Structure and Function (February 1983)
26. Estuaries and Enclosed Seas (February 1983)

PREFACE

This account of broad-leaved evergreen forests is composed of contributions covering South America, New Zealand, Australia, southeastern North America, the Mediterranean and the Middle East, India, Africa south of the Sahara and Japan. The coverage is incomplete; for instance, it proved impossible to obtain a contribution from China although Wang (1961) identified broad-leaved evergreen forests of oaks, schima and laurels as a significant vegetation component of the vegetation of China.

Furthermore, the contributions differ somewhat in the emphasis placed on various aspects of broad-leaved evergreen forests. This is because there are great differences in knowledge of these forests and the uses made of them, a reflection of geographical, historical, cultural and economic differences.

Nevertheless, it is evident that broad-leaved evergreen forests with their wealth and diversity of species and ecosystems are under attack throughout the world. Much has been destroyed and large areas are being cleared for agriculture or are being replaced by forest plantations of alien tree species. Other areas are being progressively changed by fire, domestic stock, introduced animals and the harvesting of a variety of forest products.

Compared with other kinds of forests, we know little of the synecology of evergreen broad-leaved forests nor of the autecology of individual species. Undoubtedly many species are potentially very useful. For instance, some species of eucalypts have been planted extensively outside their natural range because of their capacity for rapid growth and the usefulness of their wood. Eucalypt plantations are improving the economy of many impoverished rural areas. More research is needed about the species in broad-leaved evergreen forests and their possible benefits.

Justification for taking greater care of the remaining broad-leaved evergreen forests arises from their potential for timber production and the protection of useful and interesting species and of their component ecosystems. This justification is strengthened by growing recognition of the importance of these forests in soil and water protection and for scientific study and recreation. Increasingly evergreen broad-leaved forests are being seen as attractive landscape components and an important part of national heritages.

Hopefully this book will serve to draw attention to the need to have a greater care for broad-leaved evergreen forests and to the need for more information about them.

I wish to thank Dr. D.W. Goodall, the Editor in Chief, for inviting me to serve as editor for this volume in the series on Ecosystems of the World. I also wish to express my appreciation of the authors who have contributed to the project.

REFERENCE

Wang, C., 1961. *The Forests of China with a Survey of Grassland and Desert Vegetation.* Maria Moors Cabot Foundation, Publication 5, Harvard University, Cambridge, Mass., 313 pp.

J.D. OVINGTON
Australian National Parks and Wildlife Service
Canberra, A.C.T.

LIST OF CONTRIBUTORS

M.J.A. BULFIN
Botany Division
DSIR
Christchurch (New Zealand)

E.F. DEBAZAC
Casier, O.N.U.
Rabat Chellah (Morocco)

D.G.M. DONALD
Faculty of Forestry
University of Stellenbosch
Stellenbosch 7600 (South Africa)

J. DUGDALE
Entomology Division
DSIR
Auckland (New Zealand)

B. GOPAL
Department of Botany
University of Rajasthan
Jaipur 302004 (India)

V.M. MEHER-HOMJI
Department of Botany
University of Rajasthan
Jaipur 302004 (India)

D.F. OLSON Jr.
USDA Forest Service
Pacific Southwest Forest and Range Experiment Station
4955 Canyon Crest Drive
Riverside, Calif. 92507 (U.S.A.)

J.V. OLTREMARI
Facultad de Ingenieria Forestal
Universidad Austral de Chile
Valdivia (Chile)

J.D. OVINGTON
Australian National Parks and Wildlife Service
P.O. Box 636
Canberra, A.C.T. 2601 (Australia)

L.D. PRYOR
Department of Forestry
Australian National University
Canberra, A.C.T. (Australia)

T. SATOO
Faculty of Agriculture
University of Tokyo
Tokyo 113 (Japan)

F.M. SCHLEGEL
Facultad de Ingenieria Forestal
Universidad Austral de Chile
Valdivia (Chile)

J.M. THERON
Faculty of Forestry
University of Stellenbosch
Stellenbosch 7600 (South Africa)

T.T. VEBLEN
Department of Geography
University of Colorado
Boulder, Colo. 80309 (U.S.A.)

P. WARDLE
Botany Division
DSIR
Christchurch (New Zealand)

CONTENTS

Chapter 1

INTRODUCTION

J.D. OVINGTON

The asymmetric distribution of land and sea between the Northern and Southern Hemispheres largely determines the distribution and nature of temperate forests. Broadly, temperate needle-leaved coniferous forest and deciduous broad-leaved forest are characteristic of the continental, winter cold climates of North America and Eurasia whilst evergreen broad-leaved forest prevails in the southern temperate region where the winters are milder.

Sakai et al (1981) have drawn attention to the large differences in actual or potential freezing resistance of different tree species. They consider the extreme degree of winter hardiness, which they found to be a feature of many northern conifers and deciduous broad-leaved species, reflects a very long history of adaptation to climates with cold winters. In contrast they found a low level of cold resistance in Southern Hemisphere evergreen broad-leaved trees.

They suggest that tree species capable of withstanding freezing did not evolve in the Southern Hemisphere because the mild, oceanic winters did not provide any stimulus for this. Further, hardy northern tree species, with minor exceptions, were unable to migrate southwards across the barrier formed by the tropics. When Northern Hemisphere species have been introduced into the Southern Hemisphere, there is little evidence that, under the natural conditions there, they are more competitive than the native tree species.

The temperate broad-leaved evergreen forest occurs in widely disjunct areas of the world, and because of the different evolutionary histories of these areas they differ greatly in the species composition of their flora and fauna. Nevertheless, they show an affinity of structure and there is a tendency for the tree storey to be predominantly species of *Quercus* in the Northern Hemisphere and of *Nothofagus* or *Eucalyptus* (Australia) in the Southern Hemisphere.

Two main categories of broad-leaved evergreen forest are usually distinguished in temperate regions, namely temperate broad-leaved sclerophyll forest and temperate broad-leaved rain forest. A third intermediate category may be recognized (UNESCO, 1973), namely temperate evergreen season forest, where there is some foliage reduction but not complete leaf shedding during the unfavourable season.

TEMPERATE BROAD-LEAVED SCLEROPHYLL FOREST

Temperate broad-leaved sclerophyll forest occurs essentially in areas having a Mediterranean-type climate characterized by winter rain and summer drought. The vegetation is dominated by evergreen hardwood trees whose leaves have hard, leathery or waxy surfaces. The leaves may be relatively small for hardwoods. Conifers sometimes occur as an admixture in the overstorey. The few epiphytes present are usually lichens or moss and woody evergreen lianes may occur.

Walter (1973) considers that the ecological significance of sclerophylly is that sclerophyll plants are able to achieve a high level of gaseous exchange when the water supply is adequate, but when water is scarce during the drought season gaseous exchange is greatly reduced by shutting of the stomata. With the onset of autumn rain the plants are able to switch quickly to more rapid organic production.

On the North American continent temperate broad-leaved evergreen sclerophyll forest is of limited extent, occurring as discontinuous patches. In the southwest of the U.S.A., forests stands of evergreen trees such as *Arbutus menziesii*, *Lithocarpus densiflorus*, *Quercus agrifolia*, *Q. chrysolepis* and *Q. wislizenii* are present in the northern part of the Californian sclerophyllous zone (Shelford, 1963) and in south and central Arizona within an altitude range of about 1200 to 1900 m. In the southeast of the U.S.A., small scattered stands of evergreen trees occur from Virginia to Florida as part of the southern mixed hardwood forest. Here the evergreen trees (*Quercus myrtifolia* and *Q. virginiana*) tend to be present on the drier sites where the soils are poor in plant nutrients. In the bayhead forests, on soils which are subject to seasonal flooding and contain much organic matter, evergreens such as *Gordonia lasianthus*, *Magnolia virginiana* and *Persea borbonia* occur.

In Europe the broad-leaved sclerophyll forest of evergreen trees forms a narrow ring around the Mediterranean, the breadth and altitudinal limits being determined by climate and human influence. The tree overstorey is not rich in species and consists mainly of oaks (*Quercus coccifera*, *Q. ilex* and *Q. suber*) and *Pistacia lentiscus*. These forests have a long and continuing history of cutting and grazing and are highly modified from the natural condition. Though many have regenerated naturally some have been planted.

The European broad-leaved sclerophyll zone can be seen biogeographically as extending in a broken belt into the Middle East and to the western slopes of the Himalayas. The few remaining trees and small stands in this belt are probably the remnants of once more extensive forests. On the Indian subcontinent, stands of *Quercus ilex* on the drier sites of the Himalayas and the forests of the southern Indian Hills dominated by trees of the Lauraceae, Myrtaceae and Styracaceae can be regarded as temperate broad-leaved evergreen sclerophyll forest because of the temperate climate at the higher altitudes.

Evergreen broad-leaved sclerophyll forest extends over large areas of temperate Asia in a broad band up to 800 km wide across southern China and extending to the mountain regions of Taiwan, Vietnam, Thailand and Burma. Whilst fairly uniform in structure and appearance, the forest is rich in species with over 150 species of evergreen oaks. The different kinds of oak forest have an admixture of other species. The most important groups of overstorey species based on number of species present are *Quercus*, *Pasania*, *Castanopsis*, *Machilus*, *Magnolia*, *Cinnamomum*, *Illicium*, *Schima*, *Michelia*, *Cryptocarya*, *Manglietia* and *Altingia*. Much of this forest has been cleared for agriculture and there has been some reafforestation, often with coniferous species.

Temperate broad-leaved evergreen sclerophyll forest retains its greatest natural extent in Australia, despite large-scale clearing for agriculture and urban development. It is remarkable that the principal tree genus virtually throughout is *Eucalyptus*. This eucalypt forest extends southwards as a more or less continuous coastal forest band of varying width from the Tropic of Capricorn to South Australia and Tasmania with a second area in the southwestern part of Western Australia. Scattered stands of eucalypts also occur in the interior where soil moisture is available. Whilst the eucalypt forest is fairly uniform in general appearance, over 450 eucalypt species have been identified and the eucalypt forest contains a great diversity of ecosystems depending on the species present, climatic, soil, moisture and fire conditions and the effects of people. At the better locations with adequate soil water, the eucalypts may attain a very large size and be highly productive of timber.

Little, if any, broad-leaved evergreen sclerophyll forest occurs in South America, Africa south of the Sahara or in New Zealand.

TEMPERATE BROAD-LEAVED RAIN FOREST

Temperate broad-leaved evergreen rain forest is found in areas with a humid, virtually frost-free climate. Evergreen hardwood trees predominate although sometimes conifers may be present. Such forest tends to be multi-layered, rich in epiphytes and to have a well-developed understorey in which ferns are often abundant.

Although generally absent from the Northern Hemisphere, temperate broad-leaved rain forest of evergreen tree species occurs in Asia, in the Himalayas, eastern mainland China, the southern part of the Korean Peninsula, the lowlands of Taiwan and southern Japan. As a result of population pressure the Asian rain forests are greatly modified by shifting agriculture, grazing and cutting.

In the Indian Himalayas marked altitudinal zonation is shown in the distribution of the tree species. At the lower altitudes of the temperate climatic zone there is a varied tree cover with species such as *Alcimandra cathcartii*, *Castonopsis tribuloides*, *Machilus edulis*, *Magnolia campbelli* and *Mallotus nepalensis* present whilst at the middle and higher altitudes the oaks *Quercus lamellosa* and *Q. pachyphylla* become dominant. In the drier regions of the Himalayas in Nepal, the vegetation becomes more sclerophyllous and the dominant oaks are *Q. dilata* and *Q. incana*.

In eastern Asia much of the mixed evergreen broad-leaved temperate rain forest has been cleared, but scattered remnants occur along the China coast and in the lowland areas of offshore islands including Taiwan. Historically the rain forests are closely allied to those of the neighbouring tropical regions and there is a great variety of tree species and of epiphytes, climbers and lianes.

In Japan the northern limit of the temperate broad-leaved evergreen rain forest is determined by the gradual reduction in the warming influence of the north-flowing ocean currents. In the south the evergreen rain forest attains its maximum altitude of about 850 m above sea level. In relatively undisturbed forest there are many lianes, epiphytes and ferns and a great variety of evergreen trees and shrubs. The dominant tree species include evergreen members of the Fagaceae (*Castanopsis* spp., *Cyclobalanopsis* spp., *Quercus phylliraeoides*), the Lauraceae (*Cinnamomum* and *Machilus* spp.) and other broad-leaved evergreen species (*Camellia japonica*, *Distylium racemosum* and *Myrica rubra*). Most of the forest has been cleared for agriculture and industry and only a few greatly modified remnants are left.

In South America, the temperate broad-leaved evergreen rain forest is largely confined to southern Chile, the northern boundary being determined by increasing aridity and the southern extremity by decreasing temperature. In general these forests display a multi-layered structure and a diversity of species in the tree, shrub and ground layers. Diversity tends to decrease southwards and with increasing altitude. Trees of *Nothofagus* species (*N. betuloides*, *N. dombeyi*, *N. nitida* and *N. obliqua*) tend to be dominant and other common species of the overstorey include *Laurelia philippiana*, *L. sempervirens*, *Persea lingue*, *Tepualia stipularis* and *Weinmannia trichosperma*. Clearing for pasture and agriculture has been most extensive in the north and bamboos of the genus *Chusquea* may form a closed cover over disturbed areas without rain forest.

On the African continent, which does not extend so far south as South America, temperate evergreen rain forest is relatively poorly developed and present on separate mountain ranges where temperate conditions exist because of altitude. Despite the disjunct nature of the temperate evergreen rain forest there are similarities in floristic composition and the majority of plant species are endemic to these moist, high-altitude areas. The annual rainfall is usually greater than 2000 mm. Aspect in relation to rain-bearing winds is important and the damaging effects of any dry period may be mitigated by mist at the high altitudes. Undisturbed forest is multi-layered with epiphytes and lianes present, but in general the forest has been much reduced in extent and modified by burning, grazing and clearing.

Temperate broad-leaved evergreen rain forest in Australia is best developed in Tasmania where it occurs from sea level to 900 m. It is also present in Victoria and occurs as small patches within the coastal ranges and tablelands of New South Wales. The Tasmanian temperate evergreen rain forest is relatively undisturbed and the major hardwood tree species in order of importance are *Nothofagus cunninghamii*, *Atherosperma moschatum*, *Eucryphia lucida* and *Acacia melanoxylon*, with the conifer *Phyllocladus asplenifolius* also present. Epiphytes are common and there is a well-developed ground flora with mosses, lichens and ferns abundant.

In New Zealand temperate broad-leaved evergreen rain forest of one form or another covered most of the islands before the first human settlers arrived about 1000 years ago, the exception being some areas dominated by conifers or deciduous trees and subalpine stands of *Nothofagus*. A diversity of forest types has been recognized, and in general the forest is structurally complex with abundant epiphytes, lianes and ferns. Extensive areas of evergreen hardwood forest occur dominated by trees of *Nothofagus* species.

The New Zealand forests are unusual in that until recently they have evolved in the complete absence of people and other mammals and largely in botanical isolation. The early Polynesian settlers damaged the native forest by burning. Since the coming of Europeans large areas of native rain forest have been cleared for agriculture. This relatively recent clear-felling and the introduction of mammals, which have killed many trees and restricted regeneration, has shown the importance of these forests in protecting the soil and reducing soil erosion on steep mountain slopes. In the past half century many areas where native forest formerly grew, have been planted with commercial stands of the introduced *Pinus radiata*. These plantations are some of the most productive timber-producing stands in the world and of great significance in the New Zealand economy.

REFERENCES

Sakai, A., Paton, D.M. and Wardle, P., 1981. Freezing resistance of trees of the south temperate zone, especially subalpine species of Australasia. *Ecology*, 62: 563—570.

Shelford, V.E., 1963. *The Ecology of North America.* University of Illinois Press, Urbana, Ill., U.S.A., 610 pp.

UNESCO, 1973. *International Classification and Mapping of Vegetation. Ecology Conserv.*, 6: 93 pp.

Walter, H., 1973. *Vegetation of the Earth in Relation to Climate and the Eco-Physiological Conditions.* English Universities Press Ltd., London, 237 pp.

Chapter 2

TEMPERATE BROAD-LEAVED EVERGREEN FORESTS OF SOUTH AMERICA[1]

T.T. VEBLEN, F.M. SCHLEGEL and J.V. OLTREMARI

INTRODUCTION

The temperate broad-leaved evergreen forests of South America occur principally in Chile with only minor extensions into neighbouring Argentina. Chile is a long and narrow country extending over 4200 km from latitude 18° to 56°S. Precipitation generally increases from north to south at these latitudes on the western coast of South America and there is a gradation from a pattern of desert in the north, *matorral* scrub and sclerophyllous woodland in central Chile, to temperate rain forest in the south. The flora of Chile and adjacent regions in Argentina is relatively well known. The composition and distribution of plant communities is known only incompletely because of the complexity and diversity of the vegetation and the difficulty of access in the fiord-lined coast of the southernmost latitudes. Ecological investigations needed to elucidate the dynamic interrelations and the factors controlling the distributions of plant assemblages are only in an incipient stage. The functional processes of these forest ecosystems have not yet been investigated.

Southern Chile, or the region south of the Río Bío-Bío (*c.* 37°45′S), was phytogeographically described by Reiche (1907) who relied on personal knowledge of the region and the writings of explorer naturalists such as R.A. Philippi (1865) and F.W. Neger (1899). To these early workers and to Skottsberg (1910 and 1916), all subsequent authors on the vegetation of southern Chile are greatly indebted. Skottsberg (1910) subdivided the coastal region south of 41° into: (1) a northern species-rich Valdivian forest of evergreen *Nothofagus* and (2) a southern species-poor subantarctic forest of evergreen *Nothofagus*. He also recognized southern and northern deciduous *Nothofagus* forests at higher altitudes in the Andean Cordillera.

The Valdivian rain forest region as described by Hauman (1913) has its northern limit at the Río Bío-Bío and its southern limit at 46°S. North and south of these limits the tree and shrub strata of the forests are markedly less rich in species due to increasing aridity and diminishing temperatures, respectively. Skottsberg (1916) set the southern limit of the Valdivian rain forest region at 48° and named the forest further south the Magellanic rain forest. While the boundary between the Valdivian and Magellanic (or subantarctic) rain forests cannot be determined precisely because there is a continuum of change in species composition, these two rain forest regions differ markedly in species richness. Although the coastal region south of 48° is frequently mapped as Magellanic rain forest, on the outer coastal islands oceanic heath vegetation abounds and the evergreen *Nothofagus* forests are not continuous. Godley (1960) has described much of this vegetation as Magellanic moorland with evergreen *Nothofagus* rain forest only locally established on more favourable sites.

Schmithüsen (1956) recognized two major subdivisions of the forested region of southern Chile: (1) a northern lowland deciduous forest (*c.* 37°35′ to 40°40′S at sea level) and (2) a southern evergreen rain forest (south of 40°40′S). On the basis of decreasing diversity of tree species present from north to south, he further subdivided the evergreen rain forest into the Valdivian rain forest (40°40′ to 43°20′S at sea level), the north Patagonian rain forest (43°20′ to 47°30′S), and the subantarctic rain

[1] Manuscript completed November, 1980.

forest (south of 47°30′S). Schmithüsen (1956) regarded the forests of the mid- to high elevations at latitudes north of 40°40′ as outliers of the north Patagonian rain forest normally occurring further south.

Oberdorfer's (1960) classification of the vegetation of southern Chile generally coincides with that of Schmithüsen except for the inclusion of the so-called northern lowland deciduous forest in the Valdivian rain forest region. Lowland deciduous forest as recognized by Schmithüsen is more appropriately considered as mainly evergreen forest and included in the Valdivian rain forest unit originally delimited by Hauman (1913).

In this description of the forests of southern Chile the following phytogeographical regions are recognized (Fig. 2.1): (1) the Valdivian rain forest (*c.* 37°45′ to 43°20′S); (2) the north Patagonian rain forest (*c.* 43°20′ to 47°30′S); and (3) the Magellanic rain forest (south of *c.* 47°30′S). For convenience,

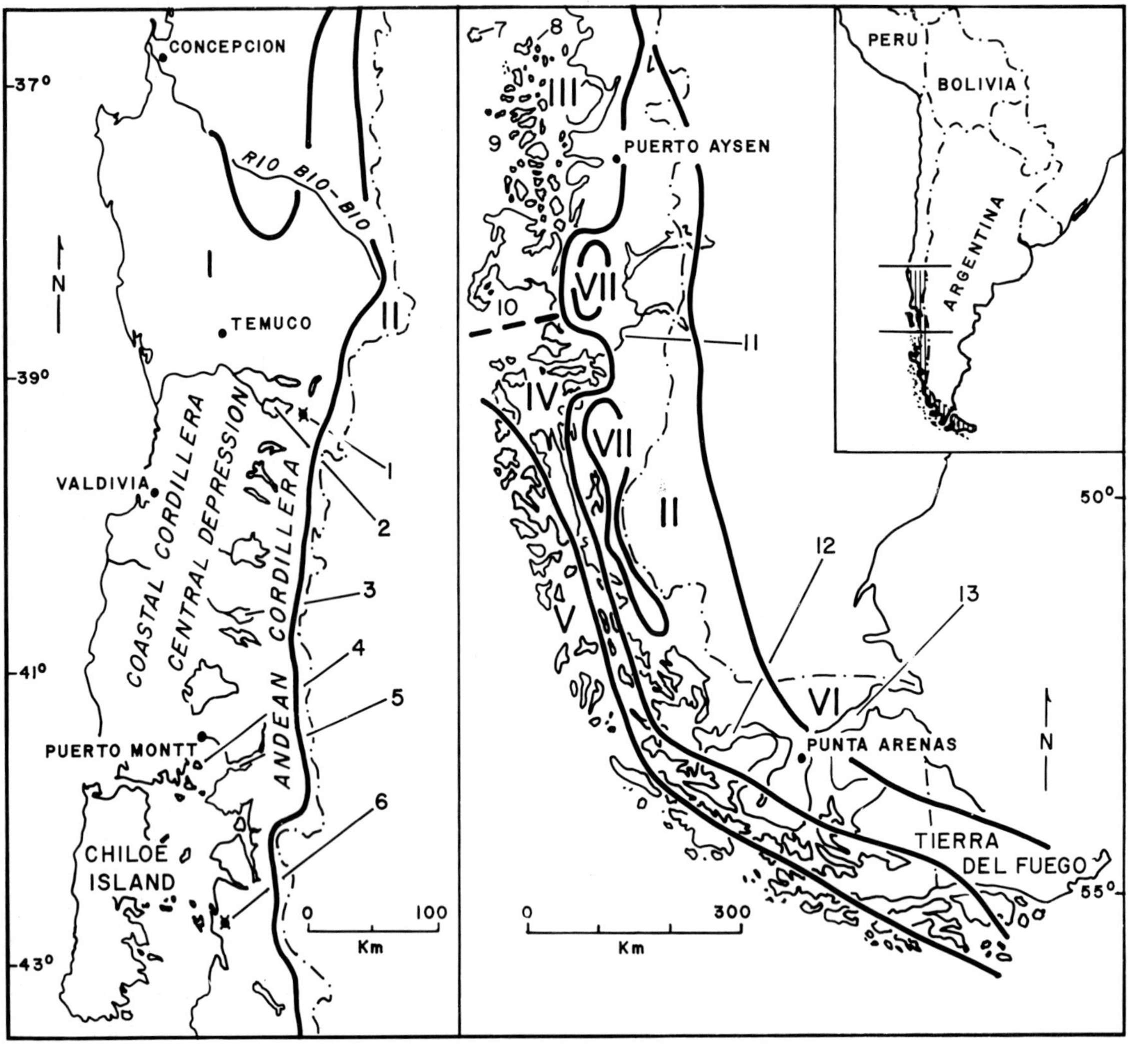

Fig. 2.1. Southern Chile. Locations: *1* = Volcán Villarrica; *2* = Lake Villarrica; *3* = Lake Puyehue; *4* = Reloncaví Sound; *5* = Estero de Reloncaví; *6* = Volcán Michinmávida; *7* = Guafo Island; *8* = Guaitecas Islands; *9* = Chonos Islands; *10* = Gulf of Penas; *11* = Río Baker; *12* = Skyring Sound; and *13* = Straits of Magellan. Vegetation regions: *I* = Valdivian rain forest; *II* = deciduous *Nothofagus* forest; *III* = north Patagonian rain forest; *IV* = Magellanic rain forest; *V* = Magellanic moorland; *VI* = Patagonian steppe; and *VII* = snow, ice, or above timberline.

the latter two regions are discussed together.

To facilitate comparison with the vegetation of other parts of the world, the forests of the west coast of southern South America are considered in relation to Mueller-Dombois and Ellenberg's (1974) physiognomic-ecological classification (Table 2.1). While for completeness all the forest types of southern Chile are included in Table 2.1, only those in the mainly evergreen formation subclass are discussed here.

TABLE 2.1

Physiognomic-ecological classification of forest types of southern Chile

I. Mainly evergreen forests
- A. Temperate broad-leaved rain forests largely without conifers
  - 1. On adequately drained sites
    - a. *Nothofagus obliqua–Laurelia sempervirens–Persea–Aextoxicon*
    - b. *Eucryphia–Aextoxicon–Laurelia philippiana*
    - c. *Eucryphia–Weinmannia–Laurelia philippiana*
  - 2. On poorly drained sites
    - a. *Nothofagus dombeyi–Drimys*–Myrtaceae
- B. Temperate broad-leaved rain forests with conifers admixed
  - 1. On adequately drained sites
    - a. *Nothofagus dombeyi–Eucryphia–Laurelia philippiana–Saxegothaea*
    - b. *Nothofagus dombeyi–Laurelia philippiana–Saxegothaea–Weinmannia*
    - c. *Nothofagus dombeyi–N. alpina–Laurelia philippiana–Saxegothaea*
    - d. *Nothofagus nitida–Saxegothaea–Podocarpus nubigenus–Weinmannia*
    - e. Evergreen *Nothofagus–Weinmannia*
    - f. *Nothofagus betuloides*
    - g. *Nothofagus dombeyi–N. pumilio*
    - h. *Nothofagus betuloides–N. pumilio*
- C. Subalpine elfin temperate broad-leaved rain forest
  - 1. Subalpine *Nothofagus betuloides*
- D. Evergreen coniferous forests with broad-leaved trees admixed
  - 1. *Araucaria araucana*
  - 2. *Fitzroya cupressoides*

II. Mainly deciduous forests
- A. Cold-deciduous forests largely without evergreen trees
  - 1. *Nothofagus pumilio*
- B. Subalpine elfin cold-deciduous forests largely without evergreen trees
  - 1. Subalpine *Nothofagus pumilio*

VALDIVIAN RAIN FOREST REGION

Physical environment

Physiographically the region from the Río Bío-Bío south to 43°20′S can be divided into the Coastal Cordillera, the Central Depression, and the Andean Cordillera. The oldest rocks in the region occur in the Coastal Cordillera, which are made up of Paleozoic and Precambrian metamorphic rocks. The Coastal Cordillera exceeds 1000 m in altitude in only a few places and has not been glaciated (Ruíz et al., 1965; Illies, 1970). The uplands of western Chiloé Island are an extension of the Coastal Cordillera but have maximum altitudes from 700 to 800 m. The Central Depression is a structural depression, or graben, filled with Quaternary glacial, fluvioglacial, aeolian, and alluvial deposits and volcanic ash. The maximum altitude of approximately 300 m is in the north and the elevation gradually diminishes southwards. The Andean Cordillera at these latitudes is a broad mountainous zone of 60 to 100 km width divided into a labyrinth of uplands and valleys.

The Andes, which are still uplifting and whose summits average about 2000 m in this region, were extensively glaciated during the Pleistocene. Most glaciated surfaces have been covered by extensive andesitic volcanic deposits of recent origin. Glaciers scoured the numerous valleys in the Chilean Lake District and filled the Central Depression with fluvioglacial sediments which underlie the younger volcanic deposits. In the southern part of the Lake District, Pleistocene glacial ice extended nearly to the foothills of the Coastal Cordillera and covered the northwestern sector of Chiloé Island. Southwards from Estero de Reloncaví on the mainland, the ice reached the sea and the resulting fiord coastline has greatly hindered access.

The Valdivian rain forest region generally has a west coast maritime climate characterized by a mild temperature range and high annual precipitation (Di Castri, 1968). To the north the climate of Central Chile is mediterranean in type with precipitation being greatest in the winter. From north to south, the total annual precipitation increases and the summer drought becomes gradually less severe (Fig. 2.2 and Table 2.2). In the southern part of this region moist conditions prevail throughout the year with the average monthly precipitation ex-

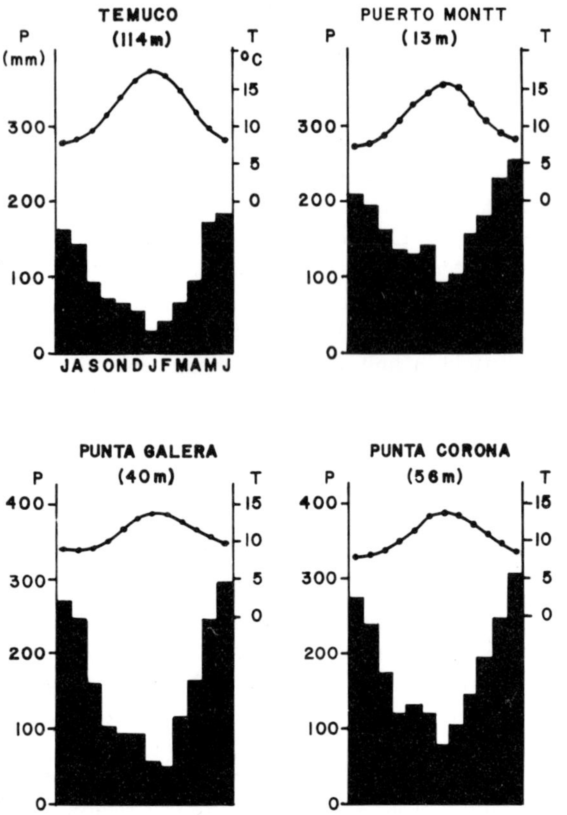

Fig. 2.2. Climatograms of selected stations from the Valdivian rain forest region. Key: T = average monthly temperature (°C); P = average monthly precipitation (mm); AT = average annual temperature (°C); and AP = average annual precipitation (mm). Stations: Temuco (38°45′S, 72°35′W), AT = 12, AP = 1190; Punta Galera (40°01′S, 73°44′W), AT = 11.2, AP = 1933; Puerto Montt (41°28′S, 72°56′W), AT = 11.1, AP = 1996; and Punta Corona (41°47′S, 73°52′W), AT = 10.5, AP = 2148. Altitude above sea level is given below the name of each station. (Source: Anonymous, 1964.)

ceeding 100 mm even during the driest summer month. The wettest zones are probably the higher elevations of the west-facing slopes of the Coastal Cordillera which receive the full effect of the moist Pacific air-masses but meteorological stations are lacking there. The west-facing slopes of the higher altitudes of the Andean Cordillera in the northern region as well as the low altitudes south of *c.* 41°30′S are very moist, receiving over 4000 mm of precipitation annually. Due to the rain-shadow influence of the Coastal Cordillera, there is less precipitation in the Central Depression and precipitation rapidly diminishes eastward from the crest of the Andean Cordillera. Above approximately 800 m in the Coastal and Andean Cordilleras much winter precipitation falls as snow. Gale-strength winds commonly occur in association with the winter cyclonic storms, and also during summer when a high-pressure cell is frequently located over southern Argentina and produces strong, warm winds.

Forest types

While the Valdivian rain forest region is better known than the more southerly regions, the existing classifications of its vegetation are inadequate for ecological description. Based on aerial photographs supplemented by limited forest inventories, these forests have been classified to determine the volume of timber present and for economic evaluation of the forest resource (Instituto Forestal, 1966a). However, in this classification forests of very distinct species composition and altitudinal distribution are considered as single types. Similarly, the well-known phytosociological studies by Schmithüsen (1956) and Oberdorfer (1960) are valuable initial efforts but fall far short of adequately describing the complex variety of plant assemblages present. These phytosociological studies are of restricted value for the definition of forest types because of the small number and small size of sample stands in relation to the size of component trees and because of the emphasis given to slight floristic differences in the understorey vegetation.

Consequently, the following classification of forest types of the Valdivian rain forest region is of necessity an original synthesis based on our experience and the information provided by the following authors: Philippi (1865), Neger (1899), Reiche (1907), Skottsberg (1910, 1916), Hauman (1913), Bernath (1937), Ljunger (1939), Looser (1948), Schmithüsen (1956, 1960), Thomasson (1959, 1963), Godley (1960, 1968), Oberdorfer (1960), Holdgate (1961), Fuenzalida (1965), Heusser (1966, 1974), Instituto Forestal (1966a), Eskuche (1968), Villagran and Soto (1974), Brun (1975) and Burschel et al. (1976). In describing forest types, it is recognized that they represent abstract classification categories somewhat artificially imposed on the continuum of vegetational changes in the Valdivian rain forest. In the moist oceanic climate of this region abrupt vegetational boundaries are rare

TABLE 2.2

Precipitation at selected stations in the Valdivian rain forest region (Data from Almeyda and Saez, 1958; and Anonymous, 1964)

Locality	Coordinates	Altitude (m)	Annual precipitation (mm)	Percentage occurring in winter (June—August)
Western slopes of the Andean Cordillera				
Lonquimay	38°26′S, 71°15′W	900	1919	47%
Cherquenco	38°39′S, 72°01′W	520	2345	42%
Pedregoso	39°11′S, 72°10′W	500	3120	40%
Trafún	39°40′S, 71°51′W	550	4627	40%
Puerto Fuí	′39°52′S, 71°53′W	540	4970	41%
Puñire	40°16′S, 72°12′W	300	3400	40%
Peulla	41°05′S, 72°02′W	190	3468	36%
Reñihue	42°35′S, 72°35′W	20	5765	34%
Chaitén	42°54′S, 72°45′W	40	3675	33%
Western side of the Coastal Cordillera and Chiloé Island				
Puerto Domínguez	38°54′S, 73°14′W	5	1480	43%
Toltén	39°13′S, 73°13′W	80	1799	44%
Niebla	39°52′S, 73°24′W	10	2254	41%
Punta Galera	40°01′S, 73°44′W	40	2125	41%
Venecia	40°15′S, 73°42′W	50	3500	40%
Maullín	41°37′S, 73°35′W	20	1946	37%
Punta Corona	41°47′S, 73°52′W	50	2444	38%
Quilán Cucao	42°38′S, 74°05′W	30	1900	34%
Central Depression and east side of Chiloé Island				
Victoria	38°14′S, 72°18′W	300	1652	46%
Temuco	38°45′S, 72°35′W	114	1345	42%
Gorbea	39°06′S, 72°40′W	150	1946	47%
Paillaco	40°03′S, 72°50′W	200	1440	45%
Osorno	40°35′S, 73°09′W	180	1330	40%
Frutillar	41°08′S, 73°01′W	130	1580	38%
Puerto Montt	41°28′S, 72°56′W	15	1960	35%
Morro Lobos	42°04′S, 73°24′W	20	2350	39%
Castro	42°29′S, 73°45′W	30	2070	43%

and usually associated with azonal or intrazonal soils. Normally there are gradual changes of species composition along altitudinal and moisture gradients.

While the region as a whole is characterized by evergreen rain forests, other forest formations also occur. In the northern part of the region, forests of the conifer *Araucaria araucana* are found in the Coastal Cordillera from 37°20′ to 38°40′S mainly from 1000 to 1400 m altitude and in the Andean Cordillera near the timberline from 37°30′ to 39°30′S. Forests of the giant conifer *Fitzroya cupressoides* occur discontinuously from approximately 39°50′ to 43°30′S in both the Coastal and Andean Cordilleras. They are present on poorly drained lowlands and in montane zones of very high precipitation mainly from 700 m to near timberline at over 1000 m altitude. Excessive cutting and poor natural regeneration has eliminated this highly valued species from all but the most remote localities (Veblen et al., 1976). Deciduous forests of *Nothofagus pumilio* and *N. antarctica* occur at the timberline in the Andean Cordillera and at the forest/steppe ecotone east of the Andes from approximately 36°S to Tierra del Fuego. While *N. pumilio* forms a relatively abrupt timberline on the western slopes of the Andes, the change from forest to steppe east of the Andes is much more gradual and is characterized by *N. antarctica* scrub. No timberline is present in the relatively low Coastal Cordillera.

Lowland forests of adequately drained sites (Table 2.3)

***Nothofagus obliqua–Laurelia sempervirens–Persea–Aextoxicon* Type 1.** These forests occur in the Central Depression between *c.* 38° and 41°S at altitudes below 400 m, and have been described by

TABLE 2.3

Lowland forests of adequately drained sites[1]

Type 1 *Nothofagus obliqua–* *Laurelia sempervirens*	Type 2 *Eucryphia–Aextoxicon–* *Laurelia philippiana*	Type 3 *Eucryphia–Weinmannia–* *Laurelia philippiana*
Emergent tree stratum		
Nothofagus obliqua	*Eucryphia cordifolia*	*Eucryphia cordifolia*
Matrix of dominant and subdominant trees		
Laurelia sempervirens *Persea lingue* *Aextoxicon punctatum*	*Aextoxicon punctatum* *Laurelia philippiana*	*Laurelia philippiana*
Tall shrub–small tree stratum		
Myrceugenia planipes *Chusquea quila* *Myrceugenella apiculata* *Gevuina avellana* *Lomatia dentata*	*Drimys winteri* *Gevuina avellana* *Myrceugenia planipes* *Amomyrtus luma* *A. meli* *Chusquea quila* *Pseudopanax laetevirens* *Myrceugenella apiculata* *Lomatia ferruginea*	*Drimys winteri* *Amomyrtus meli* *A. luma* *Pseudopanax laetevirens*
Small shrub stratum		
Rhaphithamnus spinosus *Rhamnus diffusus* *Pseudopanax valdiviensis* *Azara lanceolata*	*Lophosoria quadripinnata* *Rhaphithamnus spinosus* *Pseudopanax valdiviensis* *Ribes punctatum* *Greigia sphacelata*	[the shrub strata and liana–epiphyte stratum are undescribed but are probably very similar to those of Type 2]
Lianas and epiphytes		
Lapageria rosea *Cissus striata* *Hydrangea integerrima* *Luzuriaga radicans* *Mitraria coccinea* *Sarmienta repens* *Boquila trifoliata*	*Hydrangea integerrima* *Luzuriaga erecta* *L. radicans* *Boquila trifoliata* *Lapageria rosea* *Mitraria coccinea* *Cissus striata* *Hymenophyllum* spp. *Fascicularia bicolor*	

[1] Species are listed in decreasing order of importance in this and all subsequent tables.

Schmithüsen (1956) as lowland summergreen deciduous forest. However, the evergreen trees far exceed the single deciduous species, *Nothofagus obliqua*, in number and basal area, and the general canopy maintains its evergreen character. Lying in the zone of most severe summer drought, this type borders the sclerophyll woodlands of *Cryptocarya alba*, *Peumus boldus*, and *Quillaja saponaria* to the north and is composed of trees with scleromorphic leaves. It is found on well-drained soils typically derived from volcanic ash overlying morainic and fluvioglacial deposits. Clearing has transformed this forest zone into a landscape of pasture and cultivated fields dominated by large scattered *Nothofagus obliqua* and *Laurelia sempervirens*. Virtually the only extensive virgin stands of this type are on private estates near Lake Villarrica. There, the forest is characterized by 40 m tall *Nothofagus obliqua* with large predominant crowns extending 5 m or more above the general canopy of *Laurelia sempervirens*, *Persea lingue*, and *Aextoxicon punctatum*. The conifer *Podocarpus salignus* is often present in the semi-exploited stands in the northern part of the region but is not always characteristic of this type. Although the small shrub stratum is very sparse and movement through the forest easy, lianas are abundant, especially *Lapageria rosea*, the national flower of Chile. In disturbed stands the large climbing bamboo, *Chusquea quila*, proliferates.

***Eucryphia–Aextoxicon–Laurelia philippiana* Type 2.** Forests dominated principally by *Eucryphia cordifolia* and *Aextoxicon punctatum* and less so by *Laurelia philippiana* occur in a zone of great humidity along the Pacific coast from *c*. 38°S to the northwestern coast of Chiloé Island (Schmithüsen, 1956; Holdgate, 1961). Minor areas of this forest type also occur below altitudes of 400 m in the Andean Cordillera from *c*. 39°30′ to 41°30′S (Oberdorfer, 1960). The small tree *Drimys winteri* and several myrtaceous shrub or small tree species are typically present in these moister habitats. Similarly, hymenophyllaceous ferns and bromeliads (e.g., *Fascicularia bicolor*) are abundant epiphytes. In some instances, the main tree canopy may consist of nearly pure *Aextoxicon punctatum* while elsewhere 40 m tall *Eucryphia cordifolia* extend 5 m above the main canopy, and *Laurelia philippiana* may be present. Closer to the coast, where *Aextoxicon punctatum* tends to be more dominant, the forest is usually of considerably less stature, scarcely reaching 15 m tall. Typically the soils are well drained and derived from marine or alluvial sediments. Virgin stands of this type are rare, and the exploited stands are characterized by a proliferation of *Chusquea quila* whose culms often exceed 20 m in length.

***Eucryphia–Weinmannia–Laurelia philippiana* Type 3.** In northwestern Chiloé Island, inland from the coast and at slightly greater altitudes, *Aextoxicon punctatum* is not present but *Weinmannia trichosperma* appears, and *Laurelia philippiana* becomes more abundant (Godley, 1960; Holdgate, 1961). Although not fully described floristically, this type is known from forest inventory data to be important at altitudes up to 350 m on northern Chiloé Island. It occurs on well-drained slopes and ridges in an area where much of the land is otherwise poorly drained. The soils here are of a brown-earth type; they are not very acid, are fairly rich in soil fauna, and exhibit rapid incorporation of litter into the upper horizon (Holdgate, 1961). In this type trees of several species are twice as large as trees of the same species further north in the Andes. Examples include *Amomyrtus meli* 20 m tall and 60 cm in diameter at breast height (dbh) and *Drimys winteri* of 150 cm in dbh. This and the previous type are unique in the Valdivian rain forest region for the absence of *Nothofagus* and conifers.

Forests of poorly drained sites (Table 2.4)

***Nothofagus dombeyi–Drimys*–Myrtaceae Type 4.** This type occurs in two topographic situations both with imperfect drainage. The first is known locally as *ñadi* which refers to the flat to undulating topography on glacially derived sediments where an iron-rich hardpan impedes drainage. During winter the soil is saturated but the layer above the hardpan is thin, and since it lacks volume for moisture storage the soil is desiccated during the summer months. The second situation is the *hualve* which refers to backswamps resulting from deranged drainage patterns and immature stream systems in this tectonically active region.

The vegetation of both topographic situations is very similar, being dominated by the trees *Nothofagus dombeyi* and *Drimys winteri*, and a diverse

TABLE 2.4

Forests of poorly drained sites

Type 4 *Nothofagus dombeyi–Drimys*–Myrtaceae	Type 5 *Tepualia–Pilgerodendron*
Emergent tree stratum	
Nothofagus dombeyi	*Pilgerodendron uvifera*
Drimys winteri	*Drimys winteri*
Tall shrub–small tree stratum	
Temu divaricatum	*Tepualia stipularis*
Tepualia stipularis	*Pseudopanax laetevirens*
Myrceugenia exsucca	
Myrceugenella gayana	
M. apiculata	
Chusquea uliginosa	
Caldcluvia paniculata	
Lomatia ferruginea	
Ovidia pillo-pillo	
Embothrium coccineum	
Nothofagus antarctica	
Small shrub stratum	
Escallonia leucantha	*Desfontainia spinosa*
Baccharis lycioides	*Pernettya poeppigii*
Berberis buxifolia	*Escallonia pterocladon*
Pernettya spp.	*Baccharis magellanica*
Lophosoria quadripinnata	*Empetrum rubrum*
Ugni molinae	
Lianas and epiphytes	
Boquila trifoliata	*Philesia magellanica*
Luzuriaga radicans	*Griselinia ruscifolia*
Hymenophyllum spp.	*G. racemosa*
Luzuriaga erecta	*Luzuriaga radicans*
	L. erecta

assemblage of myrtaceous shrubs. Cyperaceae, epiphytic hymenophyllaceous ferns, and a large ground-rooted fern, *Blechnum chilense*, are also abundant (Ramírez et al., 1976). *Ñadis* occur principally in the Central Depression between Valdivia and Puerto Montt where much of this vegetation type has been burned and cleared in futile attempts to cultivate the land.

***Tepualia–Pilgerodendron* Type 5.** This type occurs on poorly drained ridge flanks, near river courses, along strands, and in hollows principally in the southern district of the Valdivian rain forest region at all altitudes. The small tree *Tepualia stipularis* may be the sole dominant or, more frequently, it is mixed with the conifer *Pilgerodendron uvifera* and less frequently with *Drimys winteri*. Known locally as *tepual* vegetation, it does not dry out during the summer months as do the *ñadis* and *hualves*. The multi-stemmed *Tepualia stipularis* characteristically forms a dense lattice 5 to 8 m above the surface of dark, organic-rich standing water, and the area may only be traversed by crawling over the horizontal trunks. It is very reminiscent of tropical mangrove vegetation. On Chiloé Island where *tepuales* are common, *Nothofagus nitida* is often found on the adjacent slightly better drained topography, and true bog vegetation lacking trees occurs on the surrounding saturated soils. Beneath the *Tepualia stipularis* are cushions of *Astelia pumila*, *Oreobolus obtusangulus*, and *Sphagnum* moss supporting scattered *Myrteola mummularia*, *Drosera uniflora*, and *Leptocarpus chilensis*. The soil is true peat, very acid with a pH as low as 3.5 (Holdgate, 1961).

Mid-elevation forests dominated by *Nothofagus dombeyi* (Table 2.5)

***Nothofagus dombeyi–Eucryphia–Laurelia philippiana–Saxegothaea* Type 6.** With increasing altitude from the area characterized by Type 2, in the Andean Cordillera *Aextoxicon punctatum* gradually phases out and the conifer *Saxegothaea conspicua* and *Nothofagus dombeyi* appear. This type is generally absent from the western slope of the Coastal Cordillera south of Valdivia and similarly in northwestern Chiloé Island it is of restricted importance.

However, at altitudes from *c*. 400 to 700 m this type is abundant in the Lake District of the Andes. It occurs on well-drained slopes on soils derived principally from volcanic ash. The emergent trees, *N. dombeyi* and *Eucryphia cordifolia*, may be over 40 m tall with diameters greater than 1.5 m; *Laurelia philippiana* and *Saxegothaea conspicua* are typically 5 to 10 m shorter. Epiphytes and lianas densely cover the tree trunks. These forests have been very much destroyed by fires, and the early successional communities following fire are characterized by *Chusquea quila* and the arborescent

TABLE 2.5

Mid-elevation forests dominated by *Nothofagus dombeyi*

Type 6 *Nothofagus dombeyi–Eucryphia–Laurelia philippiana–Saxegothaea*	Type 7 *Nothofagus dombeyi–Laurelia philippiana–Saxegothaea–Weinmannia*	Type 8 *Nothofagus dombeyi–N. alpina–Laurelia philippiana–Saxegothaea*
Emergent tree stratum		
Nothofagus dombeyi *Eucryphia cordifolia*	*Nothofagus dombeyi*	*Nothofagus dombeyi* *N. alpina*
Matrix of dominant and subdominant trees		
Laurelia philippiana *Saxegothaea conspicua*	*Saxegothaea conspicua* *Laurelia philippiana* *Weinmannia trichosperma* *Podocarpus nubigenus*	*Saxegothaea conspicua* *Laurelia philippiana* *Weinmannia trichosperma*
Tall shrub–small tree stratum		
Chusquea quila *Myrceugenella apiculata* *Amomyrtus luma* *Pseudopanax laetevirens* *Gevuina avellana* *Drimys winteri*	*Dasyphyllum diacanthoides* *Chusquea culeou* *Amomyrtus luma*	*Dasyphyllum diacanthoides* *Chusquea culeou* *Myrceugenia chrysocarpa* *Chusquea quila*
Small shrub stratum		
Rhaphithamnus spinosus *Lophosoria quadripinnata* *Ribes punctatum*	*Azara lanceolata* *Ribes punctatum*	*Azara lanceolata* *Ribes punctatum*
Lianas and epiphytes		
Luzuriaga radicans *Cissus striata* *Fascicularia bicolor* *Asplenium dareoides* *Elytropus chilensis* *Mitraria coccinea* *Hydrangea integerrima* *Hymenophyllum* spp.	*Hydrangea integerrima* *Hymenophyllum* spp. *Asteranthera ovata* *Asplenium dareoides* *Luzuriaga radicans* *Mitraria coccinea*	*Hydrangea integerrima* *Hymenophyllum* spp. *Asteranthera ovata* *Asplenium dareoides* *Luzuriaga radicans* *Griselinia racemosa*

shrub *Aristotelia chilensis*, as well as abundant small shrubs and lianas.

***Nothofagus dombeyi–Laurelia philippiana–Saxegothaea–Weinmannia* Type 7.** This type occurs in the Andean Cordillera, extending from altitudes at the upper limit of the previous type to *c*. 950 m and on similar soils. It is the most extensive forest type remaining in the Lake District of the Andes. *Weinmannia trichosperma* is not always present, being more abundant in the southern part of the region, and *Dasyphyllum diacanthoides* typically is present only in small numbers. *Nothofagus dombeyi* reaches maximum development, being up to 50 m tall and over 2.5 m in dbh. With emergent crowns often as much as 20 m in diameter and huge fluted trunk buttresses, *N. dombeyi* majestically presides over the dominant and subdominant trees which form a canopy at 30 to 35 m. The robust liana *Hydrangea integerrima* also reaches its maximum development here, attaining heights of 40 m and diameters of 35 cm. The constantly moist condition of the understorey results from the location of this type on the west-facing slopes of the Andes. At this altitude the annual precipitation exceeds 4000 mm and the dense canopy cover reduces maximum temperatures and hence evaporation. Wiry lianas, epiphytic ferns, bryophytes, and foliose lichens abound; South America's tallest moss, *Dendroligotrichum dendroides*, is common. Despite the heavy shade, the understorey is often dense with regeneration of *Laurelia philippiana* and *Dasyphyllum diacanthoides* and scattered *Amomyrtus luma*, but in terms of species diversity the shrub strata are less rich than for the lower altitude types. In the lighter spots below small gaps in the canopy the 7 m tall bamboo *Chusquea culeou* forms clumps of fifty or more culms, and where the forests are disturbed dense thickets (Fig. 2.3). This type usually occurs on deep soils rich in organic matter derived from volcanic ash deposits. At the moistest sites, *Drimys winteri* and the conifer *Podocarpus nubigenus* may be present in small numbers, more so towards the south. This forest type extends through the lowest passes to the eastern side of the Andes especially in the zone of Lake Nahuel Huapi (Thomasson, 1959; Eskuche, 1968).

***Nothofagus dombeyi–N. alpina–Laurelia philippiana–Saxegothaea* Type 8.** This type is similar to Type 7 except for the presence of the deciduous *Nothofagus alpina* in the emergent stratum and a less luxuriant development of epiphytes. Although *N. alpina* may reach 45 m in height and 2 m in dbh it is generally smaller than the evergreen *N. dombeyi*. This mixed *Nothofagus* forest occurs within the same altitudinal range as the previous type. However, it is displaced towards the north and occurs on perhaps slightly drier sites but still within a zone of high precipitation. In some stands, *Weinmannia trichosperma* and a few large *Dasyphyllum diacanthoides* as well as numerous saplings

Fig. 2.3. Proliferation of *Chusquea culeou* following burning and exploitation of a *Nothofagus dombeyi–Laurelia philippiana–Saxegothaea conspicua* forest at *c.* 700 m in the Andes of the Province of Valdivia.

and seedlings may be present. Very few virgin stands of this type remain due to the extensive logging of the highly prized *Nothofagus alpina*. Forest litter production has been measured only in this type and has been found to be 5.4 t ha^{-1} annually (Burschel et al., 1976).

TABLE 2.6

Mid- and high-elevation *Nothofagus* forests

Type 9 *Nothofagus nitida–Saxegothaea–Podocarpus nubigenus–Weinmannia*	Type 10 *Nothofagus dombeyi–N. pumilio*	Type 11 *Nothofagus betuloides–N. pumilio*	Type 12 Subalpine *Nothofagus betuloides*
Emergent tree stratum			
Nothofagus nitida	*Nothofagus dombeyi*	*Nothofagus betuloides*	*Nothofagus betuloides*
Matrix of dominant and subdominant trees			
Saxegothaea conspicua *Podocarpus nubigenus* *Weinmannia trichosperma* *Drimys winteri*	*Nothofagus pumilio*	*Nothofagus pumilio*	
Tall shrub–small tree stratum			
Lomatia ferruginea *Pseudopanax laetevirens* *Gevuina avellana* *Amomyrtus meli* *A. luma* *Ovidia pillo-pillo* *Tepualia stipularis* *Myrceugenia* spp.	*Chusquea tenuiflora* *Myrceugenia chrysocarpa*	*Chusquea tenuiflora*	*Chusquea tenuiflora* *Maytenus magellanica* *Gaultheria myrtilloides*
Small shrub stratum			
Desfontainia spinosa *Ugni molinae* *Chusquea culeou* *Pernettya mucronata*	*Drimys winteri* var. *andina* *Berberis serrato-dentata* *B. buxifolia* *Maytenus disticha*	*Drimys winteri* var. *andina* *Pernettya poeppigii* *Berberis serrato-dentata* *Maytenus disticha* *Berberis buxifolia* *Desfontainia spinosa*	*Drimys winteri* var. *andina* *Desfontainia spinosa* *Escallonia alpina* var. *alpina* *Berberis serrato-dentata* *B. linearifolia* *Maytenus disticha* *Pernettya poeppigii*
Lianas and epiphytes			
Campsidium valdivianum *Asteranthera ovata* *Luzuriaga radicans* *Philesia magellanica* *Griselinia scandens* *Fascicularia bicolor*	*Asteranthera ovata*		*Asteranthera ovata*

Mid- and high-elevation *Nothofagus* forests (Table 2.6)

***Nothofagus nitida–Saxegothaea–Podocarpus nubigenus–Weinmannia* Type 9.** This type is common at altitudes of 600 to 900 m on the western slopes of the Coastal Cordillera south of Valdivia and on western Chiloé Island. The dominant is principally *Nothofagus nitida* of 25 to 35 m in height with a mixture of the conifers *Saxegothaea conspicua* and *Podocarpus nubigenus*, and of *Weinmannia trichosperma* and *Drimys winteri* a few metres less in height. This type occurs under conditions of great and constant humidity, often with imperfect drainage, and has many species of the *tepual* vegetation (Type 5) in the understorey. On Chiloé Island, *Podocarpus nubigenus* is more important in this type and *Saxegothaea conspicua* is nearly absent. Soils here are high in organic matter with pH values ranging from 4.0 to 4.6 (Holdgate, 1961). Based on leaf morphology, *Nothofagus nitida* appears to form hybrids with *N. betuloides* at the higher altitudes and with *N. dombeyi* at lower altitudes in this type. As for the other types of the western slopes of the Coastal Cordillera and Chiloé Island, small tree and tall shrub species such as *Amomyrtus meli*, *Drimys winteri* and *Gevuina avellana* attain substantially greater dimensions than they do further north.

***Nothofagus dombeyi–N. pumilio* Type 10.** At altitudes above *c.* 1000 m and north of 40°47′S in the Andes a narrow transitional zone rarely greater than 150 m in altitudinal extension occurs between the *Nothofagus dombeyi*- or *N. dombeyi–N. alpina*-dominated forests and the pure stands of deciduous *N. pumilio* which form the timberline. Although scattered individuals of *Podocarpus nubigenus* are found at the wettest sites near the lower limit of this type, it is mostly above the altitudinal limit of conifers. *Nothofagus dombeyi* up to 1.5 m in dbh typically forms a stratum of emergent trees 25 to 30 m tall. *N. pumilio* is about 5 m shorter. The understorey is dominated by bamboo 2 to 3 m tall. This bamboo, *Chusquea tenuiflora*, is taller and more abundant beneath the evergreen *Nothofagus dombeyi* than below the deciduous trees. *Drimys winteri* var. *andina*, a dwarf variety of the tree species common on wet sites at lower altitudes, is an important constituent of the understorey. Lianas are generally absent at this altitude except for the delicate *Asteranthera ovata* which is common in the lower altitudinal range of this type. Bryophytes and lichens are common on the ground and as epiphytes on the lower 1 or 2 m of the tree trunks. These forests occur typically on thin soils high in organic matter derived from recent volcanic scoria and tuff. For several months of the year the understorey is covered by 1 to 2 m of snow.

***Nothofagus betuloides–N. pumilio* Type 11.** South of 40°47′S, the northern limit of *Nothofagus betuloides*, this type occurs in the Andes as a transitional zone between the completely evergreen forests below and the timberline-forming *N. pumilio* forests. In the northern area of its distribution it is nearly identical to the previous type except for the replacement of *N. dombeyi* by *N. betuloides* as the emergent tree species. Probably these two very similar species hybridize. Occurring near the timberline and often on the steep slopes of recently active volcanic cones, this type is frequently affected by snow avalanches and volcanic ash falls.

Subalpine *Nothofagus betuloides* Type 12. This type occurs sporadically south of 41°S on the wettest west-facing slopes where the evergreen *Nothofagus betuloides* forms the timberline. These are elfin forests of gnarled, multi-stemmed trees barely attaining 10 m in height. Physiognomically, this type is very similar to the timberline forests of *Nothofagus solandri* var. *cliffortioides* of the southern Alps of New Zealand. The understorey vegetation closely resembles that of the previous two types but is less completely dominated by *Chusquea tenuiflora*, and *Drimys winteri* var. *andina* and other shrub species are more common. Often occurring on the western flanks of recently active volcanoes, this type is found on coarse poorly developed soils derived from scoria and tuff.

THE NORTH PATAGONIAN AND MAGELLANIC RAIN FOREST REGIONS

Physical environment

The north Patagonian and Magellanic rain forest regions (south of 43°20′S) are located on heavily glaciated fiord lands and mazes of islands. Here,

TABLE 2.7

Precipitation at selected stations in the north Patagonian and Magellanic rain forest regions (data from Almeyda and Saez, 1958; and Fuenzalida, 1966)

Locality	Coordinates	Altitude (m)	Annual precipitation (mm)	Percentage occurring in winter (June–August)
Island and coastal locations				
Guafo	43°34′S, 74°45′W	140	1391	34%
Melinka	43°54′S, 73°46′W	5	4277	37%
Puerto Aysén	45°24′S, 72°42′W	10	2868	30%
Ofqui	46°44′S, 74°00′W	10	3166	16%
Cabo Raper	46°50′S, 73°35′W	40	2013	26%
San Pedro	47°43′S, 74°55′W	22	4076	24%
Guarelo	50°21′S, 75°21′W	10	7500	25%
Evangelistas	52°24′S, 75°06′W	55	2625	24%
Bahía Felix	52°58′S, 74°04′W	42	4866	23%

the western margin of the Andes dips below sea level and forms an archipelago of mountainous islands. The islands appear to have been completely covered by Pleistocene ice sheets which left surfaces of exposed bedrock; soils are mainly developed in the hollows and ravines. Inside the coastal fringe the mainland rises abruptly to the high summits and ice fields of the Andean Cordillera. Inland, where not covered by ice fields, the topography consists of glacially-oversteepened slopes, talus slopes, and unconsolidated colluvial and fluvioglacial deposits in the valley bottoms.

The climatic characteristics of the zone south of 43°20′S are summarized in Table 2.7 and Fig. 2.4 which are based on data from the few meteorological stations in this region. The west coast maritime climate gradually becomes cooler southwards with a more uniform precipitation regime. South of approximately 47°10′S a cold-temperate climate of great humidity prevails (Fuenzalida, 1965) with the diurnal and seasonal temperature ranges much reduced. Cyclonic storms moving in from the Pacific and the orographic effects of the mountainous islands produce relatively high annual rainfall (but less than in much of the Valdivian rain forest region) which is uniformly distributed throughout the year. The constantly humid conditions of the west coast contrast dramatically with the semi-arid steppe climate a few kilometres to the east on the leeward side of the Andean Cordillera. While the total precipitation over much of the coastal region is not exceptionally large, the lack of a summer dry season and the constantly low temperatures (and therefore reduced evaporation) produce very moist conditions. Precipitation may fall as snow at any season, but at sea level snow is neither deep nor long-lasting. Gale-strength winds are common throughout the year.

Forest types

To the south of the Valdivian rain forest region the vegetation becomes less rich in species due to decreasing warmth. This transition towards species-poor forests at higher latitudes is a gradual process and the southern limits of most species are only roughly known. Although Hauman (1913) and Skottsberg (1916) considered the Valdivian rain forest to extend as far south as 46° and 48°S, respectively, this seems inconsistent with the disappearance at much more northerly latitudes of such typically Valdivian species as *Aextoxicon punctatum*, *Dasyphyllum diacanthoides*, *Eucryphia cordifolia*, *Laurelia sempervirens* and *Nothofagus alpina*. Thus Schmithüsen (1956) recognized a north Patagonian rain forest in the coastal area from 43°20′ to 47°30′S which in terms of tree dominants is reminiscent of the forests described for the middle elevations of the Valdivian rain forest region. Despite the difficulty of precisely delimiting a boundary separating the northern mid-elevation forests from the southern coastal forests, important floristic and structural differences preclude the

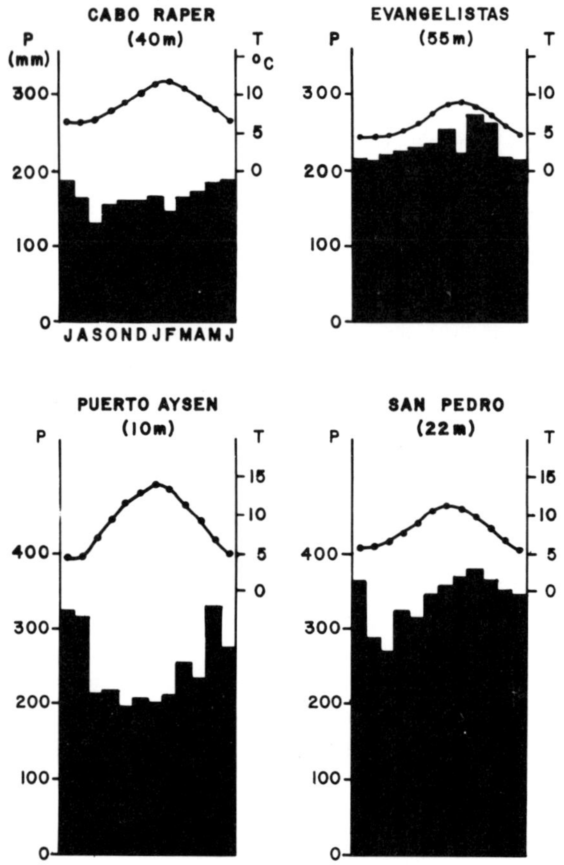

Fig. 2.4. Climatograms of selected stations from the north Patagonian and Magellanic rain forest regions. Key: Same as Fig. 2.2. Stations: Puerto Aysén (45°24′S, 72°42′W), AT = 9, AP = 2973; Cabo Raper (46°50′S, 73°35′W), AT = 8.6, AP = 1979; San Pedro (47°43′S, 74°55′W), AT = 8.2, AP = 4076; and Evangelistas Islands (52°24′S, 75°06′W), AT = 6.4, AP = 2773. Altitude above sea level is given below the name of each station. (Source: Fuenzalida, 1966.)

classification of the southern coastal forests as part of the forest types previously described for the Valdivian rain forest region. Schmithüsen (1956) sets the northern limit of the much impoverished Magellanic rain forest at 47°30′S while Skottsberg (1916) and Godley (1960) have placed the boundary at 48°S.

Extending south of 55° to within 800 km of the northern tip of the Antarctic Peninsula, the Magellanic rain forest is the southernmost forest formation of the world (Young, 1972). Due to the inhospitable climate and difficult terrain of the coastal region south of 43°20′S the vegetation is inadequately known and forest types must be very broadly defined. The following discussion is largely a synthesis of descriptions by: Dusén (1903–1906), Reiche (1907), Skottsberg (1910, 1916), Godley (1960), Heusser (1960), Holdgate (1961), Pisano (1972, 1973) and Young (1972). Recognizing the importance of non-forest vegetation and that the rain forest is characteristically confined to a narrow coastal fringe or to sheltered gullies, the term "Magellanic moorland" may best describe the western zone of the Magellanic region (see Pisano, 1983). Godley (1960) has noted that the moorland zone, which refers to bare rock, bog, or sparse covers of cushion and tussock plants, corresponds closely to the distribution of Andean diorite. This poor substratum and the constant wind of the exposed locations are probably the most critical factors in restricting tree growth.

Lowland coastal forest on adequately drained sites (Table 2.8)

Evergreen *Nothofagus–Weinmannia* Type 13. This type is similar to Type 9 described for Chiloé Island and the Coastal Cordillera and is approximately equivalent to Schmithüsen's (1956) north Patagonian rain forests, but excludes the forests of mid-elevations in the Valdivian rain forest region. An important difference from the northern distribution of Schmithüsen's north Patagonian rain forest is that mixtures of the three evergreen southern beeches — *Nothofagus dombeyi*, *N. nitida*, and *N. betuloides* — occur in the south. This type extends as far south as approximately 48°S but occurs only along the narrow coastal fringe. At 45°23′S it is dominated by *N. dombeyi* with a matrix of subdominant *Weinmannia trichosperma*, *Saxegothaea conspicua*, *Podocarpus nubigenus*, and *Drimys winteri* (Reiche, 1907). At 46°40′S, *Saxegothaea conspicua* is no longer present and there is a mixture of the three species of *Nothofagus* of about 25 m in height with *Weinmannia trichosperma* nearly as tall. *Caldcluvia paniculata*, *Drimys winteri* and *Podocarpus nubigenus* are also important, while *Laurelia philippiana* is of reduced size and abundance (Heusser, 1960; Muñoz, 1960). Throughout this type *Nothofagus nitida* is more dominant and *Podocarpus nubigenus* more abundant on soils with a tendency towards waterlogging. *Chusquea quila* continues to be abundant in the understorey at least as far

TABLE 2.8

Lowland coastal forests on adequately drained sites

Type 13 Evergreen *Nothofagus–Weinmannia*	Type 14 *Nothofagus betuloides*
Emergent tree stratum	
Nothofagus dombeyi *N. nitida* *N. betuloides*	*Nothofagus betuloides*
Matrix of dominant and subdominant trees	
Weinmannia trichosperma *Drimys winteri* *Laurelia philippiana* *Saxegothaea conspicua* *Podocarpus nubigenus*	*Drimys winteri* *Podocarpus nubigenus*
Tall shrub–small tree stratum	
Caldcluvia paniculata *Maytenus magellanica* *Lomatia ferruginea* *Pseudopanax laetevirens* *Chusquea quila* *Embothrium coccineum* Myrtaceae (in north only)	*Maytenus magellanica* *Embothrium coccineum* *Pseudopanax laetevirens*
Small shrub stratum	
Desfontainia spinosa *Pernettya mucronata* *Blechnum chilense* *B. magellanicum*	*Pernettya mucronata* *Berberis ilicifolia* *Blechnum magellanicum* *Lebetanthus myrsinites* *Desfontainia spinosa*
Lianas and epiphytes	
Hymenophyllum spp. *Asteranthera ovata* *Mitraria coccinea* *Philesia magellanica* *Griselinia racemosa* *Campsidium valdivianum*	*Hymenophyllum* spp. *Polypodium billardlieri* var. *magellanicum* *Philesia magellanica* *Serpyllopsis caespitosa*

south as 46°40′S. In the area of Río Baker *Nothofagus betuloides* predominates in the wet coastal region, and 60 to 70 km inland as conditions become drier *N. dombeyi* prevails (Reiche, 1907); *N. dombeyi* has not been reported further south than 48°S. Typical of this forest type are the tree-like ferns *Blechnum chilense* and *B. magellanicum*, and very abundant epiphytic hymenophyllaceous ferns and woody vines. The parasitic *Myzodendron punctulatum* is also common on *Nothofagus* branches.

Nothofagus betuloides **Type 14.** Along the coasts, on low-lying islands, and extending inland on slopes favoured by high precipitation from *c.* 48° to 55°S, the *N. betuloides* type grows on the better drained sites. The soils are generally characterized by a surface layer up to 20 cm thick of partially decomposed organic matter overlying a bleached, podzolic A horizon (Pisano, 1972; Young, 1972). Derived from recent fluvioglacial deposits, the soils are typically shallow. The evergreen *N. betuloides* forests are associated with soils undergoing podzolization in contrast to the surrounding bogs and heath vegetation where gleization prevails. *N. betuloides* is tolerant of moderately poor drainage and can grow on soils with peat layers up to 2 m deep. *Podocarpus nubigenus* is locally abundant along the forest borders as far south as 51°S, especially in areas of poor drainage. The shrub layer is poor in species but dense. Epiphytic ferns and mosses are very abundant particularly on the slowly decaying fallen trees. As far south as 53°30′S, the small tree *Pseudopanax laetevirens*, which also grows as a liana, is found only at the lowest altitudes along the coast where minimum temperatures are relatively high (Pisano, 1973). The same tree is an important component of the Valdivian rain forests at latitudes north of 40°S. This type extends eastward towards the semi-arid steppe as gallery forests along stream courses, while epiphytes diminish rapidly in abundance and shrubs characteristic of the steppe appear in the understorey (Pisano, 1973).

The very reduced climatic seasonality of the evergreen Magellanic rain forest region distinguishes it from the rest of southern South America. Near the coast there is no period of complete dormancy, and some plants are in flower at all times of the year (Young, 1972). However, under the constantly low temperatures tree growth is generally slow.

Forests of poorly drained sites

Tepualia–Pilgerodendron **Type 5.** The previously described Type 5 can be considered a northern outlier of the *Tepualia–Pilgerodendron* type which is widely distributed south of 43°20′S at low altitudes on poorly drained sites. It is transitional between the *Nothofagus betuloides* forests and the treeless bogs. The southern distribution of this type differs slightly in species composition from its northern equivalent; in the south the small shrubs present include *Berberis ilicifolia*, *Desfontainia spinosa*, and *Empetrum rubrum*. The dominant trees are rarely more than 5 m tall.

Non-coastal forests (Table 2.9)

***Subalpine Nothofagus betuloides* Type 15.** This elfin type, which is very similar to Type 12 (Table 2.6), forms the timberline only on the wettest slopes and borders the western sides of the Patagonian ice fields on steep slopes. It is found on thin rocky soils. The multi-stemmed trees are low, often less than 5 m and always less than 10 m tall, and are greatly deformed by the weight of the snow on steep slopes. Lianas and vascular epiphytes are generally lacking, the shrub layer is very reduced in stature and the relative poverty of species differentiates it from Type 12. The subalpine *N. betuloides* forests are transitional to the scrublands at higher altitudes in which *N. betuloides* occurs as prostrate individuals scattered among the shrubs *Chiliotrichium diffusum* and *Empetrum rubrum*.

TABLE 2.9

Non-coastal north Patagonian and Magellanic forest types

Type 15 Subalpine *Nothofagus betuloides*	Type 16 *Nothofagus betuloides–N. pumilio*
Emergent tree stratum	
Nothofagus betuloides	*Nothofagus betuloides* *N. pumilio* *Drimys winteri*
Tall shrub–small tree stratum	
	Maytenus magellanica *Embothrium coccineum*
Small shrub stratum	
Empetrum rubrum *Chiliotrichium diffusum* *Maytenus disticha* *Desfontainia spinosa*	*Berberis ilicifolia* *Lebetanthus myrsinites* *Maytenus disticha*
Lianas and epiphytes	
	Hymenophyllum spp.

***Nothofagus betuloides–N. pumilio* Type 16.** This type, which is similar to Type 11, occurs at higher altitudes on the drier slopes and towards the eastern drier zone. It is transitional between the purely evergreen subantarctic forest (Type 14) and the deciduous *N. pumilio* forests which form the timberline on all but the wettest sites. These mixed evergreen–deciduous forests are characteristically found on brown podzolic soils showing less intense and less complete signs of podzolization than the soils of the pure evergreen forests to the west (Pisano, 1973). *N. betuloides* attains heights of 20 to 25 m and *N. pumilio* a few metres less, although this relationship is reversed towards the east. Characteristically *Drimys winteri* is of very minor importance in this type. The hymenophyllaceous fern epiphytes are rarely very abundant, although herbaceous ground-rooted plants such as *Osmorrhiza obtusa*, *Ranunculus minutiflorus*, *Rubus geoides*, *Valeriana lapathifolia*, *Viola maculata*, and *V. magellanica* are often abundant.

FOREST DYNAMICS

Forest distribution and climatic change

The boundary between the forests of southern Chile and the Patagonian steppe [described elsewhere in this series (Soriano, 1982)] may be slowly shifting westwards as a consequence of a long-term trend towards a drier climate in this region. Kalela (1941) argued that on the eastern side of the Andes at the latitudes of the Valdivian rain forest, the forest/steppe boundary has been retreating to the west for the past several centuries. He inferred this westward retreat of the forest vegetation from observations of the structure and vigour of stands located near the forest/steppe ecotone. Islands of *Nothofagus antarctica* surrounded by steppe vegetation appear not to be regenerating, and to the west stands of *N. alpina* show signs of decay in the form of dead tree tops. Similarly, Kalela reported that stands of *N. dombeyi* on the eastern side of the

Andes are being invaded by the more drought-tolerant *N. obliqua*. However, an alternative explanation to recent climatic change may be tree seedling destruction by livestock. For example, during winter on the western side of the Andes at these latitudes cattle are often sheltered in the *Nothofagus* forests, where the browsing damage is more severe on the evergreen *N. dombeyi* seedlings than on the deciduous *N. obliqua* seedlings. The westward retreat of the forests before the advancing Patagonian steppe appears to have been caused, at least partially, by livestock browsing as well as the human use of fire in recent centuries (Eriksen, 1975).

In the latitudes of the north Patagonian and Magellanic rain forests, palynological studies by Auer (1958, 1960) suggest an increase in the extent of the steppe vegetation beginning *c.* 2200 years B.P. This seems to indicate a long-term drying trend. Based on palynological analyses of cores taken at *c.* 41°S in the Valdivian rain forest region, Heusser (1966, 1974) suggested post-glacial climatic fluctuations have resulted in considerable displacement of the existing forest types. Furthermore, Skottsberg (Heusser, 1960) described a relict plant community on a north-facing exposure on the south side of Skyring Sound (52°40′S). This contained *Gunnera chilensis* and *Adiantum chilense*, species normally not found south of Ultima Esperanza Sound (51°30′S) and Chiloé Island, respectively. Plants of these species may have migrated to Skyring Sound during a period of moister climate and were subsequently isolated by the effects of increasing aridity. Mercer (1970) has demonstrated a series of advances and recessions of the glaciers of Patagonia related to climatic fluctuations during the past several hundred years. Pisano (1975) suggested that these fluctuations and particularly a long-term but erratic trend towards greater aridity are reflected in the vegetation pattern of the Patagonian and Magellanic regions. As evidence, he cited: (1) the establishment of trees such as *Nothofagus betuloides*, *N. antarctica*, and *Pilgerodendron uvifera* in herbaceous tundra and bog communities; (2) the invasion of *Sphagnum* bogs by species characteristic of less humid soil conditions, such as *Empetrum rubrum* and *Pernettya mucronata*; and (3) the unstable ecotone between the *Nothofagus pumilio* forest and the Patagonian steppe vegetation. Much of this purported evidence for recent directional climatic change may reflect improvement of drainage due to natural erosion, the effects of sheep and cattle, and the influence of fire.

While fluctuations in the climate of Patagonia over the past few millenia are clearly indicated by the palynological evidence and the fluctuations of glaciers, a climate-caused contemporary retreat of the western Patagonian forests has yet to be demonstrated conclusively. Pisano (1975) suggested that post-Pleistocene climatic fluctuations in the Magellanic region have resulted in a species-poor biota which renders these ecosystems vulnerable to sometimes irreversible deterioration processes triggered by human actions. The very poor regeneration of these forests and exceedingly slow tree growth following felling or burning is probably accounted for by the low temperatures and the almost constant, strong winds. Given these conditions, the forested ecosystems of the Patagonian and Magellanic regions must be considered to be very fragile in the light of increasing forest destruction by man.

Role of *Chusquea* in the Valdivian rain forest region

The large bamboos of the mainly neotropical genus *Chusquea* play a very important role in the dynamics of forest regeneration and succession throughout the Valdivian rain forest region (Müller-Using, 1973; Veblen et al., 1977a). The five common species of *Chusquea* in the region (Table 2.10) occur from sea level to above the timberline and in virtually all forest types. While *Chusquea* is present even in virgin forests, it often becomes the dominant cover following disturbance by fire or logging. At such disturbed sites at lower altitudes *C. quila* and *C. uliginosa* (in well-drained and periodically water-logged sites, respectively) form impenetrable thickets 5 m tall often extending over several hectares. Where some supporting trees remain *C. quila* may become 20 m or more tall. At mid-elevations (*c.* 500–900 m) where forests are disturbed, the tall *C. culeou* forms dense stands with over 100 000 culms ha^{-1}. Over areas up to hundreds of square kilometres most of the population of a given *Chusquea* species will flower synchronously, produce seed, and die within a two to three year period. Such mass reproduction occurs with a frequency of eighteen years or more (Gunckel, 1948). Until the *Chusquea* flowers and

TABLE 2.10

Principal species of *Chusquea* in the Valdivian rain forest region

Species	Growth-habit	Approximate altitudinal range at 40°S	Habitat
C. quila	climbing, multi-branched culms often exceeding 20 m in length	sea level to 600 m	common in undisturbed forests on well-drained sites; dominant in secondary vegetation
C. uliginosa	climbing, multi-branched culms often exceeding 8 m in length	sea level to 350 m	typical of poorly drained sites
C. culeou	erect, separate, unbranched culms often exceeding 7 m in height	500 to 900 m	abundant in undisturbed forests on well-drained sites; forms dense thickets where the forest is disturbed
C. tenuiflora	erect, separate culms usually 2 to 3 m in height	900 to 1200 m	typically dominates the understoreys of the mixed evergreen–deciduous forests just below timberline
C. nigricans	erect, separate, short culms rarely exceeding 1.5 m in height	900 to 1300 m	typical of poorly drained basins and frost pockets; sometimes found in scrublands above timberline

dies, it tends to impede or at least greatly reduce tree regeneration. Only regeneration of the most tolerant species, such as *Aextoxicon punctatum* and *Laurelia philippiana*, is possible under the shade produced by a dense cover of *Chusquea*. Thus, tree regeneration is often precluded for lengthy intervals following logging or burning. Forest management schemes which depend on natural tree regeneration are greatly complicated by the proliferation of *Chusquea* following silvicultural interventions. Vigorous resprouting of *Chusquea* following the cutting of culms makes mechanical control very difficult, and the great extent of *Chusquea* makes chemical control costly both in financial terms and in the undesirable effects of massive chemical applications.

The sudden production of huge masses of seed means there is more seed than the existing population of seed-eaters can cope with, so that ample seed survives (Evans, 1976; Janzen, 1976). Following the gregarious flowering of *Chusquea*, local rodent populations explode, and, as the readily available *Chusquea* seed supply is exhausted, they become serious pests in cultivated fields. Serious rodent plagues are reported in association with the gregarious *Chusquea* flowering during the eighteenth and nineteenth centuries in the Valdivian rain forest region (Urban, 1934; Gunckel, 1948) and one occurred in 1975–77 on the outskirts of Valdivia. The severity of plagues has probably been exacerbated by the increase in the abundance of *Chusquea* as a consequence of forest clearing and burning. Population reduction of foxes, raptors and other predators by hunting and habitat destruction has also probably contributed to the rodent problem triggered by the flowering of *Chusquea*.

Catastrophism and *Nothofagus*

Virgin old-growth stands dominated by *Nothofagus obliqua* at low altitudes (Type 1) and the *N. dombeyi–N. alpina*-dominated forests of the middle elevations (Types 6, 7 and 8) have a similar structure (Veblen et al., 1979; Veblen et al., 1980). These forests lack *Nothofagus* individuals in the smaller size classes. Characteristically, trees of *Nothofagus* less than 80 cm in dbh are absent, indicating a total lack of regeneration of the emergent trees in undisturbed stands. The transitional zone towards the higher altitude and more open *N. pumilio* forests is exceptional, for there is some regeneration of *N. alpina* in undisturbed stands. This lack of regeneration at middle and low elevations seems to be due to the intolerance of these *Nothofagus* species to shade. The seed bed seems adequate because seedlings up to a size of c. 1 m are present. The presence of numerous individuals in all size classes from seedlings to trees over 1 m in dbh of *Aextoxicon punctatum*, *Laurelia sempervirens* and *Persea lingue* at low elevations and of *Dasyphyllum diacanthoides*, *Laurelia philippiana* and *Saxegothaea conspicua* at middle elevations suggest that these shade-tolerant species are gradually replacing the emergent *Nothofagus*.

Forests in which the dominants are not replacing themselves have been described for many areas in the Northern Hemisphere and are usually ascribed to climatic change, fire suppression, or changes in the herbivore populations (Jones, 1945; Spurr and Barnes, 1973). In the Valdivian rain forest region the wide distribution and the abundant regeneration of *Nothofagus obliqua*, *N. dombeyi*, and *N. alpina* in highly disturbed areas show that these species are well adapted to the present climate. Native herbivores capable of destroying tree regeneration are very few and fires have greatly increased as a result of the presence of people.

The 1960 catastrophic earthquake which devastated much of southern Chile provided an important clue to understanding the dynamics of the low- and mid-elevation *Nothofagus* forests in the Valdivian rain forest region.

Associated with the 1960 earthquake thousands of debris avalanches, landslides, and mudflows occurred in the Andes from 39°S to at least as far south as 42°S (Saint-Amand, 1962). In the Andes of the Province of Valdivia (39°25′–40°40′S) analysis of aerial photographs taken less than a year following the 1960 disaster showed that at least 2.8% of the entire surface area had slid (Fig. 2.5). The glacially steepened slopes of the Andes with a mantle of one to several metres of volcanic ash are particularly prone to mass movement. This volcanic ash mantle consists of successive layers of very porous pumiceous lapilli (gravel-size ash) separated by fine andesitic ash weathered to a stage high in allophane (Wright and Mella, 1963). Allophane is an amorphous hydrated alumino-silicate gel characteristic of young, well-drained soils derived from volcanic ash (Brady, 1974). It can absorb a large amount of water which is subsequently expressed when shaken under pressure, producing a thixotropic phenomenon resembling liquefaction. This noncrystalline allophane-rich layer when moist serves as a lubricant for the movement of the pumiceous lapilli and the overlying soil and vegetation, and under certain conditions whole hillsides may collapse. The frequency and extent of mass movements in the Andes was greatest along the Reloncaví fault zone extending from Volcán Villarrica (39°25′S) through Estero de Reloncaví and south of Volcán Michinmávida (42°50′S), where the shaking in 1960 must have been very intense. Mass movements were much less abundant in the Coastal Cordillera where volcanic deposits are less frequent, thinner, and of finer material forming less porous layers.

Fig. 2.5. Landslides at the eastern end of Lake Pellaifa in the Andes of the Province of Valdivia. Note the inundation of the forest which occurred when landslide debris blocked the egress of the lake and forced the water level to rise.

On bare areas produced by these slides, *Nothofagus obliqua* and *Eucryphia cordifolia* are the common colonizers at low elevations and *Nothofagus dombeyi* at mid-elevations (Veblen and Ashton, 1978). Although not as abundant over the entire region, the much exploited *N. alpina* also commonly colonizes the bare slide surfaces where a few seed trees remain. The same species which dominate the surrounding old-growth forests are important colonizers of primary bare areas.

The historical record indicates that catastrophic mass movements triggered by earthquakes similar in magnitude to that of 1960 have affected the Valdivian rain forest region several times during the past five centuries (Fonck, 1896; Guarda, 1953; Weischet, 1960). Further support for this is provided by the presence of young (60–100 years old), even-aged, single-species stands of *N. dombeyi* on older slide surfaces. They are also common in valley bottoms where floods, resulting from landslide debris blocking channels, have destroyed the original vegetation. Consequently the widespread *Nothofagus*-dominated forests of shade-tolerant subdominant trees of the low and middle altitudes of the western slopes of the Andes from *c.* 39° to 41°30′S appear to be relatively early successional phases.

Catastrophic volcanism also partially accounts for the forest structure. In the Andes of the Valdivian rain forest region, numerous active or recently active volcanoes have explosively erupted many times during historic times, significantly affecting the regional vegetation (Casertano, 1963; Veblen et al., 1977b). Ash showers from the more explosive eruptions similar to those which deposited the regionally extensive pumiceous lapilli and tuff deposits would have destroyed the existing vegetation and created conditions favourable for the establishment of *Nothofagus*. The amount of ash fall from even relatively mild eruptions such as the December 31, 1971 eruption of Volcán Villarrica is remarkable. In an experimental plot located in a *Nothofagus dombeyi–N. alpina–Laurelia philippiana–Saxegothaea conspicua* forest 25 km to the southeast of Volcán Villarrica, measurements from 96 collection boxes indicated 13.62 ± 0.06 t ha^{-1} of sand- to small gravel-sized ash fell during the few days of volcanic activity. Furthermore, fires associated with volcanism or lit by the aboriginal population may have contributed to the forest structure described previously.

The *Nothofagus*-dominated forests (below the altitude of *N. pumilio* at *c.* 1100 m) which account for the major portion of the plant cover of the Andes of the Valdivian rain forest region appear to be slowly undergoing directional changes in terms of species dominance and floristic composition. However, the frequency of catastrophic phenomena and the slowness of succession involving the replacement of the long-lived (over 400 years old) intolerant tree species prevents the development over large areas in the Andes of a vegetation without *Nothofagus* and consisting only of shade-tolerant species. This interpretation of the dynamics of *Nothofagus* forests under periodic catastrophic influences is presently restricted to the mid-elevations of the western side of the Andes from *c.* 39° to 41°30′S. Elsewhere, such as at mid to high elevations on Chiloé Island, *Nothofagus* may regenerate following treefalls in forests of *N. nitida–Saxegothaea–Podocarpus nubigenus–Weinmannia* (Godley, 1960). This contrasts with the situation in the *Nothofagus dombeyi* forests of the Andes where canopy gaps created by treefalls are rapidly occupied by abundant saplings of *Laurelia philippiana*.

The two non-*Nothofagus* forest types (Types 2 and 3) are of special interest. Except for restricted occurrences of Type 2 in the foothills of the Andes in the southern part of the Valdivian rain forest region, both types are limited to Chiloé Island and the western slopes of the Coastal Cordillera (Fig. 2.6). The extent of these two types on Chilóe Island is poorly known, while the structure and composition of the non-*Nothofagus* vegetation at *c.* 40°S on the western side of the Coastal Cordillera is reliably known (Philippi, 1865; Ardiles and Maldonado, 1977). There, *N. nitida* is present from approximately 600 to 850 m, below which *Nothofagus* species are generally absent except for a few *N. obliqua* and *N. dombeyi* near stream courses below 150 m. The vegetation below 600 m is dominated by the same shade-tolerant trees which are the subdominants of the *Nothofagus* forests of the low and middle elevations of the Andes at this latitude. The Coastal Cordillera has been much less affected by volcanism and mass movements than the Andes. While the absence of *Nothofagus* may be due to catastrophic phenomena being of lesser importance in the Coastal Cordillera, the milder temperature ranges and probably greater humidity

Fig. 2.6. The non-*Nothofagus* Valdivian rain forest at *c.* 40°S on the western slopes of the Coastal Cordillera. The gray area on the summit is *Fitzroya cupressoides* forest which has been burned.

characteristic of these Pacific slopes may also be significant.

FOREST USE AND MANAGEMENT

Past use

The aboriginal population of southern Chile made relatively little direct use of the forests, only utilizing small quantities of wood in the construction of shelters and canoes and for spears and other armaments. The population of the Valdivian rain forest region lived mainly along the edges of forest and water bodies in small impermanent settlements. Hunting, fishing, and collecting in the riparian habitats were far more important than the rudimentary agriculture they practiced. Aboriginal impact on the vegetation was principally through the use of fire. The Mapuches (or Araucanians), famous for their warlike habits, set fire to the vegetation as both a defensive and aggressive tactic, thus altering considerable areas of forest.

The Spaniards began their conquest of southern Chile in 1552 but fierce resistance by the Mapuches forced them to retreat to the area north of the Río Bío-Bío in 1599 (Guarda, 1953). For the next two centuries, the Mapuches successfully resisted Spanish attempts to conquer the territory from 37° to 40°S. Further south, in the territory of the more tractable Huilliches, the Spaniards were able to maintain permanent forts at Valdivia, Calbuco (near present-day Puerto Montt), and on Chiloé Island for defence against the Dutch and English corsairs. During the colonial period, the Spaniards maintained only a weak sovereignty over the interior of the southern Lake District. The area between 37° and 40°S became known as the Frontera and remained under Mapuche control until 1888.

The exploitation of the Valdivian rain forests south of 40° began with the establishment of Spanish forts. Wood was used in house construction and particularly in the shipyards of seventeenth century southern Chile. Chilean woods, most notably *Fitzroya cupressoides*, were the basis of the colonial ship-building industry along the entire Pacific coast of South America from Lima, Peru, to the south. The massive clearing and burning of the forests of southern Chile began in the mid-nineteenth century when colonists, primarily of German origin, began settling in the interior of the southern Lake District (Pérez, 1958). Conversion of the forested areas to agricultural use was the principal motive for the burning, but much more forest land than could be used for agriculture was destroyed. With the settlement and agricultural development of southern Chile, there was a concomitant growth in the use of forest products such as firewood, timber for construction, and bark for tannins. The timber of the Valdivian rain forests was of little commercial importance until 1912 when the railroad was completed to link the south with the populous markets of Central Chile. The present-day exploitation pattern began with the completion of trunk lines into the Andes in 1930.

Settlement and forest exploitation of the north Patagonian rain forest region is largely a twentieth century phenomenon. For strategic, as well as economic, reasons the government encouraged the colonization of this region by concessions of national lands to colonists (Saelzer, 1973). Since 1930 there has been extensive forest clearing in the Patagonian rain forest region and especially in the adjacent deciduous *Nothofagus pumilio* zone. The forest destruction and establishment of very extensive sheep and cattle raising industries have resulted in severe erosion and sedimentation. During this period, the forests of the island archipelago, notably on the Guaitecas Islands, have been burned to facilitate the exploitation of the highly prized conifer *Pilgerodendron uvifera.* Settlement of southernmost Chile, the area of the Magellanic rain forest, is very sparse and recent. The difficult access and the continual inclement weather have discouraged agriculture and livestock raising and, consequently, much of the Magellanic rain forest remains intact.

Present use

Public ownership of forest lands includes both Forest Reserves and National Parks. Although the area of publicly owned forest land is large—nearly 11 million hectares—the accessible areas of greatest commercial value are largely privately owned (Thelen and Miller, 1976). Over half of the forest land in the north Patagonian and Magellanic rain forest regions is state-owned. In this southern area, timber production is a minor activity, and vast tracts of ice and treeless lands are included in the Forest Reserves and National Parks. State-owned forest land accounts for a very minor proportion of forest land in the Valdivian rain forest region where commercial forestry is much more important.

In general, the native forest of southern Chile has not been subject to forest management practices. Used primarily for saw wood and railroad ties, the more valuable species and individuals of best form have been selectively cut and no measures have been taken to assure adequate regeneration of the desirable species. While fires have destroyed much more forest, selective cutting has seriously degraded the remaining forest resource. Virgin stands containing the most valuable species, such as *Nothofagus alpina*, are now found only in the most remote and inaccessible areas.

In the Valdivian rain forest region slow and labor-intensive tree-felling using the axe and hand-saw was normal until just a few years ago, but increasingly the chain saw is being used. Felling and the cutting of logs to 3.6 m in length take place during winter, and extraction begins during the spring as the weather improves. Oxen are still the most important means of extracting the logs, but a few larger enterprises also use tractors. Saw wood production occurs primarily in small, temporary sawmills which are frequently moved to minimize the transport distance from the exploitation site to where the wood is processed. Processing is by large circular saws powered in most cases by steam engines using the wood residues from the cutting process for fuel. Only a few larger enterprises use electric or diesel motors. Under the crude processing methods utilized, no more than 30% of the volume of the wood cut is utilized in the final product. Transportation of the saw wood to market is by rail and increasingly by truck. Approximately 60% of the saw wood produced from native species, particularly the better quality timber, is exported to Argentina.

The presence in the Valdivian rain forest region of some 400 000 ha of secondary forests (*renovales*) principally of *Nothofagus alpina* and *N. obliqua* represents an important economic potential (De Camino et al., 1974). These discontinuous and heterogeneous secondary forests are derived from both coppicing and seed regeneration following exploitation by means of small clear-cuts or group selection which leave a drastically opened-up forest canopy. Rarely do these stands extend over more than a few hectares. Due to the value of the wood of these two species of *Nothofagus* and to their relatively rapid growth, considerable interest exists to manage these stands. The Chilean Forest Service is currently applying such techniques as thinnings and enrichment planting to secondary forests of this kind.

Plantations of exotic species

At the beginning of the present century, Monterey pine (*Pinus radiata*) was introduced to Chile. Native to central coastal California and a few islands off the coast of northwestern Mexico and California, this species adapted very well in Chile. While *Eucalyptus* spp. and *Pseudotsuga menziesii* are also economically important, plantation forestry in Chile is overwhelmingly dominated by *Pinus radiata*. Plantations of Monterey pine are found in the coastal area from 33° to 40°S, although the greatest concentrations are near 37°S.

The very rapid growth of *P. radiata* has spurred the planting of large areas of this species; the area planted increased from 260 685 ha in 1965 to approximately 450 000 ha in 1976 (Instituto Forestal, 1966b). The extent to which *P. radiata* has replaced native species as the raw material of the forestry industry is evident from Table 2.11. Despite the overwhelming importance of Monterey pine for pulp and saw wood, native species continue to be more important for use as railroad ties and fence posts and for the manufacture of plywood and veneer. The major increases in consumption of *P. radiata* for saw wood and pulp from 1963 to 1974 reflect the growth of a modern forestry industry which has made Chile one of the leading exporters of forest products in Latin America.

CONSERVATION STATUS

The protective functions of the forest

Given that southern Chile is one of the rainiest regions of the world and that it is characterized by extremely rugged topography, the soil and watershed protection afforded by the Chilean rain forests is of paramount importance. Successful, long-term realization of several proposed large-scale hydroelectric projects—from Lake Pirihueico to Río Baker—will require recognition of the protective function of the surrounding forests. The great susceptibility of the Andes of southern Chile to catastrophic mass movements of soil and the importance of the forest cover in diminishing the frequency and extent of mass movements should be taken into account in all land-use planning in the Andean zone.

TABLE 2.11

Consumption of raw material by the Chilean forestry industry: changes from 1963 to 1974 (volume in 1000 m^3; solids without bark) (Sources: Instituto Forestal, 1968; and FAO/SIDA, 1976)

Product	*Pinus radiata*		Native species	
	1963	1974	1963	1974
Saw wood	477.2	2300.0	1152.4	600.0
Pulp	641.4	1888.5	–	–
Fiber boards	32.0	72.5	–	–
Particle boards	19.8	28.4	–	5.2
Plywood	–	–	17.3	22.4
Veneer	–	–	2.7	2.7
Railroad ties	–	–	43.2	156.0
Fence posts	–	5.5	4.1	114.5
Total	1170.4	4294.9	1219.7	900.8

The provision of wildlife habitat is another important function of the Chilean rain forests. In contrast to the species diversity and variety of life forms characteristic of the rain forests, the mammalian fauna is inconspicuous, rarely being observed or heard in the virgin forests. A few birds are common in the Valdivian rain forests, the most important of which are the *chucao* (*Scelorchilus rubecula rubecula*) and *huet-huet* (*Pteroptochos tarnii*) on the forest floor and *carpintero negro* (*Campephilus magellanicus*) high above in the tree canopy.

As a consequence of habitat destruction and hunting, many native bird and mammal species in southern Chile are in danger of extinction. For example, the *huemul* (*Hippocamelus bisulcus*) —Chile's only large deer—formerly ranged from the latitude of Santiago to the Straits of Magellan, but today only an estimated 500 to 1500 survive in nearly inaccessible timberline areas of the southern Andes (Miller et al., 1973). The population of the small Chilean deer, the *pudú* (*Pudu pudu*), has been greatly reduced due to destruction of its habitat—the lowland Valdivian rain forest—and hunting and persecution by domestic dogs. Nevertheless, their population still probably numbers in the thousands. The endangered river otter (*Lutra provocax*) has been severely restricted to small remnant populations in the Andean streams of the Valdivian rain forest region. Similarly, it is likely that habitat destruction in the Valdivian rain forest region is driving to extinction the poorly known marsupials *Dromiciops* and *Rhyncholestes* (Taber et al., 1974). Much of the native avian fauna is in a similarly precarious status. Wildlife protection legislation, which includes total protection for fourteen mammal and twenty-eight bird species, dates from 1929–1931. However, the fines and other sanctions against violators are so mild as to be ineffective and are inadequately enforced (Von Furstenberg, 1969).

One of the more alarming developments, which is likely to affect adversely both the native fauna and vegetation of southern Chile, is the introduction and spread of the European red deer (*Cervus elaphus*). The red deer has been recently introduced at several points in the Valdivian rain forest region as hunting game to attract international sportsmen. Even though the inhibitory effects of red deer on forest regeneration in New Zealand are well known to the Chilean natural resource managers, the continued release of red deer is being officially sanctioned. The justification given for this indifference is the presence of the Chilean puma (*Felis concolor*) which allegedly will control the red deer population. However, the puma is commonly considered to be a threat to man and his livestock, and as a consequence of hunting pressure only small vulnerable populations remain (Taber et al., 1974). Thus, red deer may soon join the European hare and English rabbit as an exotic species inflicting serious economic damage in the forestry and agricultural sectors of the Chilean economy.

National Parks

The Chilean National Park system includes 50 parks, 35 of which are located in the region of the rain forests (Table 2.12). The National Park System was initiated in 1926, primarily out of an interest to preserve areas of attraction for tourism, which is reflected in the relatively early establishment of parks in the well-populated Valdivian rain forest region. In the less populated north Patagonian and Magellanic rain forest regions, where preservation objectives have played a more important role, park establishment has taken place largely in the past two decades.

The large number of National Parks as well as the amount of designated surface area would imply preservation of a great diversity of virgin ecosystems. However, as is the case in many Latin American countries, the selection of an area as a National Park was based on inadequate knowledge of the features of the ecosystems involved and lacked overall planning. Consequently, many types of ecosystems are unprotected whereas others are represented in several parks. Considering the scarcity of the financial resources and trained personnel required for proper park management, a recent study proposes a substantial reduction in the number of National Parks (Thelen and Miller, 1976). These authors recommend maintaining three National Parks in the Valdivian rain forest region,

TABLE 2.12

The National Parks of southern Chile (after Corp. Nacl. Forestal, 1973); parks marked with an asterisk (*) are recognized by the International Union for the Conservation of Nature and Natural Resources

Park	Surface (ha)	Year of establishment
Valdivian rain forest region[1]		
Ralco	no data	1972
Nahuelbuta*	5432	1939
Contulmo	82	1941
Tolhuaca*	3500	1935
Cerro Ñielol	81	1939
Los Paraguas–Conguillio*	58 000	1940
Huerquehue*	3900	1967
Villarrica*	13 780	1940
Pirihueico	no data	1945
Los Alerzales*	1230	1967
Barra del Río Bueno	619	1949
Puyehue*	117 000	1941
Vicente Pérez Rosales*	135 175	1926
North Patagonian rain forest region		
Lago Rosselot	12 390	1968
Lago Las Torres	15 280	1969
Isla Guamblin*	10 625	1967
Cinco Hermanas	227	1964
Río Simpson	41 160	1967
Dos Lagunas	180	1967
Puerto Chacabuco	221	1964
Quitralco	10 900	1967
Los Huemules	12 500	1967
Bahía Erasmo*	28 320	1967
Laguna San Rafael	1 350 123	1967
Magellanic rain forest region		
El Guayaneco*	30 428	1967
Bernardo O'Higgins*	1 761 000	1970
Torres del Paine*	161 635	1959
Monte Balmaceda*	7900	1966
Pali-Aike*	3000	1970
Los Pinguinos*	97	1966
Laguna de los Cisnes*	25	1966
Hernando de Magallanes*	800 000	1970
Alberto de Agostini*	800 000	1965
Cabo de Hornos*	63 093	1945

[1] Within each region, parks are listed in order from north to south.

one in the north Patagonian forest region, and three in the Magellanic rain forest region. They also suggest combining some of the adjacent National Parks and expansion or creation of new parks to include representative ecosystems not presently protected.

The reduction in the total number of National Parks and the concentration of scarce management resources are intended to alleviate some of the serious problems currently afflicting the park system. For example, many parks lack clearly defined boundaries, and consequently illegal agriculture, livestock raising, and forest exploitation continue to encroach on the ecosystems for which protection is intended. One of the most severe problems facing park managers is the inadequacy of budgets. In a developing country, investment in park management is widely perceived as a non-essential or unproductive use of scarce public funds. Considering the great economic importance of tourism in southern Chile, this is clearly a short-sighted and unjustified policy.

In addition to the lack of appreciation of the economic value of the Chilean National Parks, there is little public understanding or support of nature preservation. Within this context, the establishment of environmental education and nature interpretation programs within the National Parks is a critical need (Oltremari, 1975). On the positive side, Chilean authorities have taken decisive action to assure preservation of their two giant conifers *Araucaria araucana* and *Fitzroya cupressoides* by declaring them natural monuments. Despite the considerable economic importance of both *Araucaria araucana* and *Fitzroya cupressoides* for timber exports, their exploitation has been totally prohibited since 1975 and 1977, respectively. In the long run, their economic importance in attracting tourism, particularly that of *Araucaria araucana*, will probably exceed the revenues that would have been derived from the exploitation of the remaining stands.

CONCLUSIONS

While the imminent demise of the neotropical rain forests is widely perceived, the endangered status of the temperate rain forests of South America is little appreciated. Although large-scale destruction due to burning in southern Chile has abated, the remaining areas of virgin rain forest are of very restricted extent and not fully protected. To a certain extent, the availability of products from *Pinus radiata* plantations has relieved some of the pressures on the native forests. However, the risk of epidemic disease or pest attack associated with a monoculture planted on this scale cannot be ignored. Similarly, detrimental soil change after several rotations is a real threat. Presently, both disease and declining soil fertility are problems in the *P. radiata* plantations. Research in both of these areas is of crucial importance to the long-term productivity of plantation forestry.

Proposals to exploit large areas of native forest in southern Chile for pulp production are currently under consideration. Such proposals promise heavy investments in infrastructure, the creation of thousands of jobs, and significant export earnings. Consequently, it is likely that such projects will become realities in the near future. It is crucial that they be planned and executed within the context of multiple use of the forest resource. The rational management of those areas of native forest selected for exploitation will depend on an adequate knowledge of the ecological and silvical characteristics of Chilean trees. An understanding of the functioning of the forest ecosystems as well as of the watershed and habitat protection afforded by these forests is also essential. The need for basic and applied research in these areas is paramount.

ACKNOWLEDGEMENTS

For sharing with us their knowledge of the forests and forestry activities in Chile we are deeply indebted to F. Saelzer, B. Olivares, E. Morales, and B. Escobar. We are grateful to A. Veblen for preparing the figures. Many of our observations were made while we were engaged in fieldwork financed by the Facultad de Ingeniería Forestal of the Universidad Austral de Chile, the Facultad de Ciencias Forestales of the Universidad de Chile (Santiago), the American Geographical Society, the Corporación Nacional Forestal de Chile, and the Smithsonian–Peace Corps Environmental Program.

REFERENCES

Almeyda A., E. and Saez S., F., 1958. *Recopilación de Datos Climáticos de Chile y Mapas Sinópticos Respectivos.* Ministerio de Agricultura, Santiago, 195 pp.

Anonymous, 1964. *Climatología en Chile, Fascículo I, Valores Normales de 36 Estaciones Seleccionadas, Período 1916–1945.* Proyecto Hidrometeorológico, Naciones Unidas—Gobierno de Chile, Santiago, 51 pp.

Ardiles A., R. and Maldonado R., E., 1977. *Contribución al Estudio del Bosque Nativo Perennifolio de la Cordillera de la Costa, en Tres Niveles Altitudinales.* Thesis, Universidad Austral de Chile, Valdivia, 83 pp.

Auer, V., 1958. The Pleistocene of Fuego–Patagonia. Part II: The history of the flora and vegetation. *Ann. Acad. Sci. Fenn., III, Geol. Geogr.*, 50; 1–239.

Auer, V., 1960. The Quaternary history of Fuego–Patagonia. *Proc. R. Soc. Lond., Ser. B*, 152; 507–516.

Bernath, E.L., 1937. Coniferous forest trees of Chile. *Trop. Woods*, 52; 19–26.

Brady, N.C., 1974. *The Nature and Property of Soils.* MacMillan, New York, N.Y., 8th ed., 639 pp.

Brun, R., 1975. Estructura y potencialidad de distintos tipos de bosque nativo en el sur de Chile. *Bosque* (Valdivia, Chile), 1: 6–17.

Burschel, P., Gallegos, C., Martínez, O. and Moll, W., 1976. Composición y dinámica de un bosque virgen mixto de raulí y coigue. *Bosque* (Valdivia, Chile), 1: 55–74.£nm

Casertano, L., 1963. General characteristics of active Andean volcanoes and a summary of their activities during recent centuries. *Bull. Seismol. Soc. Am.*, 53: 1415–1433.

De Camino, R., Smith, B., Benavides, M., and Rodas, J., 1974. Los renovales de bosque nativo como recurso forestal: Antecedentes para la discusión del problema.In: *Situación Actual y Posibilidades Futuras del Manejo de los Renovales en Chile. Actas del Primer Seminario.* CONAF-FAO, Santiago, pp. 25–40.

Di Castri, F., 1968. Esquisse écologique du Chili. In: D. Deboutteville and E. Rapoport (Editors), *Biologie de l'Amérique Australe, IV. Documents Biogéographiques et Ecologiques.* C.N.R.S., Paris, pp. 6–52.

Dusén, P., 1903-1906. The vegetation of western Patagonia. *Rep. Princeton Univ. Exped. Patagonia, 1896–1899*, 8(Pt. 1): 1–34.

Eriksen, W., 1975. Disruptions in the ecosystems of the steppe and forest regions of Patagonia by climate and man. *Appl. Sci. Devel.*, 6: 127–142.

Eskuche, U., 1968. Fisionomía y sociología de los bosques de *Nothofagus dombeyi* en la región de Nahuel Huapi. *Vegetatio*, 16: 192–203.

Evans, G.C., 1976. A sack of uncut diamonds: the study of ecosystems and the future resources of mankind. *J. Ecol.*, 64: 1–39.

FAO/SIDA, 1976. *Seminario sobre Ocupación Forestal en América Latina.* Food and Agricultural Organization, Rome, 520 pp.

Fonck, F., 1896. *Viajes de Fray Francisco Menéndez a La Cordillera.* Niemeyer, Valparaiso, 528 pp.

Fuenzalida V., H., 1965. Biogeografia. In: *Geografía Económica de Chile.* Corporación de Fomento de Producción, Santiago, 2nd ed., pp. 228–267.

Fuenzalida V., H., 1966. Climatología. In: *Geografía Económica de Chile, Primer Apéndice.* Corporación de Fomento de Producción, Santiago, pp. 31–44.

Godley, E.J., 1960. The botany of southern Chile in relation to New Zealand and the subantarctic. *Proc. R. Soc. Lond., Ser. B*, 152: 457–475.

Godley, E.J., 1968. A plant list from the Cordillera de San Pedro, Chiloé. *Rev. Univ.* (*Univ. Católica Chile*), 53: 65–77.

Guarda G., F., 1953. *Historia de Valdivia 1552–1952.* Imprenta Cultural, Santiago, 360 pp.

Gunckel L., H., 1948. La floración de la quila y del colihue en la Araucania. *Cienc. Invest.* (*Buenos Aires*), 4: 91–95.

Hauman M., L., 1913. La florêt valdivienne et ses limites. *Rec. Inst. Bot. Léo Errera* (Bruxelles), IX: 346–408 (reprinted in *Inst. Bot. Farmocol., Univ. Buenos Aires*, 1916: 1–91).

Heusser, C.J., 1960. Late–Pleistocene environments of the Laguna de San Rafael area, Chile. *Geogr. Rev.*, 50: 555–577.

Heusser, C.J., 1966. Late-Pleistocene pollen diagrams from the province of Llanquihue, southern Chile. *Proc. Am. Philos. Soc.*, 110: 269–305.

Heusser, C.J., 1974. Vegetation and climate of the southern Chilean Lake District during and since the last interglaciation. *Quaternary Res.*, 4: 290–315.

Holdgate, M.W., 1961. Vegetation and soils in the south Chilean Islands. *J. Ecol.*, 49: 559–580.

Illies, H., 1970. *Geología de los Alrededores de Valdivia y Volcanismo y Tectónica en Márgenes del Pacífico en Chile Meridional.* Instituto de Geografía, Universidad Austral de Chile, Valdivia, 64 pp.

Instituto Forestal, 1966a. Classificación preliminar del bosque nativo de Chile. *Inf. Técn.* (*Santiago*), 27: 94 pp.

Instituto Forestal, 1966b. Inventario de las plantaciones forestales de la zona centro-sur de Chile. *Inf. Técn.* (*Santiago*), 24: 53 pp.

Instituto Forestal, 1968. Estadísticas forestales 1966–1967. *Inf. Técn.* (*Santiago*), 31: 29 pp.

Janzen, D.H., 1976. Why bamboos wait so long to flower. *Annu. Rev. Ecol. Syst.*, 7: 347–391.

Jones, E.W., 1945. The structure and reproduction of the virgin forests of the North Temperate Zone. *New Phytol.*, 44: 130–148.

Kalela, E.K., 1941. Über die Holzarten und die durch die klimatischen Verhältnisse verursachten Holzartenwechsel in den Wäldern Ostpatagoniens. *Ann. Acad. Sci. Fenn., Ser. A, Biol.*, 4(2): 1–151.

Ljunger, E., 1939. A forest section through the Andes of northern Patagonia. *Sven. Bot. Tidskr.*, 33: 321–337.

Looser, G., 1948. The ferns of southern Chile. *Am. Fern J.*, 38: 33–44; 71–87.

Mercer, J.H., 1970. Variations of some Patagonian glaciers since the late-glacial: II. *Am. J. Sci.*, 269: 1–25.

Miller, S., Rottman, J. and Taber, R.D., 1973. Dwindling and endangered ungulates of Chile: *Vicugna*, *Lama*, *Hippocamelus*, and *Pudú*. In: *Trans. 38th North American Wildlife and Natural Resources Conference, 1973*, pp. 55–68.

Mueller-Dombois, D. and Ellenberg, H., 1974. *Aims and Methods of Vegetation Ecology.* Wiley, New York, N.Y., 547 pp.

Müller-Using, B., 1973. Untersuchungen über die Verjüngung von *Nothofagus alpina* (Poepp. et Endl.) Oerst. und ihrer wichtigsten Begleitbaumarten in der chilenischen Anden- und Küstenkordillere. Thesis, University of Munich, Munich, 229 pp.

Muñoz P., C., 1960. *Preliminary List of Plants Collected for the Expedition to Laguna de San Rafael, Province of Aysén, 1959*. Technical Report, American Geographical Society Southern Chile Expedition, 16 pp.

Neger, F.W., 1899. Informe sobre las observaciones botánicas efectuadas en la cordillera de Villarrica en el verano 1896–1897. *An. Univ. Chile*, 90: 903–967.

Oberdorfer, E., 1960. *Pflanzensoziologische Studien in Chile—Ein Vergleich mit Europa*. J. Cramer, Weinheim, 208 pp.

Oltremari, J., 1975. La interpretación y el desarrollo de los parques nacionales. *Bosque* (Valdivia, Chile), 1: 28–32.

Pérez R., V., 1958. *Recuerdos del Pasado (1814–1860)*. Carlos de Vidts, Santiago, 463 pp.

Philippi, R.A., 1865. Excursión botánica en Valdivia desde los Cuncos en el departamento de la Unión, a través de la Cordillera de la Costa, hasta la mar. *An. Univ. Chile*, 27: 289–351.

Pisano V., E., 1972. Comunidades vegetales del área del Bahía Morris, Isla Capitán Aracena, Tierra del Fuego (Parque Nacional Hernando de Magellanes). *An. Inst. Patagonia*, 3: 103–130.

Pisano V., E., 1973. Fitogeografía de la Península de Brunswick, Magallanes. I. Comunidades meso-higromórficas e higromórficas. *An. Inst. Patagonia*, 4: 141–206.

Pisano V., E., 1975. Características de la biota magallánica derivadas de factores especiales. *An. Inst. Patagonia*, 6: 123–137.

Pisano V., E., 1983. The Magellanic tundra complex. In: A.J.P. Gore (Editor), *Mires: Swamp, Bog, Fen and Moor, B. Regional Studies*. Ecosystems of the World, 4. Elsevier, Amsterdam, pp. 289–323.

Ramírez, C., Steubing, L. and Alberdi, M., 1976. La familia Hymenophyllaceae (Pteridophyta) en el fundo San Martín, Valdivia–Chile. I. Taxonomía y Ecología. *Medio Ambiente (Valdivia, Chile)*, 2: 21–28.

Reiche, K.E., 1907. Grundzüge der Pflanzenverbreitung in Chile. In: A. Engler and O. Drude (Editors), *Die Vegetation der Erde, 8*. Engelmann, Leipzig, pp. 1–374.

Ruíz F., C., Corvalán, J. and Aguirre L., L., 1965. Geología. In: *Geografía Económica de Chile*. Corporación de Fomento de Producción, Santiago, 2nd ed., pp. 35–97.

Saelzer B., F., 1973. *La Evolución de la Legislación Forestal Chilena*. Borneck, Valdivia, 49 pp.

Saint-Amand, P., 1962. The great earthquake of May 1960 in Chile. *Ann. Rep. Smithson. Inst.*, 1962: 337–363.

Schmithüsen, J., 1956. Die raümliche Ordnung der chilenischen Vegetation. *Bonn. Geogr. Abh.*, 17: 1–89.

Schmithüsen, J., 1960. Die Nadelhölzer in den Waldgesellschaften der südlichen Anden. *Vegetatio*, 10: 313–327.

Skottsberg, C., 1910. Übersicht über die wichtigsten Pflanzenformationen Südamerikas S. von 41°, ihre geographische Verbreitung und Beziehungen zum Klima. *K. Sven. Vetenskapsakad. Handl.*, 46(3).

Skottsberg, C., 1916. Die Vegetationsverhältnisse langs der Cordillera de los Andes S. von 41°S. Br. Ein Beitrag zur Kenntnis der Vegetation in Chiloé, Westpatagonien und Feuerland. *K. Sven. Vetenskapsakad. Handl.*, 56.

Soriano, A., 1982. Deserts and semi-deserts of Patagonia. In: N.E. West (Editor), *Temperate Deserts and Semi-Deserts*. Ecosystems of the World, 5. Elsevier, Amsterdam, pp. 423–460.

Spurr, S.H. and Barnes, B.V., 1973. *Forest Ecology*. Ronald Press, New York, N.Y., 2nd ed., 571 pp.

Taber, R.D., Miller, S. and Rottman, J., 1974. *Rare and Endangered Mammals of Chile*. Unpublished report to the World Wildlife Fund, 50 pp.

Thelen, K. and Miller, K.R., 1976. *Planificación de Sistemas de Areas Silvestres*. CONAF–FAO, Santiago, 62 pp.

Thomasson, K., 1959. Nahuel Huapi: plankton of some lakes in an Argentine national park, with notes on terrestrial vegetation. *Acta Phytogeogr. Suec.*, 42: 1–83.

Thomasson, K., 1963. Araucanian lakes: plankton studies in north Patagonia with notes on terrestrial vegetation. *Acta Phytogeogr. Seuc.*, 47: 1–139.

Urban, O., 1934. *Botánica de las Plantas Endémicas de Chile*. Sociedad de Imprenta Concepción, Concepción, 292 pp.

Veblen, T.T. and Ashton, D.H., 1978. Catastrophic influences on the vegetation of the Valdivian Andes. *Vegetatio*, 36: 149–167.

Veblen, T.T., Delmastro, R.J. and Schlatter, J.E., 1976. The conservation of *Fitzroya cupressoides* and its environment in southern Chile. *Environ. Conserv.*, 3: 291–302.

Veblen, T.T., Ashton, D.H., Schlegel, F.M. and Veblen, A.T., 1977a. Distribution and dominance of species in the understorey of a mixed evergreen-deciduous *Nothofagus* forest in south-central Chile. *J. Ecol.*, 65: 815–830.

Veblen, T.T., Ashton, D.H., Schlegel, F.M. and Veblen, A.T., 1977b. Plant succession in a timberline depressed by vulcanism in south-central Chile. *J. Biogeogr.*, 4: 275–294.

Veblen, T.T., Ashton, D.H. and Schlegel, F.M., 1979. Tree regeneration strategies in a lowland *Nothofagus*-dominated forest in south-central Chile. *J. Biogeogr.*, 6: 329–340.

Veblen, T.T., Schlegel, F.M. and Escobar R., B., 1980. Structure and dynamics of old-growth *Nothofagus* forests in the Valdivian Andes, Chile. *J. Ecol.*, 68: 1–31.

Villagran M., C. and Soto Q., C., 1974. Estudio preliminar de la vegetación boscosa del Parque Nacional Vicente Pérez Rosales. *An. Museo Hist. Nat. Valparaiso*, 7: 125–151.

Von Furstenberg, P., 1969. Perspectivas económicas de un manejo de las áreas silvestres de Chile. *Publ. Cient. Univ. Aust. Chile*, 13: 21–33.

Weischet, W., 1960. Contribuciones al estudio de las transformaciones geográficas en la parte septentrional del sur de Chile por efecto del sismo del 22 de mayo de 1960. *Inst. Geol., Univ. Chile, Publ.*, 15: 1–40.

Wright, C. and Mella, A., 1963. Modifications to the soil pattern of south-central Chile resulting from seismic and associated phenomena during the period May to August 1960. *Bull. Seismol. Soc. Am.*, 53: 1367–1402.

Young, S.B., 1972. Subantarctic rain forest of Magellanic Chile: Distribution, composition, and age and growth rate studies of common forest trees. In: G. A. Llana, *Antarctic Terrestrial Biology*. *Antarct. Res. Ser.*, 20: 307–332.

Chapter 3

TEMPERATE BROAD-LEAVED EVERGREEN FORESTS OF NEW ZEALAND[1]

P. WARDLE
with M.J.A. BULFIN and J. DUGDALE

MAIN CLASSES OF NATIVE FOREST

The appellation "temperate broad-leaved evergreen" may be applied to most New Zealand forest. Exceptions such as forests dominated by conifers, subalpine *Nothofagus* (beech) stands, and minor enclaves of deciduous trees are so closely related floristically and ecologically to the prevailing forest vegetation that it would be inappropriate to exclude them from discussion. This chapter has to attempt, perforce, a précis of the forest ecology of New Zealand.

Botanists and silviculturalists traditionally recognize two broad categories of native forest, which the pioneer ecologist L. Cockayne named "subtropical" and "subantarctic", thereby making phytogeographic assumptions which are now hardly tenable. Current usage refers to "podocarp–hardwood" and "beech" forests; the first of these terms is inadequate and "conifer–broadleaf" is used here.

Conifer–broadleaf forest (Figs. 3.1 and 3.2)

In its typical development in mild lowland areas, this forest is multi-storeyed. Conifers where present form the tallest storey, usually as well-spaced, large-crowned trees, but they can also form continuous canopies. Most of the tree species are podocarps, in the genera *Podocarpus*, *Dacrycarpus*, *Dacrydium* and *Phyllocladus*, the tallest species reaching heights of over 40 m, or exceptionally 60 m. There are also two species of *Libocedrus* (Cupressaceae) and, north of 38°S, the massive kauri (*Agathis australis*).

Hardwoods and some of the less-tall podocarps form the next storey at 15 to 25 m, which is usually the main canopy. The genera *Elaeocarpus*, *Metrosideros*, *Weinmannia* and in northern districts, *Beilschmiedia*, *Knightia*, *Laurelia*, *Litsea* and *Nestegis* each contribute one or two species. A host of small trees, mainly belonging to the genera *Dysoxylum*, *Griselinia*, *Melicytus*, *Myrsine*, *Pittosporum*, *Pseudopanax*, *Pseudowintera*, and *Schefflera*, among others, form a subcanopy and fill gaps. Tree ferns (species of *Cyathea* and *Dicksonia*) and, in mild districts, the palm *Rhopalostylis sapida* are usually abundant and contribute much to the subtropical appearance of vegetation (Fig. 3.3).

Shrubs belong mainly to the genus *Coprosma*, but also include species of *Alseuosmia*, *Geniostoma*, *Melicope*, *Myrsine*, *Neomyrtus*, *Pseudopanax* and others. Many of them are twiggy and small-leaved. Large ferns, such as *Asplenium*, *Blechnum*, *Hypolepis*, *Lastreopsis*, *Polystichum*, *Pneumatopteris*, *Todea*, constitute a usually well-developed ground layer. Herbaceous vascular plants, which are relatively unimportant, include species of *Dianella*, *Libertia*, *Microlaena*, *Parietaria*, *Uncinia* and (north of 41°S) *Elatostema*, *Australina* and *Nertera* are creeping plants which share the lowest stratum with filmy ferns (Hymenophyllaceae) and bryophytes.

Dependent synusia can be exceptionally well developed. Epiphytes include not only bryophytes and filmy ferns, but larger ferns (e.g. *Asplenium*, *Pyrrosia*), and vascular plants such as large tussock-forming lilies (*Astelia* and *Collospermum*), several orchids, and the shrubs *Griselinia lucida*, *Senecio kirkii* and *Pittosporum cornifolium*. Several trees normally begin life as epiphytic shrubs which

[1] Manuscript completed in January 1981.

Fig. 3.1. Forest beside a coastal lagoon in Westland showing the upper storey of podocarps, and a lower storey of broad-leaved trees.

eventually encase their host tree in a false trunk of coalesced, descending roots. This fig-like habit is best seen in *Metrosideros robusta*, which as a mature tree shares the highest forest storey with its coniferous hosts (Fig. 3.4). The fibrous trunks of tree ferns are also a favoured site for tree seedlings, especially those of *Weinmannia racemosa*. Lianes are abundant, and in some forests form much of the canopy. They belong to the genera *Clematis*, *Freycinetia*, *Metrosideros*, *Muehlenbeckia*, *Parsonsia*, *Ripogonum*, *Rubus* and *Tetrapathaea*, and also include the ferns *Lygodium* and root-climbing species of *Blechnum*, *Phymatosorus* and *Rumohra*.

Despite the luxuriance of lowland conifer–broadleaf forest, most genera contain only one to two species; only *Blechnum*, *Coprosma*, *Hymenophyllum*, *Pseudopanax* s. l. and *Uncinia* have more than ten forest species.

In higher, colder, or drier regions, the mixed forest becomes less rich floristically and is not so structurally complex. As an extreme example, there are seral stands of *Metrosideros umbellata* and *Weinmannia racemosa* on recent moraine and landslides, in which the density of the canopy permits only a sparse undergrowth of ferns, shrubs and scattered seedlings of the podocarps and other trees which will dominate the next phase of the succession.

Beech forest

Pure subalpine stands of *Nothofagus solandri* var. *cliffortioides* provide an extreme contrast to the lowland forests just described (Fig. 3.5). Other than the dominant tree, which forms a dense canopy at 10 to 15 m, there may be only scattered *Coprosma* shrubs, corticolous lichens and bryo-

Fig. 3.2. Undergrowth of conifer–broadleaf forest in southern Canterbury, showing a trunk of *Podocarpus spicatus*, a *Cyathea dealbata* tree fern, the shrubby araliad *Schefflera digitata*, and the ground fern *Polystichum vestitum*. (Photo G.C. Kelly.)

Fig. 3.4. Lowland forest in northern Westland, with a large tree of *Metrosideros robusta* in the foreground. Most other trees are *Weinmannia racemosa*. (Photo G.C. Kelly.)

Fig. 3.3. Coastal forest in northwest Nelson with the palm *Rhopalostylis sapida* and the pandanaceous liane *Freycinetia banksii* in the foreground. (Photo G.C. Kelly.)

Fig. 3.5. Interior of a pure forest of *Nothofagus solandri* var. *cliffortioides* with undergrowth of the same species, in inland Canterbury. (Photo G.C. Kelly.)

phytes, bryophyte cushions (*Dicranoloma*, *Leucobryum*) on the forest floor, and two species of loranthaceous parasites, *Alepis flavida* and *Peraxilla tetrapetala*. On moist concave slopes, however, there can be a dense understorey of the ferns *Hypolepis millefolium* and *Polystichum vestitum*, the latter being summergreen.

There are four species of *Nothofagus*, all small-leaved and evergreen. The number of accompanying species increases along gradients towards moister or warmer climates. Many forests that are overwhelmingly dominated by beech have a sprinkling of plants characteristic of the various storeys of conifer–broadleaf forest. For example, 30 m tall forest of *Nothofagus fusca* can have *Dacrydium cupressinum* as a scattered overstorey, *Podocarpus ferrugineus* in the main canopy, a sparse subcanopy with *Weinmannia racemosa*, and a ground fern layer of *Blechnum discolor*, with other small trees, shrubs and vines well-represented, including *Carpodetus serratus*, *Coprosma* spp., *Pseudopanax* spp., *Pseudowintera colorata*, *Myrsine divaricata* and *Rubus cissoides*. Finally, some lowland areas support thoroughly mixed stands known as "beech–podocarp forests", a designation that reflects content of merchantable timber rather than botanical composition (Fig. 3.6).

Fig. 3.6. Interior of conifer–beech–broadleaf forest in inland Nelson, showing pendant foliage of *Dacrydium cupressinum* in the middle distance, large tussocks of the sedge *Gahnia xanthocarpa*, and the fern *Gleichenia cunninghamii* in the right foreground. (Photo G.C. Kelly.)

DISTRIBUTION OF FOREST IN RELATION TO BROAD CLIMATIC PATTERNS (Table 3.1 and Figs. 3.7–3.9)

Before their area had been drastically reduced through the activities of people, native forests covered most of New Zealand, and still occupy about 23% of the country's 270 000 km^2. They extend to the Kermadec Islands, which at 29°S form the extreme north of the New Zealand botanical and political region, and to the Chatham Islands 800 km east of the South Island, while the southernmost outliers grow in sheltered coves in the Auckland Islands at 51°S.

Evergreen hardwood forest also extends to the timberline, which has an altitude of 1400 m on the most northern high mountains at 38°S, and decreases to around 850 m at 46°S (Fig. 3.10). In inland regions of the South Island, the climate has distinctly continental attributes, and the timberline rises to 1500 m.

Eastern and inland parts of the South Island, and to a lesser extent those of the North Island, lie in distinct rain shadows, and there is little natural forest where annual rainfall is less than 800 mm. Where föhn westerlies and low humidities prevail, general forest cover can cease even by the 1600 mm isohyet. Existing relict stands, subfossil wood, and charcoal prove that the present limits of forest are largely determined by fire and that forest extended to a much lower isohyet prior to human settlement which began about 900 A.D.

Broadly speaking, conifer–broadleaf forests prevail at low altitudes, towards the north of the country, under the more reliable rainfall regimes, and on the better soils, whereas beech forests have the opposite distribution. To this generalization there is every conceivable exception. Conifer–broadleaf forest alone grows south of Foveaux Strait, while *Nothofagus truncata* can be found at sea level on the North Auckland peninsula, shedding its leaves into estuarine mangrove swamp (L.B. Moore, pers. comm.).

Nothofagus solandri var. *cliffortioides* or *N. menziesii* constitute most of the subalpine forests, forming remarkably level, abrupt timberlines

TABLE 3.1

Selected climatic data (from Killick, 1976)

Station (see Fig. 3.7)	Altitude (m)	Mean rainfall (mm)					Mean temperatures (°C)				
		year	wettest month		driest month		annual	January	July	annual max.	annual min.
Kaitaia	80	1430	165	Aug.	79	Jan.	15.5	19.5	11.5	28	2
Wellington	126	1271	140	Jul.	81	Jan.	12.5	16.5	8.5	26	1
Hokitika	39	2940	277	Nov.	231	Jan.	11.3	15.0	7.0	26	−2
Invercargill	0	1042	107	Jun.	66	Aug.	9.6	13.5	4.5	28	−5
Christchurch	7	658	84	May	46	Oct.	11.7	17.0	6.0	32	−4
Tekapo	683	606	56	Sep.	43	Feb.	9.1	15.5	1.5	30	−10
Alexandra	141	339	46	Jan.	15	Jul./Aug.	10.6	17.0	2.5	32	−8

against alpine grasslands (Wardle, 1973). However, on Mount Egmont, on the western flanks of the Southern Alps and the adjoining piedmont area between 42°30′ and 43°40′S, and on Stewart Island, beech is altogether absent. Instead, depauperate conifer–broadleaf forest extends into the subalpine belt. Its conifer storey consists of *Podocarpus hallii* (accompanied by *Libocedrus bidwillii*, except in Stewart Island), while the main canopy consists of *Weinmannia racemosa*, *Metrosideros umbellata* (except on Mount Egmont), or in certain circumstances, small trees such as *Dracophyllum traversii*, *Griselinia littoralis* and *Olearia ilicifolia*. Upwards, this forest grades into very dense subalpine scrub dominated by composites (*Olearia*, *Senecio*), small podocarps (e.g. *Dacrydium biforme*, *Phyllocladus alpinus*, *Podocarpus nivalis*), epacrids (*Archeria*, *Dracophyllum*) and various other genera.

There are other districts (Volcanic Plateau, the ranges on either side of Manawatu Gorge, Seaward Kaikoura Range, Banks Peninsula, southeastern Otago), where beech forest is scattered or at least interrupted by conifer–broadleaf stands. Here also, subalpine forest is most likely to be conifer–broadleaf. Furthermore, forest remnants and subfossil material show that the deforested, low rainfall region in the east of the South Island once supported both forest categories. Remnants dominated by *Podocarpus hallii* still grow in Central Otago (Fig. 3.11) where rainfall is about 700 mm yr^{-1} (Wells, 1972), which may be compared with the predominance of beech in the forests of western Fiordland, under rainfalls exceeding 6000 mm. As exceptions to the rule that beech forest occupies poor soils, *Nothofagus fusca* is at its best on deep, fertile colluvium, whereas *Agathis australis* in the north of the North Island and *Dacrydium cupressinum* in the west of the South Island dominate on infertile gley podzols.

Clearly, the distribution of beech and conifer–broadleaf forests cannot be simply related to environmental factors, but as will be seen later historical factors are deeply involved.

FACTORS INFLUENCING DISTRIBUTION OF FOREST

Soils

Yellow-brown earths of moderately low fertility supported most of the original forest cover of the North Island and eastern part of the South Island. Today, nearly all of these have been cleared to pasture, except on the steeper and higher ranges. There were also forested yellow-brown loams, and red-brown granular loams and clays in the North Island, which likewise have mostly been cleared. In Auckland province, forest characterized by *Agathis australis* growing on podzolized yellow-brown earths and podzols was once far more extensive than at present. Forest also occupies yellow-brown pumice soils on the central plateau of the North Island.

In the west and south of the South Island and Stewart Island, the predominant soils are podzolized yellow-brown earths, podzols and gley podzols, and their forest cover is still largely intact (Leamy and Fieldes, 1976).

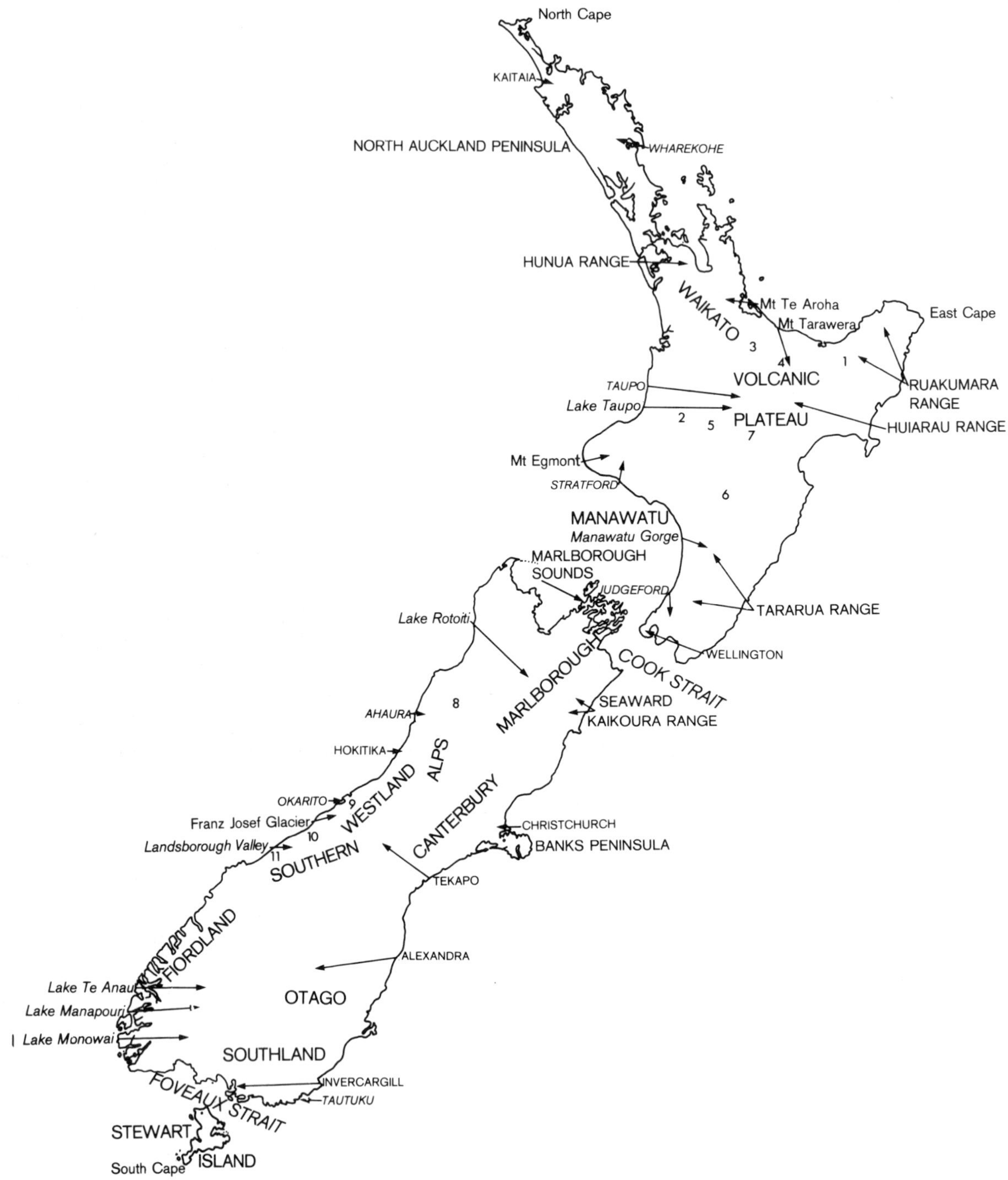

Fig. 3.7. Locality map of New Zealand. TEKAPO: climatological stations for data in Table 3.1. OKARITO: type localities for soils in Table 3.2. Numbers refer to forest type localities in Table 3.3.

Table 3.2 shows analyses of representative soils. Yellow-brown earths, yellow-brown loams and yellow-brown pumice soils are of moderately low fertility, and the podzolized yellow-brown earths are even less fertile. Fertile soils supporting forest occur only locally and include recent alluviums and

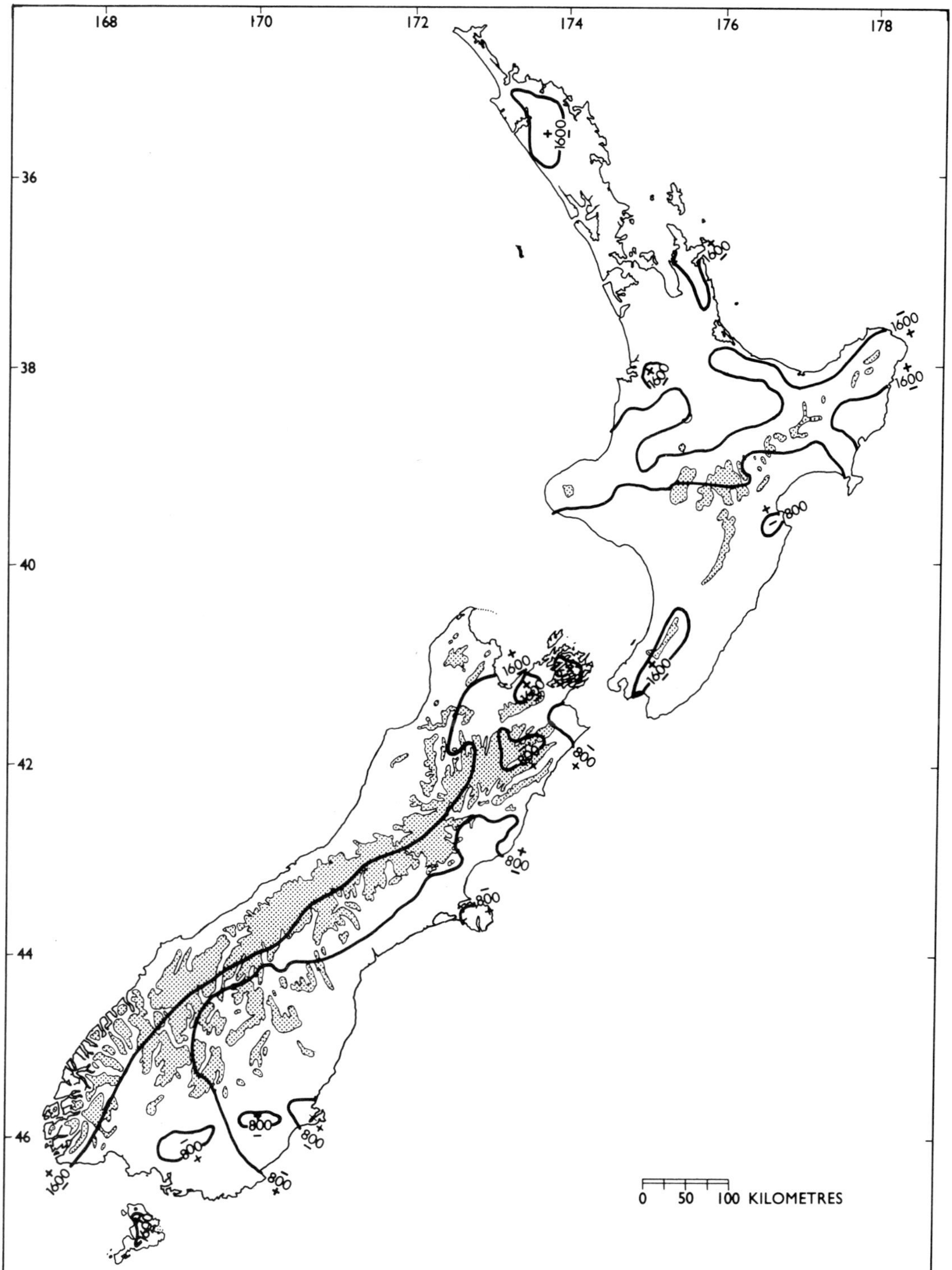

Fig. 3.8. Latitude and longitude; land above 900 m; isohyets (mm).

Fig. 3.9. Distribution of the main classes of forest in New Zealand. Forest with *Agathis australis* prominent: *1*, at present; *2*, in 1840 A.D. Conifer–broadleaf forest: *3*, at present; *4*, in 1840 A.D. Beech and conifer–beech–broadleaf forest: *5*, at present; *6*, in 1840 A.D.

Fig. 3.10. Montane and subalpine *Nothofagus* forest near the Lewis Pass between Nelson and Canterbury, showing the level timberline. (Photo G.C. Kelly.)

colluviums, and soils derived from basalt and limestone. The least fertile are gley podzols developed in areas of high rainfall on Late Pleistocene moraine hills and gravel outwash terraces in the west of the South Island, and podzols developed under *Agathis australis* in the North Auckland peninsula.

Though some species show a broad amplitude of tolerance to soil conditions, on the whole forest composition is a reliable indication of fertility. Among lowland conifers *Dacrycarpus dacrydioides*, *Podocarpus spicatus* and *P. totara* indicate recent and therefore usually fertile soils, whereas *Agathis australis*, *Phyllocladus* spp. and *Podocarpus hallii* are most abundant on poor soils. Where *Beilschmiedia tawa* and *Weinmannia racemosa* overlap

Fig. 3.11. Remnant of *Podocarpus hallii* woodland in montane grassland in inland Otago.

in distribution, the former is on more fertile sites than the latter. Ferns especially can indicate edaphic gradients. Thus, the undergrowth can be formed by *Asplenium bulbiferum* on recent alluvium and colluvium, by *Blechnum discolor* on the yellow-brown earths of stable slopes, and by *Gleichenia cunninghamii* on strongly podzolized yellow-brown earths on broad ridge crests. Adams (1976) demonstrated that the distribution pattern of *Nothofagus* species is correlated with decreasing soil fertility in the following order: *N. fusca* > *N. menziesii* > *N. truncata* > *N. solandri var. cliffortioides*. This relationship is broadly reflected in foliar nutrient concentrations, especially of nitrogen and phosphorus.

Soils which do not support forest are of four categories:

(1) Those outside the rainfall or temperature limits for forest growth.

(2) Swamps. Although stands of *Dacrycarpus dacrydioides* can have water lying between the trees almost continuously, such conditions apparently arise from secondary flooding, after the trees are well established.

(3) Extremely well-developed gley podzols, ultrabasic parent materials, and shallow, leached soils developed on hard granitic and quartzose rocks support heathland or related grass and sedge communities. At best, the vegetation is low forest of small *Dacrydium* spp., *Leptospermum scoparium*, *Phyllocladus alpinus*, etc., accompanied by stunted representatives of neighbouring tall forest, such as *Nothofagus solandri* var. *cliffortioides*, *Metrosideros umbellata* and *Weinmannia racemosa*. Heath communities are highly flammable, so that not only have they been extensively modified by fire in areas of human activity, but they have served to carry fire into tall forest, thereby often supplanting the latter on poor soils. The Auckland "gumlands", so named because of the prevalence of commercially valuable buried *Agathis* resin, support *Leptospermum sco-*

TABLE 3.2

Analysis of representative soils (from N.Z. Soil Bureau, 1968)

Soil and classification (see Fig. 3.7)	Forest	Depth (cm)	Horizon	pH	Percentage dry weight of C in fine fraction	C/N ratio	Cation exchange			NH_2SO_4 mg $(100\ g)^{-1}$	TRUOG mg $(100\ g)^{-1}$
							CEC (me/100 g)	TEB	BS (%)		
Judgeford silt loam; yellow-brown earth	podocarps and *Beilschmiedia tawa*	8–20	A_{12}	5.9	4.2	14	15.5	6.6	43	2	0.5
		42–55	B_{22}	5.9	1.0	13	8.5	0.7	8	5	0.2
		75–85	C	5.8	0.7	12	6.0	0	0	13	0.7
Stratford coarse sandy loam; yellow-brown loam	podocarp–broadleaf	8–16	A_{12}	5.6	9.0	14	23.4	6.7	29	38	1.0
		18–30	(B)	6.0	4.3	13	13.9	2.4	17	36	0.5
		40–52	BC	6.4	3.0	14	11.5	1.2	10	43	0.2
Taupo sandy silt; yellow-brown pumice	podocarp	0–8	A_1	5.5	10.1	19	31.3	9.6	30	5	0.5
		20–25	(B)	6.2	1.4	14	6.9	1.9	28		
		42–50	C	6.6	0.2	10	3.4	1.9	56		
Tautuku silt loam; podzolized yellow-brown earth	*Dacrydium cupressinum*–*Weinmannia racemosa*	0–5	A_2	4.1	31.3	86	41.9	4.4	11	2	12
		27–30	B_{2fe}	5.0	1.2	17	25.8	1.2	5		
		30–40	C_{11}							9	0.2
Wharekohe silt; kauri podzol	*Agathis australis*	8–25	A_{22}	5.5	3.5	50	9.0	0.6	7	1	0.2
		48–72	B_{22}	4.9	1.3	16	41.6	2.5	6	1	0.2
		85–115	C_{1g}	4.8	0.7	12	35.1	20	6	1	0.2
Okarito peaty loam; deep gley podzol	*Dacrydium cupressinum*–*D. colensoi*	5–12	A_1	4.7	19.7	33	30.8	5.3	14	4	0.5
		13–19	A_2G	5.1	5.2	43	15.3	0.6	4	1	0.2
		40–44	B_h	5.0	4.0	44	21.2	0.5	2	27	1

Note: CEC = cation exchange capacity; TEB = total exchangeable bases; BS = base saturation; NH_2SO_4 = total inorganic phosphorus extracted in 1.0 molar H_2SO_4; TRUOG = plant available phosphorus extracted in 0.002 molar H_2SO_4.

parium-dominated vegetation which is frequently burnt, and which probably replaced *Agathis* forest burnt by the Maoris.

(4) Very recent soils, for example on flood plains, landslide scars and volcanic areas, where forest has not had time to develop. The succession to forest is often delayed by harsh conditions, such as lack of water retention by the soil or exposure to frost and wind.

Rainfall

At present, forest margins in the drier districts are mainly anthropogenic, and adjoin a wide variety of vegetation: *Pteridium* fernland, *Leptospermum* shrubland, tussock grassland, successional forest, alien communities such as those dominated by *Rubus fruticosus* and *Ulex europaeus*, and of course artificial vegetation maintained by agriculture. It seems that prior to human intervention a zone of microphyllous woodland, consisting of species such as *Coprosma virescens*, *Discaria toumatou*, *Leptospermum ericoides*, *Olearia lineata* and *Sophora microphylla*, grew under moisture regimes intermediate between those supporting forest and semi-arid grassland.

Wind

Most of New Zealand experiences frequent, strong winds. Zonal westerlies prevail on mountain ridges and western headlands, through passes, and through Cook and Foveaux Straits, producing marked shearing of vegetation. Sea breezes also wind-train trees on eastern coasts, and produce orographic fog belts which favour mesophilous forests on coastal hills which receive quite low rainfall.

East of the axial ranges, föhn winds cause high evapotranspiration, and as precipitation decreases downwind, forest remnants become progressively more restricted to leeward slopes. In contrast, the western lowlands of the South Island are the least windy part of New Zealand, owing to a cushioning effect along the Southern Alps.

Temperature

Above the forest limits, whether they are formed abruptly by *Nothofagus* or by subalpine scrub gradually diminishing in height and extent, there is chiefly grassland dominated by large tussocks of *Chionochloa* spp. Herbfields and low scrub are also important. In many places, especially on the drier mountains, fire has removed the upper forests which have been replaced by shrubland or downward extensions of grassland.

Valley floors in mountainous parts of the South Island often lack forest. Sharp winter temperature inversions prevail in such places. In one instance, valley floor temperatures falling below −10°C have been demonstrated to kill transplanted and sown seedlings of *Nothofagus solandri* var. *cliffortioides* which dominates adjacent slopes (Wardle, 1971). Similar frosty enclaves of grassland exist even at low altitudes, but are gradually encroached upon by forest.

ENVIRONMENTAL FACTORS DAMAGING OR DISTURBING TO FOREST

Drought

The forested parts of New Zealand experience generous rainfall. On the basis of monthly means, rainfall appears well distributed, and though in some districts there is a distinct tendency for seasonal maxima and minima, these by no means constitute well-marked wet or dry seasons. Despite this, most of the country is subject to irregular, intense droughts, which can be very damaging in the warmer part of the year. Most localities experience rainless periods lasting several weeks, while in some eastern districts of the South Island severe droughts can endure for more than a year and the rainfall remains below average for years at a time. These conditions can be greatly exacerbated by föhn winds. Death of canopy trees has been observed, either as a direct effect of such droughts or through secondary causes.

During droughts, fire risk can become intense. Beginning in early Maori times, extensive tracts of forest have been lost from eastern and inland regions of both main islands by burning. Despite an active policy of preventing and extinguishing fires, the destruction continues from time to time. In the last decade, stands of both beech and conifer–broadleaf forest have been accidentally burnt during dry, föhn weather in areas where annual rainfall exceeds 2000 mm.

Temperature extremes

New Zealand temperatures (42°C is the highest recorded screen maximum) are probably never high enough to directly injure plants in the forest, though they are doubtless one of the factors inhibiting colonization of bare ground. Both winter and unseasonal frosting, on the other hand, occur from time to time, and temporarily can damage trees growing well within their range of climatic tolerance. Examples are frosting of the *Nothofagus* early summer flush of shoots and flowers, and of the new growth of the deciduous *Fuchsia excorticata*. Severe die-back of shoots in exposed trees of *Beilschmiedia tawa* has been attributed to winter frost. The southern limits of many forest species, such as *Dodonaea viscosa*, *Freycinetia banksii*, *Metrosideros robusta* and *Rhopalostylis sapida*, correspond closely to minimum isotherms. These species are known to be sensitive to winter frost when cultivated outside their natural range. Further, experimentally determined freezing resistance of native trees and shrubs generally correlates well with their natural distribution; for leaves this ranges from −4°C or warmer for broad-leaved species not extending south of 39°S, to −18°C to −25°C for the hardiest high-altitude shrubby conifers (Sakai and Wardle, 1978). "Frost desiccation" or winter drying of exposed foliage is a common timberline phenomenon in New Zealand as elsewhere.

Snow

Snow is only a regular component of the forest environment in the subalpine belt, and even here the winter snow pack is thin and shortlived by temperate-zone standards. Exceptional falls result in canopy breakage and avalanches which can initiate regeneration cycles. On regular avalanche tracks and places of deep snow accumulation, the forest is replaced by shorter, more tolerant vegetation.

Relatively light snowfalls extend down to sea level most winters south of 45°S, but usually melt within a few days, whereas north of 42°S, snow is almost unknown at low altitudes. However, the eastern foot-hills of the central ranges in the South Island occasionally receive freak falls when up to 2 m of wet snow can fall in one to two days. These can cause devastating breakage in both native and exotic forests.

Wind

Most of New Zealand experiences occasional cyclonic gales, which leave broad swathes of fallen trees, sometimes covering many hundreds of hectares. Extensive even-aged beech stands have resulted from mass release of seedlings after such episodes. Wind-throws in conifer–broadleaf forest are usually more localized, and the resultant groups of even-aged conifers seem to develop through a preliminary stage of hardwood dominance.

Onshore westerly gales, when not accompanied by rain, can cause salt-burning and defoliation several kilometres inland, especially among the faster-growing, softer-leaved hardwoods such as *Melicytus ramiflorus*. Wind-scorching has also affected actively growing shoots of *Dacrycarpus dacrydioides* far from the sea.

Topographic changes and primary succession

Since New Zealand is a tectonically active region with high relief, new surfaces such as landslide scars, recent river terraces, and prograding dunes constantly provide sites for plant succession. One of the best dated primary successions is on moraine and gravel outwash deposited during the fluctuating retreat of the Franz Josef Glacier (Wardle, 1980a)

Fig. 3.12. Franz Josef Glacier, with successional shrubland developing on outwash terraces and fans.

(Fig. 3.12). Shrub seedlings become established among pioneer herbs and mosses, leading to dominance, first of *Carmichaelia grandiflora*, and by 30 to 40 years, of *Olearia avicenniaefolia*. By this time, there are many hardwood seedlings and forest ferns in the undergrowth, and by 100 years or so *Metrosideros umbellata* and *Weinmannia racemosa* can form forest, which in turn harbours seedlings of podocarps. At 1000 years, these podocarps can become an overstorey of massive trees, but smaller tree species and ferns (e.g. *Asplenium bulbiferum*, *Melicytus ramiflorus*) still indicate recent, relatively fertile soil. In this area the youngest surfaces supporting "climax" conifer–broadleaf forest, characterized by maximum complexity and an understorey of *Blechnum discolor*, are dated at 5000 years, and their soils are mapped as yellow-brown earths. As soils become extremely gleyed and podzolized, they come to support dense stands of *Dacrydium cupressinum*, and ultimately heath and bog vegetation. This chronosequence has been assigned a span of 22 000 years (Stevens and Walker, 1970). At first, nitrogen, organic carbon and organic phosphorus increase to a steady state, but after 12 000 years there is loss of these elements, together with non-occluded inorganic phosphorus and exchangeable cations.

The mineralogy of the sequence has also been studied (Mokma et al., 1973). Iron oxide is progressively leached from the A horizon and redeposited in the B horizon; the sand, coarse-silt and medium-silt fractions become relatively enriched in quartz through weathering of the feldspars, mica and chlorite; the mica weathers to vermiculite and montmorillonite.

Successions to beech forest can be much simpler. In a subalpine situation, for example, seedlings of *Nothofagus solandri* became successfully estab lished on river gravel deposited four years earlier and could be expected to lead to a mature stand within 200 years (Wardle, 1972).

Fire

Fires, presumably caused by volcanic eruptions and lightning, are known to have occurred in New Zealand throughout the Quaternary era. Only since the advent of the Maori people of Polynesian race about 900 A.D. has fire become both frequent and extensive. This may explain much of the striking difference between the aspect of Australian and New Zealand vegetation. Australia has many lignotuberous trees which resprout after fire; New Zealand has none. Few New Zealand trees possess any significant ability to resprout after fire (exceptions are *Griselinia littoralis* and, in some degree, *Weinmannia racemosa*). The Australian *Nothofagus cunninghamii* shows marked production of epicormic shoots, whereas the closely related *N. menziesii* of New Zealand produces few. Australia is replete in species with serotinous fruit; New Zealand has only *Leptospermum scoparium* which is either conspecific with or closely related to Australian plants.

In consequence, the species of beech or mixed forests exhibit little tolerance of fire, and as they have retreated they have been replaced, not by fire-adapted forest equivalent to the Australian sclerophyll forest, but by miscellaneous shrubland, grassland and fernland communities. In drier regions, burnt *Nothofagus* forests have tended not to regenerate over large areas, although there can be vigorous spread from the margins of surviving remnants. Conifer–broadleaf forest appears to spread more widely because of the better seed dispersal of most of its species, but the successions are slow and mainly confined to mesic sites.

In humid districts not influenced by föhn winds, virgin native forest does not burn readily. In most areas, successful clearing has only followed logging or partial felling, although fire is said to have destroyed some large virgin stands of *Agathis australis*. Mesic burnt areas can be rapidly invaded by small hardwood species, especially *Aristotelia serrata*, and tree ferns can multiply from surviving rhizomes to form dense stands. In time, the re-establishment of a hardwood canopy provides the conditions for conifers to re-enter (Cameron, 1960). On sites burnt repeatedly, *Pteridium aquilinum* can become firmly established, and is only very slowly supplanted by woody vegetation. Shrubland of *Leptospermum scoparium* is also extensive, especially on dry hillsides. Like *Pteridium*, it becomes entrenched with repeated burning, and in the European era, has tended to supplant *Pteridium* when the latter has been weakened through trampling by stock. *Leptospermum scoparium* forms a dense shrubland at first, but eventually becomes taller and more open, to provide excellent shelter for seedlings of forest species. Woodlands of *L. ericoides* have also been induced by fire.

Volcanicity

Most of the central plateau of the North Island has a blanket of tephra, often containing charred logs which attest to forest destruction during eruptions. The plateau has supported a discontinuous forest cover since it was first seen by Europeans around 1840 A.D., but only some of the open areas can be directly attributed to recent eruptions (such as that of Mount Tarawera in 1886), or current volcanic and thermal activity. The last really extensive eruption was from the basin of Lake Taupo about 1800 B.P., and most of the area where tephra fell thickly remains free of forest. Fire, rather than the direct effects of volcanic eruption, seems to have prevented forest establishment. Especially after the arrival of the Maoris, flammable pioneer vegetation of *Pteridium* and *Leptospermum* and frost-hollow communities of *Dracophyllum* spp. and grasses have been maintained and extended (Nicholls, 1963).

Dense stands of podocarps on Taupo tephra have been regarded as even-aged, and successional to forests in which very large, scattered podocarp trees overtop a main canopy of hardwood species (McKelvey, 1963). However, Herbert (1980) has demonstrated that their apparent uniformity conceals an age range of about 500 years, on the basis of growth ring studies. Trees of *Podocarpus spicatus* were up to 676 years old, whereas the predominant *Dacrydium cupressinum* was mainly between 400 and 500 years. Herbert suggested that the podocarps have gradually invaded a "nurse" stand of hardwoods over a long period. It is also noteworthy that over 1000 years elapsed between the Taupo eruption and the establishment of the oldest living trees. Further, the eruption is unlikely to have destroyed the forests completely, for the existence of isolated patches of *Nothofagus* is more likely to represent survival than subsequent immigration (McKelvey, 1973).

PRINCIPLES OF FOREST CLASSIFICATION

New Zealand's generally youthful landscape of high relief, diverse geology and sharp climatic gradients supports forest communities which vary widely in composition, structure and stage of development, despite the rather narrow floristic base. In theory, New Zealand ecologists take an "individualistic" view of the plant community, perceiving numerous combinations, each depending on overlapping and ecological tolerances of component species (e.g. Wells and Mark, 1966). Yet, it is necessary to classify the vegetation, even if only for reasons of forest management, and such classifications are inevitably complex. As examples, Nicholls (1976) listed 182 forest types (excluding some degraded communities) for the North Island; Wardle (1977) has 33 forest types in a district covering 1200 km^2 in Westland; while J. Wardle (1970) has 23 types characterized by dominance or codominance of a single species, *Nothofagus solandri.*

Cockayne (1926) gave brief descriptions of the structure and floristics of the forest communities that he intuitively recognized throughout New Zealand. More recent accounts cover only single districts in any detail, the most comprehensive being McKelvey's (1963) partly quantitative study of the forest types of the west Taupo area. The National Forest Survey between 1945 and 1956 yielded a classification based mainly on timber content, dominant trees and terrain, although the types recognized correspond in varying degrees to communities based on floristic and ecological criteria. Selected Forest Survey data are given in Table 3.3.

Rather than try to do justice to the multitude of communities, the following section summarizes the ecological range of a few prominent tree species, and briefly describes some of the communities in which they grow.

ECOLOGY OF MAJOR TREE SPECIES

Agathis australis (Araucariaceae)

The kauri is the largest, though not the tallest, New Zealand tree, with a massive columnar trunk supporting a broad, heavy-limbed, but rather sparse canopy. Diameters of up to 7 m are known.

The natural range lies north of 38°S, where mature trees form groves or grow scattered among conifer–broadleaf and broadleaf forest, in which *Beilschmiedia tawa*, *B. tarairi* and *Weinmannia silvicola* are the main dominants, according to site. *Agathis* groves are concentrated on ridges.

TABLE 3.3

Mean timber volumes and tree densities, for selected forest types (from unpublished data of National Forest Survey; N.Z. Forest Service)

Forest type:	P1	P3H1	PH2	P1H1	P5	PH1	P1	P5	P6	P16H2	P3	B1	PB1	PB3	P1	P19	P8
Locality: (see Fig. 3.7)	1	1	2	3	3	4	5	5	5	5	6	7	8	8	9	10	11
Average merchantable volume (m^3 ha^{-1})																	
Dacrydium cupressinum	395.7	26.0	37.4	61.2	457.8	42.4	176.4	3.3	180.6	18.6	9.1		28.6	96.6	289.8	48.9	48.2
Dacrydium colensoi														0.4	1.7		
Dacrycarpus dacrydioides	62.2	4.8		0.4		3.2	52.6	0.1	0.7				20.6	2.0	0.4	47.3	377.9
Podocarpus ferrugineus	20.1	1.6	3.6	71.3	7.9		13.5	36.9	48.5	0.6	0.7	0.3	10.4	4.4	18.5	16.3	7.8
Podocarpus spicatus	126.0	6.4			13.8		140.9	32.0	117.4		2.7	<0.1	0.2			16.3	
Podocarpus totara	32.6	2.5		0.4			70.4										0.6
Podocarpus hallii		<0.1						29.6	(67.1)		12.0	<0.1	0.7	0.2	0.6	(5.0)	
Phyllocladus trichomanoides				1.0	0.3		<0.1	0.4	22.1								
Phyllocladus glaucus								0.5	<0.1								
Phyllocladus alpinus								0.2							<0.1		
Libocedrus bidwillii											10.7		0.5	1.1		<0.1	0.6
Nothofagus menziesii	0.3											42.2	6.4	0.7			6.0
Nothofagus fusca	2.7	0.2										38.5	13.9	3.8			
Nothofagus truncata													1.1	10.5			
Nothofagus solandri												0.2	<0.1	1.3			
Beilschmiedia tawa	11.4	13.2	28.7	65.3	13.5	35.8	<0.1		0.6	12.1							
Elaeocarpus dentatus	0.3	0.5	2.0	3.1		0.9	1.8	0.2	0.8	0.6		<0.1		<0.1			
Elaeocarpus hookerianus		0.1					<0.1	1.8	1.8		0.4	<0.1	0.1	0.1		<0.1	<0.1

Knightia excelsa	<0.1	1.2	1.8	2.0		1.1	0.2			0.3							
Nestegis spp.	0.9	0.1		0.2			0.2	<0.1	0.7		0.2	0.1					
Weinmannia racemosa			0.1	0.4		0.3							0.1	<0.1	<0.1	3.0	0.6
Metrosideros umbellata														<0.1	2.0		
Ixerba brexioides				0.1													
Quintinia acutifolia									0.5	<0.1				<0.1	<0.1		
Myrsine spp.									<0.1								
Dysoxylum spectabile			<0.1	<0.1		0.6											
Persoonia toru							<0.1	0.1									
Laurelia novae-zelandiae			3.4	0.2		3.7											
Litsea calicaris			1.1	2.5		2.0											
Total	652.7	57.1	78.5	144.7	493.4	90.4	456.6	105.6	441.5	32.3	36.2	81.7	83.2	121.7	313.6	137.1	442.1
Number of trees ha^{-1} with dbh > 12 in. (30.5 cm)																	
merchantable softwood	141	5	7	12	99	7	99	67	96	5	20		27	54	185	49	202
merchantable hardwood	17	17	27	49	10	32	2	2	2	10		52	30	25		3	6
total including unmerchantable	175	52	104	82	109	72	124	94	140	40	*	*	106	141	224	117	282

P1 = Dense mixed podocarps.
P3H1 = Scattered podocarps/mainly *Beilschmiedia tawa.*
PH2 = Scattered *Dacrydium cupressinum*/*Beilschmiedia tawa* and *Weinmannia racemosa.*
P1H1 = *Dacrydium cupressinum*/*Beilschmiedia tawa.*
P5 = Dense *Dacrydium cupressinum*/sparse *Beilschmiedia tawa.*
PH1 = Scattered *Dacrydium cupressinum* and *Dacrycarpus dacrydioides*/dense *Beilschmiedia tawa.*
P1 = Dense mixed podocarps.
P5= Dense mixed podocarps of small diameter.
P6 = Dense podocarps/*Weinmannia racemosa.*
P16H2 = Sparse *Dacrydium cupressinum* and *Metrosideros robusta*/*Beilschmiedia tawa.*
P3 = *Libocedrus bidwillii, Dacrydium cupressinum, Podocarpus hallii.*
B1 = Almost pure *Nothofagus fusca* and *N. menziesii.*
PB1 = *Dacrydium cupressinum* and *Dacrycarpus dacrydioides*/*Nothofagus fusca* and *N. menziesii.*
PB3 = Small *Dacrydium cupressinum*/*Quintinia, Weinmannia racemosa*, etc.
P1 = Dense *Dacrydium cupressinum.*
P19 = Mixed podocarps/*Weinmannia racemosa*, etc.
P8 = Dense *Dacrycarpus dacrydioides.*
*Data not available.

Podocarps and hardwoods grow within them, but the latter only poorly and sparsely where the kauri trees are closely spaced. Undergrowth in the groves is dominated by the large tussocky herbs *Astelia trinervia* and *Gahnia xanthocarpa*. Other species characteristically associated with kauri include *Dracophyllum latifolium*, the epiphytic shrub *Senecio kirkii* and a scrambling fern, *Lygodium articulatum*.

Dacrydium cupressinum (Podocarpaceae)

D. cupressinum ranges literally from North Cape to South Cape. Only on the eastern side of the South Island is it absent over large areas. The altitudinal limits decrease from north to south from 950 m to about 300 or 400 m. It is most prevalent as well-spaced or grouped overstorey trees of large girth in forests with a main canopy dominated by hardwoods such as *Beilschmiedia tawa* and *Weinmannia racemosa*. It has a similar role in most conifer–beech–broadleaf stands, and trees can also be scattered through mesic beech forests.

D. cupressinum is associated with *Dacrycarpus dacrydioides*, *Podocarpus ferrugineus*, *P. totara* and *P. spicatus* in dense conifer stands, which until recently were extensive in the central North Island. At least some of these stands are seral in origin, following fire and volcanic eruption. On the infertile gley podzols of Westland, there are nearly pure stands of *Dacrydium cupressinum*, with *Podocarpus ferrugineus* usually as the only other canopy species (Fig. 3.13). Slender, coppicing *Weinmannia racemosa* and *Quintinia acutifolia* form a sparse subcanopy, while the shrub storey consists mainly of these two hardwoods and *Neomyrtus pedunculata*, although tree ferns, *Phyllocladus alpinus* and other shrubs may be locally common. The forest floor is generally moss-covered. The presence of the medium-sized podocarp *Dacrydium colensoi*, mainly in wet hollows, indicates a relationship to heath-forest.

Fig. 3.13. Logging in dense *Dacrydium cupressinum* forest near Okarito, Westland.

Beilschmiedia tawa (Lauraceae)

B. tawa is the prevailing tall hardwood in North Island low- to mid-altitude forests on soils of moderate fertility. It also grows in the Marlborough Sounds, and very locally on the coastal hills of eastern Marlborough to 42°20′S. In warm lowland and coastal sites, stands with *B. tawa* include some of the most luxuriant examples of conifer–broadleaf forest. The conifer overstorey usually includes *Dacrydium cupressinum* and sometimes also *Podocarpus spicatus* and *P. totara*. Other tall hardwoods are *Elaeocarpus dentatus*, *Knightia excelsa*, *Litsea calicaris*, *Metrosideros robusta* and *Vitex lucens*. Typical subordinate species are the trees *Alectryon excelsus*, *Dysoxylum spectabilis*, *Hedycarya arborea*, *Melicytus ramiflorus*, *Pseudopanax arboreus* and *Pseudowintera axillaris*, various tree ferns including *Cyathea dealbata* and *Dicksonia fibrosa*, profuse lianes including *Freycinetia banksii*, *Metrosideros* spp., *Ripogonum scandens*, and the fern *Blechnum filiforme*, the palm *Rhopalostylis sapida*, shrubs such as *Coprosma australis*, *Geniostoma ligustrifolium* and *Macropiper excelsum*, and large ground ferns. *Collospermum hastatum* is an abundant epiphyte. Further species associated with *Beilschmiedia tawa* in northern districts include *Agathis australis*, *Alseuosmia macrophylla*, *Ixerba brexioides* and *Quintinia serrata*.

Floristically simpler stands dominated by podocarps and *Beilschmiedia tawa* grow in cool, moist upland sites on the Volcanic Plateau, to an altitude of nearly 900 m. In these, *Elaeocarpus dentatus*, *Nestegis* spp., *Weinmannia racemosa* and (in the north) *Ixerba* are associated tall hardwoods. Principal lower-storey species include *Blechnum discolor*, *Cyathea smithii*, *Dicksonia squarrosa*, *Hedycarya*, *Melicytus ramiflorus* and *Schefflera digitata*.

Seedlings, saplings and young trees of *Beilschmiedia tawa* are usually abundant in undisturbed forest, whereas those of the large conifers other than *Podocarpus ferrugineus* are scarce.

Nothofagus solandri (Fagaceae)

N. solandri is a wide-ranging species, with two recognized varieties. Var. *solandri* is found mainly on warm, dry, lowland sites from East Cape to mid-Canterbury. Var. *cliffortioides* replaces it at higher altitudes, in southern districts, and on poor soils under high rainfall, and itself encompasses distinct ecotypes. Where the two varieties overlap, there are mixed stands containing many trees with intermediate characteristics. Var. *cliffortioides* stands of quite remarkable floristic poverty form much of the subalpine forest east of the Main Divide in the South Island, and are described on pp. 34–35. At the other extreme, either variety can codominate in complex conifer–beech–broadleaf stands. A stand in eastern Marlborough, for example, has *Dacrydium cupressinum* and other conifers as an overstorey to a main canopy of *Nothofagus solandri* with scattered trees of *Elaeocarpus dentatus*. Lower strata include the small trees *Carpodetus serratus* and *Myrsine australis*, the tree fern *Cyathea smithii*, and the shrubs *Coprosma australis*, *C. rhamnoides* and *Cyathodes fasciculatus*, and a dense ground layer of *Blechnum discolor*.

On poorly drained soils in the northwest and southwest of the South Island and on the wetter North Island mountains, *Nothofagus solandri* var. *cliffortioides* is associated with conifers such as *Dacrydium biforme*, *Libocedrus bidwillii* and *Phyllocladus alpinus* that also tolerate wet infertile soils. *Coprosma foetidissima*, *Cyathodes juniperina*, *Myrsine divaricata* and *Pseudopanax linearis* are typical understorey shrubs, while moss usually forms a continuous ground cover. On shallow soils on granitic and quartzose rocks, those communities grade into heath-forest and scrub dominated by small *Dacrydium* spp., *Leptospermum scoparium* and stunted *Nothofagus solandri* var. *cliffortioides*.

Nothofagus fusca (Fagaceae)

In contrast to the preceding species, *N. fusca* is the most demanding beech in New Zealand. Its range extends discontinuously from East Cape and Mount Te Aroha to Lake Monowai. It ascends to the upper limit of the montane belt, and descends to sea level from Cook Strait southwards. In frosty inland valleys in regions of fairly high rainfall, *N. fusca* forms one of the most impressive native forest communities. Pure stands on colluvial fans and recent terraces have only sparse understoreys, containing scattered small trees (e.g. *Carpodetus serratus*, *Griselinia littoralis*, *Pseudopanax crassifolius*), small-leaved *Coprosma* spp., the liane *Rubus cissoides* and the tufted fern *Polystichum vestitum*.

Deep litter and humus of *Nothofagus fusca* leaves cover the ground. On adjacent, less favourable sites, *N. fusca* tends to be accompanied or replaced by *N. solandri* or *N. menziesii.*

N. fusca is also characteristic of conifer–beech–broadleaf forest, especially in the northwest of the South Island. As usual, conifers form a scattered overstorey with *Dacrydium cupressinum* and *Podocarpus ferrugineus* as the main species. Beeches form the main canopy, with *Nothofagus fusca* predominating on the deeper, more freely drained sites, but it is usually mixed with *N. truncata* and *N. menziesii*, the former increasing in dominance on the upper slopes, ridges and spurs and the latter likewise on frosty flats and low terraces. *N. solandri* can also be present on gradients towards gley podzol soils. On the optimal sites, *Weinmannia racemosa* and in places *Quintinia acutifolia* form a tall subcanopy. Small trees and shrubs include *Coprosma rhamnoides*, *Myrsine divaricata*, *Neomyrtus pedunculata* and *Pseudowintera colorata.* The ground layer is usually well developed, with *Blechnum discolor* or *Microlaena avenacea* as the main species, according to site.

Where *Nothofagus fusca* and *N. solandri* grow in proximity, secondary stands developed after fire include many hybrid trees (Cockayne, 1926).

Weinmannia racemosa (Cunoniaceae)

This is the most abundant and widely distributed of the larger hardwoods, its range extending from the south of Stewart Island to the Waikato district at 38°S. In the north it is replaced by the closely related *W. silvicola.* Being favoured by high rainfall, it is rare to absent in the dry eastern districts of both islands. In the west and south of the South Island *W. racemosa* is abundant to sea level, but in the North Island tends to be replaced at low altitudes by *Beilschmiedia tawa*, especially on the more fertile soils. The upper limits of *Weinmannia racemosa*, which decline from 1100 m to 700 m in passing southwards along its range, provide a good marker for the montane/subalpine transition.

On lower hill slopes in Westland, *W. racemosa* is the main tall tree in a luxuriant aspect of conifer–broadleaf forest. Accompanying canopy species can include *Dacrydium cupressinum*, *Elaeocarpus dentatus*, *E. hookerianus*, *Metrosideros umbellata*, *Podocarpus ferrugineus*, *P. hallii* and *Quintinia acutifolia.* Small trees and shrubs include, among others, *Coprosma lucida*, *Hedycarya arborea*, *Myrsine australis*, *Neomyrtus pedunculata*, *Pseudopanax crassifolius*, *Pseudowintera colorata*, and the tree ferns *Cyathea smithii* and *Dicksonia squarrosa. Blechnum discolor* and *B. capense* form most of the herbaceous understorey. The lianes *Metrosideros* spp. and *Ripogonum scandens* are abundant, while epiphytes include Hymenophyllaceae, *Astelia solandri* and *Phymatosorus diversifolius.*

At higher altitudes, *Weinmannia racemosa* dominates in forests that are simpler floristically and structurally. On recent surfaces it is frequently codominant with *Metrosideros umbellata* in even-aged stands. It is also an important species in conifer–beech–broadleaf forests.

Griselinia littoralis (Cornaceae)

This tree seldom attains a height of more than 10 m, but the trunk is irregular and disproportionately thick. *G. littoralis* is distributed through the three main islands but is generally a minor species in dense, tall low-altitude forests. Though mostly epiphytic in origin, it may develop into a massive tree, laden with epiphytes. It occurs sparsely in the subcanopies of most montane beech forests, locally dominating its stratum on moist, rocky sites. It is often abundant, and in places dominant, in upper montane and lower subalpine stands of conifer–broadleaf forests.

G. littoralis features most prominently in forest patches scattered through the grassland landscapes in the east of the South Island. Before the advent of the Maoris, it seems it dominated the main canopy *in lieu* of taller hardwoods in these drier areas, and some stands still have well-developed overstoreys of conifers; e.g. *Podocarpus totara*, *P. spicatus*, and at higher altitudes, *P. hallii* and *Libocedrus bidwillii.* However, other stands have few or no conifers, and appear to be secondary after fires of the "Maori" era. *Griselinia littoralis* actively coppices after fire, unlike most native trees. Stands often also border on or interdigitate with *Nothofagus* forest, in a manner that suggests old fire patterns. Other small hardwood trees, such as *Carpodetus serratus* and *Myrsine divaricata*, also contribute to the canopy of these low, broad-leaved forests, and in the absence of severe browsing there is usually a dense fern layer, typically formed by *Polystichum vestitum.*

Other important trees

Podocarpus spicatus and *P. totara*; these massive podocarps are widespread in regions of moderate rainfall, ascending well into the montane belt, but in wet districts rapid leaching of soils largely confines them to recent alluvium and tephra deposits. They are replaced by *P. hallii* on poorer soils and at higher altitudes.

Dacrycarpus dacrydioides; formerly one of the most abundant podocarps, this tree is characteristic of moist, fertile soils, and can form nearly pure stands on recent alluvial flats and bordering lakes and swamps. The community is commonly designated "swamp forest", but although water lies between the trees in many stands, this results usually from flooding after the forest is established. *D. dacrydioides* also grows in mixed podocarp stands on moist, fertile soils.

Phyllocladus trichomanoides; a podocarp up to 25 m tall, found mainly in conifer–broadleaf stands on well-drained soils in the northern part of the North Island. It tends to be replaced by *P. glaucus* on wet ground and at higher altitudes. *P. alpinus* is widely distributed as a subalpine shrub, but also grows as a small tree at low altitudes in forests on gley podzol soils in the west of the South Island, and in frosty inland valleys in both main islands.

Libocedrus bidwillii (Cupressaceae); a medium-sized conifer that often shares the overstorey with *Podocarpus hallii* in high-altitude conifer–broadleaf forests. *L. plumosa* has a very limited distribution at low altitudes in the north of the North Island and extreme northwest of the South Island.

Laurelia novae-zelandiae (Monimiaceae); a tall hardwood with buttressed lower trunk, growing on moist gullies or swampy lowlands in the North Island and northernmost parts of the South Island.

Litsea calicaris (Lauraceae); locally an important tree although restricted to low-altitude conifer–broadleaf forest of the northern half of the North Island.

Beilschmiedia tarairi (Lauraceae); locally dominates the broad-leaved canopy or shares it with *B. tawa* in lowland forests north of 38°S.

Nothofagus menziesii (Fagaceae); most of the subalpine beech forests along and west of the Main Divide in the South Island, as well as those of the Tararua, Huiarau and Raukumara Ranges of the North Island, are dominated by this species. At lower altitudes in the same districts it is widely codominant in conifer–beech–broadleaf stands. Although taxonomically apart from the other beech species in New Zealand, it shares their main ecological and silvicultural attributes. *N. menziesii* resembles *N. fusca* in growing best on moist, sheltered relatively fertile sites, but through its cold tolerance it replaces the latter at high altitudes, in valleys subject to temperature inversions, and in southern districts of the South Island. Locally, *N. menziesii* can substitute for the seemingly hardier and less demanding *N. solandri* var. *cliffortioides*, both in relict patches of forest bordering the semi-arid basins of Central Otago and on extremely infertile soils.

Nothofagus truncata (Fagaceae); similar to *N. fusca* in appearance, but tolerating drier sites and somewhat poorer soils. Outliers of *Nothofagus* in the North Auckland Peninsula are of this species whereas in the South Island it reaches south only to 42°40′S, except for recently discovered occurrences near Haast, that extend the previously known southern limit by 250 km (June, 1977).

Knightia excelsa (Proteaceae); a tree of fastigiate habit, growing in warm districts to as far south as the Marlborough Sounds. In virgin forest it is found only locally, mainly on ridge crests, but can dominate in secondary forests developed after fires.

Metrosideros robusta (Myrtaceae); important in lowland and montane forests of the North Island, usually as trees which have originated as epiphytes on large conifers, but also as a pioneer on landslide scars and volcanic debris. It extends along the west coast of the South Island to 43°S.

Metrosideros umbellata (Myrtaceae); next to *Weinmannia racemosa*, this is the main tall hardwood of *Nothofagus*-free areas in the south and west of the South Island and on Stewart Island, dominating mainly on ridge crests, landslide scars, and exposed slopes. Groves of *Metrosideros umbellata* on the Auckland Islands form the southernmost outpost of true forest in New Zealand.

Metrosideros excelsa (Myrtaceae); the celebrated *pohutukawa*, abundant in remnants of coastal forest and on coastal cliffs and rocky shores north of 39°S. It also grows on the inland shores of some of the lakes of the Volcanic Plateau.

Elaeocarpus dentatus (Elaeocarpaceae); a

hardwood, typically with a short bole and spreading crown, occurring widely in lowland and lower montane forests south to 43°40′S.

Elaeocarpus hookerianus (Elaeocarpaceae); a widely distributed tree that tends to replace *E. dentatus* on poorly drained sites, at higher altitudes, and in the southern half of the South Island.

Aristotelia serrata (Elaeocarpaceae); a semi-deciduous small tree or large shrub that rapidly colonizes mesic sites laid bare by fire, felling, windthrow, or landslide. The most abundant of a group of small hardwoods with pioneering tendencies that includes also *Carpodetus serratus*, *Fuchsia excorticata*, *Hoheria populnea*, *H. sexstylosa*, *Melicytus lanceolatus*, *M. ramiflorus*, *Myrsine* spp., *Pennantia corymbosa*, *Pittosporum* spp., *Pseudopanax* spp., *Schefflera digitata*, and the taller, larger-leaved species of *Coprosma*.

Plagianthus betulinus (Malvaceae); the largest deciduous tree in New Zealand, attaining heights of 10 to 15 m. It grows mainly on fertile alluvial flats, and can dominate small areas in frosty inland valleys.

Hoheria glabrata (Malvaceae); likewise deciduous, this small tree dominates moist subalpine talus in the west of the South Island.

Quintinia spp. (Escalloniaceae); medium-sized hardwoods, found in hilly forests north of 39°30′S, with one species (*Q. acutifolia*) abundant on the western side of the South Island to 43°30′S.

Nestegis spp. (Oleaceae); important hardwoods in hilly forests in the North Island, and extending to the northeast of the South Island. Four species are recognized.

Vitex lucens (Verbenaceae); north of 39°30′S, on fertile lowland soils. Today it is seen mainly in small forest remnants, or as isolated trees surviving forest clearance.

Dysoxylum spectabile (Meliaceae); a low-crowned hardwood which often dominates on coastal slopes in the North Island and extreme north of the South Island.

REPRODUCTIVE BIOLOGY, PHENOLOGY, DEVELOPMENT

A recent review (Godley, 1979) contains information which extends the following summaries.

Pollination

Among wind-pollinated trees, the widely dominant Podocarpaceae are mainly dioecious, while *Agathis australis* and *Nothofagus* are monoecious, the latter possessing bright red stamens despite its anemophily. Most insect-pollinated species have inconspicuously small, greenish flowers (e.g., *Beilschmiedia*, *Griselinia*, *Plagianthus* and *Pseudowintera*). However, some trees possess white flowers that are conspicuous though their size (e.g. *Hoheria* spp., *Ixerba brexioides*) or massing (e.g. *Pennantia corymbosa*, *Weinmannia* spp.). Three species of tree *Metrosideros* have bright red flowers, and two of *Sophora* have yellow flowers. These are largely bird-pollinated, as are the less conspicuously coloured flowers of *Fuchsia excorticata*, *Knightia excelsa* and *Vitex lucens*.

In the undergrowth, *Coprosma* species are anemophilous and dioecious, while most other species have inconspicuous, usually greenish, insect-pollinated flowers. Several species of *Pittosporum* have reddish-black flowers. Those of *Pratia physaloides*, the very rare *Clianthus puniceus*, and the endemic genera *Alseuosmia* and *Rhabdothamnus* are distinct in having larger flowers coloured in various shades of red.

Small, white or greenish flowers also predominate among lianes and epiphytes, though conspicuous displays are provided by the large white dioecious flowers of *Clematis paniculata*, the bright red flowers of two climbing species of *Metrosideros*, and the red or yellow flowers of three species of loranthaceous mistletoe.

New Zealand flowers are generally only weakly scented, although *Alseuosmia*, *Cordyline australis*, *Geniostoma*, *Melicytus*, *Parsonsia*, some *Pittosporum* spp., and the epiphytic orchid *Earina* are exceptions.

Dispersal of seed

Most New Zealand trees, including the podocarps, have fleshy drupes, berries, receptacles or arils that adapt the seed for dispersal by birds (Beveridge, 1964). These fruit are generally small, but *Beilschmiedia* spp., *Corynocarpus laevigata* and *Podocarpus ferrugineus* have drupes 2 to 4 cm long. These are eaten mainly by the large native pigeon *Hemiphaga novae-seelandiae* and were also gathered by the Maoris before the introduction of European fruit trees. *Metrosideros* and *Weinmannia* shed very light seed from more or less dehiscent capsules. *Agathis*, *Hoheria*, *Laurelia* and *Libocedrus* have plumed or winged seeds, while *Heimerliodendron*, *Ixerba*, and *Pittosporum* have sticky seeds liable to adhere to birds. *Nothofagus* seeds lack obvious dispersal aids, although the nuts, weighing about 5 mg, are slightly winged and smaller than in most Fagaceae. They are eaten by birds or rodents, and under special circumstances this could aid dispersal.

Few native tree species occur beyond the New Zealand archipelago. *Sophora microphylla* however, because of buoyant seeds, their tolerance of prolonged immersion in salt water, and an ability to establish on banks of estuaries, has dispersed as far east as Gough Island in the South Atlantic Ocean (Sykes and Godley, 1968). Trees which range northwards into the tropical Pacific include the berried *Melicytus ramiflorus* and the wind-distributed *Dodonaea viscosa*, while a few species, mainly seral shrubs such as *Leptospermum scoparium*, occur in Australia.

Small fleshy fruit also predominate in other synusiae. The creeping *Nertera* spp. on the forest floor and *Luzuriaga parviflora* which grows on the bases of trees, have drupes or berries possibly reflecting the importance of ground birds in the original fauna. The fruits of the sedge *Uncinia*, which differs from *Carex* in possessing an elongated, hooked rachilla, cling tenaciously to fur and feather, and species such as *Uncinia uncinata* have increased greatly since the advent of mammals to the forests. All orchids and pteridophytes, and the lianes *Clematis* and *Parsonsia*, are anemochorous.

The seeds of *Libertia* (Iridaceae) are released from a dehiscent capsule and have no obvious facility for long-distance dispersal, yet the smallest, most cryptic species, *L. pulchella*, grows on the floor of damp, mossy, montane forests in New Guinea, eastern Australia and Tasmania, as well as in New Zealand.

Phenology

The New Zealand forest would present an unchanging appearance through the year, were it not for the flowering of the relatively few showy species, and the local presence of the winter-deciduous *Fuchsia excorticata*, *Plagianthus betulinus*, and the subalpine *Hoheria glabrata*. In lowland conifer–broadleaf forest, the more conspicuous flowering sequences are as follows:

Sophora: July to October, according to locality and genotype

Clematis: October–November

Pennantia, *Weinmannia*: November–December

Metrosideros excelsa: November–January

M. robusta: December–January

M. umbellata: January–February, though with some trees in full bloom from November to March.

M. fulgens: February–July

While early summer flowering prevails among the less conspicuous species, there is no month without some flowering.

Nothofagus has a regular flush of new shoots, bearing the leaves and male and female flowers, which begins during October in the lowlands, reaching timberline by late December (J. Wardle, 1970). Most New Zealand trees commence growth as soon as spring temperatures rise enough (e.g. to 17–20°C in *Agathis australis* according to Bieleski, 1959; see also Bussell, 1968). However, *Nothofagus solandri* can undergo true winter dormancy which is unbroken by warm temperatures alone (Wardle and Campbell, 1976).

"Berries" ripen mainly between late summer and early winter, although no month has a complete dearth. Wind-dispersed seeds ripen mainly in late summer. The woody capsules of *Metrosideros umbellata* have been reported to develop for more than a year before they release their seeds, while for the drupes of *Coprosma lucida*, seventeen months elapse between fertilization and ripening (McEwan, 1966).

Periodicity in flowering and fruiting

A flowering periodicity of around three years occurs in *Nothofagus* although the interval and intensity vary, the incidence varies from district to district, and the periodicity is less marked in *N. menziesii* than in other species. At the higher altitudes, the flowers are subject to damage by late frosts, and even in the event of a heavy seed crop, adverse conditions can sharply reduce successful germination and establishment of seedlings. Because of these factors, fifteen years have separated the last two "bumper" seedling crops in a subalpine stand of *N. solandri* var. *cliffortioides*.

Weaker and less regular periodicity affects seed fall in *Dacrydium cupressinum* (Franklin, 1968) and other podocarps (Beveridge, 1964, 1973), while the flowering display of *Metrosideros* varies markedly from year to year. There is little information for other forest species.

Seed and germination (with M.J.A. Bulfin)

The seeds of most forest species are small, even minute, and lack significant amounts of endosperm or storage cotyledons. Seeds with more storage tissue are found in those species having relatively large, one-seeded succulent fruit (e.g. *Beilschmiedia, Corynocarpus, Elaeocarpus, Podocarpus ferrugineus*).

Initial viability of apparently sound seed has proved to be high in species tested, but often only a low percentage of seed is sound (e.g. 13% in a sample of *Metrosideros umbellata*). J. Wardle (1970) found that the percentage of sound seed of *Nothofagus solandri* decreases markedly in poor seed years and towards the timberline. The viability of some seeds, e.g. those of *Metrosideros* spp., *Nothofagus* spp., and *Agathis australis* (Mirams, 1957), decreases almost to zero within a year when lying in the soil or stored at ambient temperatures and humidities. Other species retain high viability longer, e.g. at least three years in *Leptospermum ericoides* and four years in *Podocarpus ferrugineus* and *P. spicatus* (Beveridge, 1973). The life of *Nothofagus* seed has been prolonged greatly by storage at 6% moisture content and 5–10°C (Bibby and Williams, 1953); that of *Agathis* can be similarly extended.

The seeds of *Griselinia*, which have green embryos, germinate immediately after they fall, and for most other species studied, germination takes place within a month of moistening. Autumn germination soon after seedfall is the rule in the northern genera *Agathis*, *Beilschmiedia* and *Litsea*, and also in *Nothofagus truncata* in at least some districts. Otherwise, *Nothofagus* species respond to low temperature pre-treatment, to the extent that usually little germination takes place during the autumn after seedfall, whereas there is mass germination in the spring (J. Wardle, 1970). Podocarps germinate in early summer, at least six months after seedfall. *Coprosma* species tend to have a long dormancy, which can be broken by low-temperature treatment. *Lophomyrtus obcordata* has germinated at 0°C, while *Pseudopanax crassifolius* can germinate over the range 2.5 to 30°C. Light is necessary for the germination of many species, including *Leptospermum* spp. and *Metrosideros* spp.

Some seeds are surrounded by hard endocarps or testas, which can prevent entry of water and delay germination indefinitely; this condition is found among plants with both small (e.g. *Aristotelia*, *Ascarina*), and relatively large fruits or seed (e.g. *Elaeocarpus*, *Vitex lucens*; Godley, 1971). The seed coat of *Sophora*, like that of many other legumes, is thin but impervious, and germination can be hastened by scarifying.

Ecology of tree seedlings

Germination is generally epigeal or nearly so, and seedlings are mostly small and delicate, in keeping with the seeds from which they emerge. High mortality has been demonstrated for *Nothofagus solandri*. Seedlings with noticeably greater persistence are produced by *Beilschmiedia tawa* (Cameron, 1963), *Podocarpus ferrugineus* and probably other large-seeded species.

The following classification of seedlings according to their establishment and growth in response to habitat depends mainly on casual observations.

(1) *Aristotelia serrata* and *Carpodetus serratus* are small hardwoods that establish and grow in well-lighted situations, being especially vigorous in recently burnt or logged forests and on moist landslide scars and talus. *Fuchsia excorticata* is characteristic of rather more mesic sites, especially stream banks and gullies. In virgin forest, these species occur only ephemerally, usually on bare ground exposed by uprooted trees. Seedlings of *Metrosideros umbellata* also establish on bare surfaces, but being slow-growing, are restricted by competition to harsh sites such as moraine crests and landslide scars, where the species can form nearly pure seral stands.

Other small trees are also effective pioneers in disturbed areas, but, being more shade-tolerant, are also common within primary forest. Examples are *Hedycarya arborea*, *Pseudopanax crassifolius* and *Pseudowintera colorata*.

(2) A pronounced feature of lowland conifer–

broadleaf forest is the "strangling" habit—that is, where seedlings commence life as epiphytes and send down roots which eventually coalesce as a false trunk encasing the host. *Metrosideros robusta*, commencing life in the crowns of large podocarps, is the most spectacular example. The seedlings of *Weinmannia racemosa* have a predilection for the trunks of tree ferns. Other trees which often have epiphytic seedlings include *Griselinia littoralis*, *Metrosideros umbellata*, *Pseudopanax arboreus*, *P. colensoi* and *P. edgerleyi*, while many other species occasionally do so. The habit is less frequent in moist beech and subalpine forests, and absent in drier regions.

(3) Seedlings of *Nothofagus* establish best under slight shading, and show steadily increasing growth up to the pole stage with maximum annual extensions of well over 50 cm attainable. Mounds of mineral soil brought up by wind-thrown trees provide especially favourable sites for regeneration within the forest. Seedlings establish also in fully lighted situations that are not too harsh, but height growth is inhibited until they become self-shading.

Seedlings establish also under closed canopy, mainly where the roots do not have to penetrate thick moss or humus to reach mineral soil, and where there is no dense understorey. They can persist at least twelve years showing minimal growth but grow rapidly when released through opening of the canopy to produce dense thickets of competing saplings.

(4) Seedlings of conifers, like beech, establish best where they can reach mineral soil through thin litter or moss, but commonly also begin life on rotting logs and stumps. As a rule, they are not found in the open, although in wet climates seedlings of *Dacrydium cupressinum* and other species can become established on bared podzolic topsoil where there is little competition. The best establishment of most species, if not all, is to be found under successional communities of *Leptospermum* or broad-leaved species such as *Weinmannia racemosa*. Seedlings also occur in mature conifer–broadleaf forests, but many stands show an apparent scarcity of conifer seedlings other than those of *Podocarpus ferrugineus*. This absence raises questions concerning the regenerative capacity and successional status of the stands. Bieleski (1959) found that *Leptospermum* provided optimal light conditions for seedlings of *Agathis australis*. The failure of seedlings to become established in mature *Agathis* forest is unlikely to be due to low light, for the compensation point of the seedlings lies near 1% of full daylight. Bieleski suggests that root competition may be inhibiting.

Burke (1974) found that, while young seedlings of *Podocarpus hallii* are frequent close to the parent tree, older seedlings with 40 to 60 growth rings were most abundant at 40 to 50 m distant. He suggested that autoallelopathy may be involved.

Growth of young conifers is also best in seral communities (e.g. 15–30 cm y^{-1} for *Dacrydium cupressinum*). Growth is very slow in high forests, but is apparently persistent despite periods of stagnation and die back; *D. cupressinum* can take 100 to 150 years to reach pole size (Franklin, 1968).

(5) The seedlings of *Beilschmiedia tawa* are more effectively shade-tolerant than those of most New Zealand species. They are robust, deep-rooting, and in the event of the shoot dying can resprout repeatedly from ground level (Cameron, 1963). Many species can produce coppice shoots when the top of saplings, or even canopy trees, is suppressed or damaged. *Beilschmiedia*, *Quintinia* and *Weinmannia* are notably vigorous coppicers. Even *Nothofagus* and conifers can coppice occasionally. Ability to coppice is also pronounced in many of the smaller hardwoods, e.g. *Coprosma australis*, *Fuchsia excorticata* and *Melicytus ramiflorus*.

(6) When growing in the shade, seedlings of several species produce weak side branches which root adventitiously to produce thickets, so multiplying the effect of each successful establishment. In lowland forest, *Quintinia acutifolia* shows this best, but it is much more widespread among species of high altitudes and in heath-forests on poor soils. Similar layering is also characteristic of some shrubs, especially *Neomyrtus pedunculata*.

Juvenile forms

Some New Zealand trees grow from relatively large-leaved juveniles to smaller-leaved adults (e.g. *Elaeocarpus dentatus*, *Pseudopanax crassifolius*), while in others, compound juvenile leaves become simple as the tree matures (e.g. *Pseudopanax simplex*, *Weinmannia racemosa*). Very commonly juveniles with tiny leaves and thin, divaricating branchlets change, often abruptly, to adults of more conventional appearance. This happens in species of widely disparate genera, such as *Pennantia corymbosa*, *Plagianthus betulinus* and *Podocarpus spicatus*. Moreover, some species showing this metamorphosis are closely related to species which give no indication of it, for example *Elaeocarpus hookerianus* as compared with *E. dentatus*. Many shrubs, likewise in diverse families, retain the microphyllous, divaricating habit through life.

The phenomenon is so widespread in New Zealand while apparently uncommon elsewhere, that it seems unusual selection pressures are operating. One may be the erratic incidence of intense droughts in an otherwise benign rainfall regime. Shrubs and juvenile trees would be expected to be especially vulnerable, and it may be that small expendable leaves are a better adaptation than either larger evergreen leaves or the deciduousness which is a typical response to seasonally recurring moisture or temperature stress. The poor representation in New Zealand of Leguminales, Euphorbiaceae, and the other groups which contain most of the world's xeromorphic species may be another factor, while Greenwood and Atkinson (1977) suggested that browsing by moas (extinct ratite birds) should be considered.

ANIMALS

Invertebrates (contributed by J. Dugdale)

Because of the instability of the forest landscape there is, or was, a plentitude of seral sites suitable for those indigenous insect species which adapted so well throughout New Zealand to the replacement of indigenous forest by exotic pasture. The transient erosion sites were the original home of *Wiseana* (Lepidoptera: Hepialidae, porina moths), *Costelytra* (Coleoptera: Scarabaeidae, grass grub), and the genera *Eudonia*, *Orocrambus* and *Scoparia* (Lepidoptera: Pyralidae, sod webworms). Even some cicadas, such as the *Kikihia muta* group, have also adapted successfully.

These are the exceptions. The most striking features of the insects of the New Zealand temperate forests are: (1) the immense complexity of the fauna (sixteen orders of terrestrial insects, with some 3000 species in Coleoptera alone) associated with fallen plant material, fungi and the bryophyte flora; and (2) the fact that less than 10% of this endemic, diverse, irreplaceable fauna survives replacement of the indigenous plant cover by exotic plant cover.

The complexity of the fauna, and its extreme sensitivity to any radical change of plant cover suggests that (1) immigration rates may be generally low, and (2) the fauna has co-evolved over an extremely long time with the forest. Certain host–plant relationships suggest a long period of co-evolution, such as the weevils associated with dead *Agathis australis*, whose relationships lie with weevils that elsewhere are only associated with Araucariaceae.

New Zealand forest insects can be conveniently divided into feeders on live plant tissue (phytophages) and those that feed on dead plant tissue (saprophages) usually in association with fungi, yeasts and bacteria. Southwood (1973) showed for phytophages in prehistoric England and in Hawaii that the more abundant the host, the more species of phytophages are specific to that host. For the New Zealand forest phytophages, no such correlation could be demonstrated, except for *Nothofagus*; curiously, good accordance was found for the alpine zone (Dugdale, 1977). The arboreal insect fauna in New Zealand forests is impoverished; the absence of such arboreal phytophage groups as Anthelidae, Lymantriidae, Notodontidae and Saturniidae (Lepidoptera), Pergidae and Tenthredinidae (Hymenoptera), and virtual lack of canopy Chrysomelidae (Coleoptera) in New Zealand is well known; all these groups are abundant in Australia.

Not so the saprophages; as well as the sixteen orders of insects mentioned above, there are abundant molluscs, several crustacean orders (Cladocera, Ostracoda, Isopoda and Amphipoda amongst others), primitive spiders and primitive harvestmen (Cyphophthalmi). The forest floor also supports such relicts as *Peripatus*, the huge, earwig-like dipluran *Heterojapyx* and earthworms that can attain 140 cm in length. A feature shared with Australia is the presence of a sizeable element of litter-inhabiting Lepidoptera; over 150 species, or about 11% of the total indigenous Lepidoptera, have litter-feeding larvae. Peculiar to New Zealand (or not recorded in Australia) is a smaller percentage of moths with larvae that feed in dead wood. The dominant litter-feeding family in both countries is Oecophoridae; in New Zealand *Cremnogenes* Meyrick (77 spp.) dominates.

Before people arrived, New Zealand was rodent-free, and the rodent niche was filled by large flightless stenopelmatid (henicid) grasshoppers, or wetas. Now, under pressure from three *Rattus* and one *Mus* species, only those groups that have deep burrows, or which can modify hepialid and large cerambycid galleries in trees have maintained themselves. The others have either been exterminated, or are present as fortuitous, localized populations, chiefly on small, rodent-free islands. The tree wetas, *Hemideina* spp., have withstood predation (G.W. Ramsay, pers. comm.), and still destroy a large proportion of fallen tree seed (Bieleski, 1959; Beveridge, 1964).

Epidemics of pathogenic insects are generally infrequent. There is a regular cycle of epidemics of an oecophorid moth, *Proteodes carnifex*, on *Nothofagus solandri* var. *cliffortioides*. All *Nothofagus* species are subject to invasion by three *Platypus* (Coleoptera: Scolytidae) species, which introduce fungi pathogenic to stressed trees or to trees subjected to unusually heavy attack (Milligan, 1972). Whether such epidemics lead to changes in composition of the beech canopy is a matter for speculation, for example replacement of *Nothofagus solandri* var. *cliffortioides* by the *Proteodes*-immune *N. menziesii*. The only other regular defoliator epidemics known are those of *Epiphryne undosata* on the subalpine tree *Hoheria glabrata*.

The New Zealand forest is alive with nematoceran Diptera. The green "gloom" is relieved by abundant, usually polyphagous, geometrid moths with cryptic-coloured forewings but bright yellowish hindwings; such flash colouration is effective in eluding pursuers. Other insects have striking green patterns, blending well with the abundant bryophytes and lichens. The large, pervasive bryophyte flora has a rich assortment of phytophages, including the relict moths with jaws, Micropterigidae, as well as some 200 species of weevils, several chrysomelids, and a significant number of moths. Ferns too have a relatively rich and specific insect fauna. Unlike Australia, butterflies are rare; the lycaenid genus *Helleia* is present wherever the lianoid *Muehlenbeckia* is present; the satyrine *Dodonidia* is locally present where *Gahnia* sedges or *Chionochloa* grasses occur in the forest, but its geographic range is much smaller than that of its hosts.

The extremely rich litter fauna contrasting with the impoverished phytophage fauna may be a reflection of Pleistocene events. Relict areas of forest, even if reduced almost to scrub during a glaciation, would still preserve a relatively even, high-humidity environment on the forest floor. This would permit the litter fauna to survive "summers" when little new growth was put out, or there was no flowering or fruiting, or sunshine was limited. The New Zealand forest floor fauna may be regarded as a Tertiary fauna, surviving little changes and possibly increasing as new forest environments were formed by uplift. In contrast the arboreal phytophage fauna, particularly of warmth-loving hosts, has been greatly pruned by Pleistocene glaciations.

Birds

The land birds of New Zealand are relatively few in species, but high in endemicity. Ratites, including the giant moas (Dinornithidae), which died out during the earlier centuries of Maori occupation, and the still extant kiwi *Apteryx* must have arrived before the isolation of New Zealand in the Late Cretaceous. The now very localized wattle birds (three monotypic genera in the endemic family Callaeatidae), the extinct "thrushes" (*Turnagra*), the nearly extinct ground parrot or *kakapo* (*Strigops habroptilus*), and the New Zealand wrens (Acanthisittidae) are also archaic.

Other forest birds, endemic at the specific or generic level include, among others, the *kaka* parrot (*Nestor meridionalis*), parakeets (*Cyanoramphus*), a large pigeon (*Hemiphaga novae-seelandiae*), an owl (*Ninox novaeseelandiae*), a falcon (*Falco novaeseelandiae*), honey-eaters (Meliphagidae), and the flightless rails *Gallirallus* and *Notornis*. Only a few species, such as the kingfisher (*Halcyon sancta*) and migratory cuckoo occur beyond New Zealand. However, much of the avifauna is clearly derived from Australia, and self-introductions from the west are still continuing. The white-eye *Zosterops lateralis* arrived early in the nineteenth century as did several bird species of open country.

The native avifauna has suffered grievously since European settlement through reduction of their habitat and introduction of predators. According to Williams (1973), 36 species are primarily associated with forest. Though few of these have become extinct, several are highly endangered and fewer than a dozen can be found in fair abundance through most of the remaining forests. Ten introduced species, mainly songbirds, are also found mainly in lowland forest and scrub, but among these, only the European blackbird (*Turdus merula*) is thoroughly well established in virgin forest, the others frequenting margins and disturbed areas.

Most forest birds are insectivorous and frugivorous. Insects are taken on the wing by the fantail *Rhipidura fuliginosa*, from foliage, bark and litter by many other species, and excavated from dead trees by the kaka parrot. Seed-eaters include parakeets, *kaka* and introduced finches (Beveridge, 1964). Nectar is an important food for several birds. The New Zealand pigeon, *kokako* (*Callaeas cinerea*), *kakapo*, and *takahe* (*Notornis mantellii*) eat foliage and so, apparently, did moas. The role of the latter in forest ecology would be interesting to know, particularly in view of the changes wrought by recently introduced herbivores.

Mammals

Two species of bat, one endemic at the family level, *Mystacina tuberculata*, and the other at the specific, are the only native land mammals, and both are now scarce. The Maoris, whose arrival dates from around 900 A.D., brought *Rattus exulans* and a dog, and the former at least probably influenced the existing flora and fauna. For example, the tuatara "lizard" (*Sphenodon*) disappeared from the main islands during the Maori era.

Effective European contact began with Cook's first voyage in 1769, and from this time there was continual introduction of mammals from Europe and elsewhere beginning with Norway rats (*Rattus norvegicus*), cats, pigs and goats. Plagues of *R. exulans* continued in the South Island until the late 1880s but this rat appears to have largely disappeared following the spread of mice (*Mus musculus*), ship rats (*Rattus rattus*), and stoats (*Mustela erminea*). It occurs in Fiordland but otherwise is now confined to a number of off-shore islands, as is the case with the tuatara and several species of native forest birds. The recent introduction of *Rattus rattus* to Big South Cape Island, southwest of Stewart Island, resulted in the extinction of four populations of native birds, major reductions in the populations of four more species, and the extinction of a population of native bats (Atkinson and Bell, 1973). This episode may parallel major reductions and extinctions of native birds which occurred on the main islands in the middle and latter parts of the last century when *R. rattus* was spreading through the country (Atkinson, 1973). Rats and mice also eat significant quantities of podocarp, beech, and other tree seed.

In the latter part of the nineteenth century, several species of deer were naturalized, and there are now few forest areas without at least one species. Red deer (*Cervus elaphus scoticus*) are widespread; species of more or less restricted distribution are wapiti (*C. elaphus nelsoni*), sika deer (*C. nippon*), white-tailed deer (*Odocoileus virginianus*), fallow deer (*Dama dama*) and sambar (*Cervus unicolor*); while moose (*Alces alces*) possibly still survive in Fiordland.

Goats are found in many forest margins and openings, while chamois (*Rupicapra*), although typically alpine animals, stray down to low altitudes in forests on the flanks of the Southern Alps. Feral cattle were well established in some forests up to the earlier years of the present century, but are now rare. Domestic cattle, on the other hand, still penetrate many forest margins, and unfortunately this includes many biologically and scenically valuable forest remnants which are nominally protected as reserves.

Australian possums (*Trichosurus vulpecula*) were introduced from 1837 onwards, in the mistaken belief that they could sustain a fur trade without adverse effects on the forest. They are still spreading inexorably from numerous points of introduction.

Also in the nineteenth century, mustelids were introduced to control rabbits which were devastating the native grasslands. These, especially stoats (*Mustela erminea*), spread into the forests, and together with feral cats are commonly believed to be an important factor in the decline of native birds.

SYMBIOSIS

Nothofagus, like *Fagus*, has ectomycorrhizas (Morrison, 1956). The rest of the vascular flora in forests, except the sedges and rushes, is normally in symbiosis with vesicular–arbuscular endomycorrhizal fungi, but in *Leptospermum* ectomycorrhizal symbionts are common alternatives (McNabb, 1968) and they have even been recorded on ferns in special circumstances (Cooper, 1976).

Griselinia littoralis was the first plant shown to depend on endomycorrhizas for uptake of phosphorus from its natural soil (Baylis, 1959). Root morphology and growth rate mainly determine the degree of this dependence. It is complete in the podocarps and *Agathis* in which the ultimate rootlets are contracted into nodules that accommodate the symbionts, and also in species like *Coprosma robusta*, *Griselinia littoralis* and *Weinmannia racemosa*, in which root hairs are lacking or poorly developed. It is less in species with hairy roots especially if their growth is slow. Examples are *Metrosideros umbellata* and the ferns (Baylis, 1975).

Sophora is the only legume in New Zealand forests and is confined to dry open forests, forest margins, river banks and lake shores. *Coriaria arborea*, which often has a place in successions leading to forest, has efficient nitrogen-fixing nodules (Harris and Morrison, 1958).

FOREST HEALTH AND STABILITY

Many New Zealand forests give an impression of instability, resulting from widespread death of canopy trees, or from apparent scarcity of young plants of important species. In some instances this can be attributed to predators or pathogens; in others, the cause of the imbalance remains obscure, even where harmful organisms are present.

Understorey damage

Deer, goats and more locally cattle, pigs and chamois have drastically modified the understorey in much New Zealand forest, especially on the more fertile soils where palatable species such as the fern *Asplenium bulbiferum* or shrubs such as *Coprosma foetidissima* or *Pseudopanax colensoi* are abundant. The immediate effect is denudation, trampling and superficial erosion of the forest floor, and eventually spread of unpalatable or browse-tolerant plants such as *Microlaena avenacea*, *Pseudowintera colorata* and *Uncinia uncinata*. Failure of regeneration can lead to opening of the canopy, and eventual deterioration to an incomplete tree cover that cannot adequately protect steep slopes.

On less fertile soils, understorey species are mostly unpalatable, and, although species such as *Blechnum discolor* and the suckering tree *Quintinia acutifolia* may increase through elimination of competitors, change of the forest structure is less obvious.

Failure of regeneration

Nothofagus seedlings are only moderately palatable to deer, yet beech forest is peculiarly vulnerable to browsing pressures. This is partly because there are few other available understorey plants in the drier forests, and even in the wetter forests, palatable species such as *Astelia nervosa*, *Polystichum vestitum* and *Pseudopanax colensoi* are quickly eliminated. Furthermore, beech forests are often heavily grazed during severe weather because they provide shelter for large numbers of animals sustained by adjoining expanses of grasslands during fine weather. Commercial hunting of deer from helicopters has greatly reduced deer numbers during the last ten years, and although this has resulted in a general improvement in the condition of mountain vegetation, response has been tardy in those beech forests providing refuge for the remaining animals.

J. Wardle (1970) has shown that browsing history is reflected in the age structure of beech forest, with periods of high pressure resulting in deficient age classes.

The paucity of conifer seedlings in many mature conifer–broadleaf and conifer–beech–broadleaf stands is not easy to explain. The nature of the problem can best be illustrated by reference to *Dacrydium cupressinum* in central Westland. This tree is numerically most successful on infertile gley podzols where it forms stands with a mosaic structure. Some patches of the mosaic contain trees of all ages, others are even-aged, and at least some of the latter are initiated through wind fall. If a sequence beginning with these infertile soils and ending with soils developed on recent talus and alluvium is examined, it can be seen that *D. cupressinum* trees become larger and more widely spaced and the proportion of young trees and saplings decreases nearly to zero, as the vigour and density of hardwoods and understorey plants increases. Table 3.4 illustrates this difference in stand structure. Subalpine populations of *Libocedrus bidwillii* on the western flanks of the Southern Alps also

TABLE 3.4

Representation of size-classes of *Dacrydium cupressinum* in: I, nine samples in conifer–broadleaf stands on yellow-brown earths on hill slopes near Franz Josef Glacier; and II, six samples in denser *D. cupressinum* stands on flat or gently rolling podzolized sites in the same district

	< 1.2 m* tall	Diameter (cm)*									
		0–10	–20	–30	–41	–51	–61	–71	–81	–91	91
I	60	40	8	3	3**	15	21	23	11	18	19
II	107	70	13	9	17	7	4	10	6	3	8

*Measured in inches, converted to nearest centimetres.
**Class includes two very suppressed trees.

show a striking lack of young trees and saplings (Wardle, 1978).

On the basis of growth ring counts, the "regeneration gap", although it varies greatly in breadth, seems to be centred on the seventeenth and eighteenth centuries. It does not seem to be accounted for by rapid early growth and quick passage through smaller size classes, and obviously predates introduction of mammals. Nor do successional or disruptive processes appear to provide a complete explanation, despite the undoubted vigour of conifers in successions. To some ecologists (e.g. Holloway, 1954), climatic explanations have seemed an attractive hypothesis, in view of the fluctuations indicated by glacial variations and other evidence. Wardle (1963) suggested that, during certain periods, conifers may be able to establish on sites which under other circumstances are denied to them by excessive competition.

Possible biological explanations include the effect of extinction of moas on forest structure (it is noticeable that browsing mammals can favour conifer regeneration by removing competition), and the influence of rodents on seed supply. The effect of rodents is debateable since seedlings of *Podocarpus ferrugineus* are usually more abundant than those of other conifers, even though their large seeds are often eaten by rats.

While broadly similar "regeneration gaps" have been described over most of New Zealand, Barton (1972) suggested that in the Hunua Range near Auckland, only *Agathis australis* and *Phyllocladus trichomanoides* follow this pattern. These are species with a decidedly northern distribution. Here other podocarps, and also *Nothofagus truncata*, were mainly established during the 1600–1850 A.D. period.

Canopy death

The death of dominant trees has caused much concern during the last thirty years. In many instances, this has followed the establishment of brushtail possums, which browse and undoubtedly weaken trees, although the final demise may result from a combination of adverse events. Species killed include *Metrosideros excelsa*, *M. robusta*, *M. umbellata* and *Weinmannia racemosa*. The last two species dominate large areas of forest which protect steep headwaters, and in some localities death of the canopy combined with modification of the understorey and destruction of seedlings by deer have proved disastrous. Rather inexplicably, it is mature trees which are killed by possums, whereas neighbouring young seral stands remain healthy. Recently, Veblen and Stewart (1980) compared the structure and regeneration of *Metrosideros umbellata*- and *Weinmannia racemosa*-dominated forest containing deer and possums with undamaged forest on a neighbouring offshore island, while Fitzgerald and Wardle (1979) described the food of possums in similar forest from faecal analyses over a year.

Smaller trees are also vulnerable to possums; *Fuchsia excorticata* in particular has been rapidly eliminated over large areas, and this is a matter of concern where it has provided much of the nectar source for native birds. On the other hand, fuchsia and other palatable trees are still common in some areas where possums have been present for many years. It may be that possum numbers stabilize at a lower level after an initial population explosion, and in turn this may allow some forest recovery.

So far, possums have had little serious effect on beech forests, apart from the extinction over wide areas of the red- or yellow-flowered *Peraxilla* mistletoes which provided a striking seasonal display in the otherwise sombre forest.

Podocarpus hallii and *P. totara* have also died over wide areas. While possums may be a cause, the mortality has also been pronounced in places where possums are absent. In Westland, the buds of moribund trees have been found to harbour the caterpillar of *Ctenopseustis obliquana*.

Widespread defoliation by caterpillars periodically occurs in *Nothofagus* species, and can afflict extensive stands of *N. solandri*, but most trees recover. On the other hand, attack by species of *Platypus*, a small boring beetle, very often leads to tree death, probably through entry of pathogenic fungi into the holes. Severe epidemics seem to follow partial opening of stands through logging, snow breakage, avalanche or windfall. *Platypus* may have also caused the extensive death of beech trees which followed the Murchison earthquake in 1929. Destruction of the canopy by *Platypus* has serious implications where, at the same time, deer are killing seedlings.

Following a severe drought in 1969 and 1970, many forest plants died in the Manawatu district, including mature trees of *Beilschmiedia tawa*, *Dacrydium cupressinum* and *Podocarpus dacrydioides* (Atkinson and Greenwood, 1972). Similarly, many *Nothofagus* trees died as a result of the drought that afflicted Canterbury and Westland in 1970 and 1971.

Beilschmiedia tawa is prone to severe die-back of the canopy, especially when exposure increases following partial logging. MacKenzie and Gadgil (1973) have produced good evidence that frost is responsible. Salt burning after storms can defoliate and cause severe die-back in hardwoods in coastal districts, but they usually recover.

Other pathogens

Numerous pathogens, mainly insects and fungi (Gilmour, 1966; Milligan, 1972) have been described from the New Zealand forest, although their significance, if any, is not well understood. Probably the most obvious, because of the conspicuous sooty moulds that they support, are coccids, which are especially prevalent on *Metrosideros umbellata* and *Nothofagus solandri*. *N. menziesii*, like its close relatives in Australia and Patagonia, is host to the strawberry fungus (*Cyttaria* spp.) which produces stem galls.

Phytophthora, although recognized as a lethal pathogen of exotic conifers in New Zealand (Newhook, 1960) has only been demonstrated as a sporadic pathogen in native forest; *Armillaria* is possibly a more serious root parasite.

REGENERATION PATTERNS IN FOREST

Subalpine stands of *Nothofagus solandri*

This is the simplest forest community in New Zealand, and the best understood following the study by J. Wardle (1970). The periodic seed years lead to a persistent carpet of seedlings, broken only where the canopy is particularly dense, the litter too deep, or the mineral soil exposed and subject to frost-heave and wash. Death of canopy trees releases thickets of saplings, which become thinned through competition.

Such a regeneration pattern should lead to a mosaic with the total population forming a "reversed J-curve" when numbers of plants are plotted against age. In practice, some age- and size-classes are nearly always over- or under-represented, and indeed, forests can consist of one or more discrete age-groups over very large areas. The reasons for this can include windfall, snow breakage, avalanche and insect attack. These open the canopy and release seedlings. Browsing mammals, on the other hand, cause discontinuities by preventing recruitment to the canopy.

Hot fires destroy both canopy and seedlings, and the forest regenerates usually through gradual invasion of seral vegetation from the margins of surviving beech forest, often hastened by concentric or downwind spread from isolated trees. Primary succession on to bare surfaces such as slips and recent river terraces is likewise mainly through marginal spread from adjacent forest.

Stands of *Nothofagus solandri* near the upper altitudinal or lower rainfall limits tend to contain few seedlings. They are likely to be vulnerable to minor climatic fluctuations and mammalian browsing, because of inability to repair any gradual or catastrophic reduction of their canopies.

Wetter beech and conifer–beech–broadleaf forests

In *Nothofagus* forests in high-rainfall districts, including subalpine *N. menziesii* stands (Wardle et al., 1970) and mid-altitude *N. fusca* stands (June and Ogden, 1978), dense undergrowth restricts beech seedlings to bare or moss-covered ground, logs and stumps.

In low-altitude conifer–beech–broadleaf forests,

there is even heavier competition from tree ferns, ground ferns, shrubs and suckering trees such as *Quintinia acutifolia* and *Weinmannia racemosa*. Silviculturally, these forests are best managed as even-aged stands of beech, and to obtain regeneration after logging it is necessary to remove any dense ground vegetation immediately before a good seed fall. Under natural conditions, the persistence of beech in these mixed forests probably depends on the chance establishment of seedlings where competition is temporarily reduced, for example on logs and mineral soil exposed by windthrown trees. Beeches are also at a comparative advantage on harsher sites such as ridge crests, and in successions such as on slips and burns.

In some stands transitional to conifer–broadleaf forest, mainly at low altitudes, beech populations consist mainly of mature and decadent trees, and may be relict from a more favourable climatic regime. This has been noted in eastern Otago (Simpson and Thompson, 1928), coastal Fiordland, north Westland, and near Auckland (Barton, 1972).

Conifer–broadleaf forest

The conifers forming the tallest stratum of mixed forest establish well in seral forests, and in the case of some species also on poor soils where there is reduced competition. However, on the whole regeneration is sparse in mature forest, although it has been observed that mortality of established seedlings is very low, particularly in *Dacrydium cupressinum* (Franklin, 1968). In contrast, the main hardwoods usually regenerate freely, whether as terrestrially based seedlings (e.g. *Beilschmiedia* spp.), as low epiphytes (*Weinmannia racemosa*), or high epiphytes (*Metrosideros robusta*). In some species, notably *Quintinia acutifolia*, efficiency of regeneration is enhanced by vegetative multiplication. Consequently, some ecologists have seen the conifers as a transitory (Cockayne, 1928) or relict (Robbins, 1962) element in what is destined to become a hardwood climax. A view now prevalent is that presence of the conifers is generally the result of establishment during favourable climatic periods.

Nevertheless, cyclical processes of regeneration have been postulated, by Bieleski (1959) for *Agathis* and Cameron (1954) for podocarps. More recently, Beveridge (1973), has described cyclical regeneration in a forest dominated by podocarps and *Weinmannia racemosa* on the volcanic plateau, while stressing the general paucity of podocarps between 10 and 30 cm trunk diameter. The sequence involved is: (1) death of a large podocarp; (2) development of a colony of tree ferns; (3) replacement of tree ferns by hardwoods, especially *Weinmannia* beginning as epiphytic seedlings on tree fern trunks; and (4) establishment of podocarp seedlings, some of which grow into the canopy as the *Weinmannia* crown thins and dies.

In the dense *Dacrydium cupressinum* stands on gley podzols in Westland, the even-aged groups initiated by windthrow also can develop through a phase of hardwood dominance (Poole, 1937). Gale damage may also have initiated regeneration cycles in *Libocedrus bidwilli* (Clayton-Greene, 1977) and *Agathis australis* (A.E. Beveridge, pers. comm.)

ECOSYSTEM FUNCTIONING

Most work has concerned regeneration and growth of dominant trees, as discussed previously. Litter fall has been measured in several forests. In lowland stands of *Nothofagus solandri* (Bagnall, 1972) and *N. truncata* (Miller, 1963) it amounted to between 4.9 and 6 t ha^{-1} yr^{-1}, with peak falls in winter or spring. In subalpine *N. solandri* stands it ranged from 3.7 t at 1050 m to 3 t at timberline (J. Wardle, 1970) with leaf fall mainly during summer and autumn. In lowland conifer–broadleaf forest, litter fall totalled 6.9 t ha^{-1} yr^{-1}, 51% being contributed by the canopy (Daniel, 1975).

Rowe (1975) found that subalpine stands of *Nothofagus solandri* intercepted and returned to the atmosphere nearly 40% of the total summer rainfall of 250 to 275 mm, the remainder nearly all reaching the ground as throughfall rather than stem flow.

Campbell (1974) found that large *Nothofagus* trees have a distinct individual influence on pedogenesis. In an Ahaura soil in Westland assumed to be 18 000 to 22 300 years old, pH values are above 4.5 more than 8 m from living *N. fusca* trees, and mica, chlorite and feldspar weather to a 2:1 to 2:2 Al-intergrade, kaolinite, allophane and gibbsite. Close to trees and decaying stumps, pH is below 4.5, phosphorus retention is higher, allophane and gibbsite are dissolved, and the intergrade is converted to vermiculite and then to montmorillinite. This is seen as a stage in the genesis of gley podzol soils on older surfaces.

McColl (1974) compared arthropod numbers and species diversity in the litter, humus and mineral A horizon in several forest types in the Reefton district. Both values were much greater in conifer–beech–broadleaf and beech stands than in heath-forest on gley podzol or an exotic *Pinus* plantation. This finding is supported by Forster and Wilton (1973) who found that, of seventy species of spider inhabiting *Nothofagus* forest in Westland, only 10% were able to survive in contiguous *Pinus* stands.

PHYTOGEOGRAPHY AND HISTORY

Cretaceous and Tertiary origins (based on Raven and Axelrod, 1972; Fleming, 1975).

Palaeogeographic reconstructions based on plate tectonics now give a reasonably clear picture of the origins of the New Zealand forest flora. Genera such as *Nothofagus* and *Podocarpus*, which have closely related modern species in Australia, New Zealand and South America, probably evolved and dispersed when these land masses were still contiguous in their Gondwana configuration, and were present in New Zealand in the Upper Cretaceous. The Tasman Sea began to expand about this time, and through the early part of the Tertiary era warm temperate to subtropical rain forest prevailed on both sides of the Sea. In New Zealand, *Nothofagus* of the "brassii" type, which is now confined to New Caledonia and New Guinea, dominates most of the pollen record, but the "menziesii" and "fusca" types are also represented, together with other hardwoods, podocarps and Araucariaceae.

During the Tertiary, sclerophyllous vegetation spread at the expense of rain forest in Australia, whereas in New Zealand the latter continued to prevail, developing an increasingly modern aspect as archaic genera were gradually replaced by modern genera, of mainly Malesian affinity (e.g. *Alectryon*, *Elaeocarpus*, *Macropiper*, *Pittosporum*) which presumably dispersed across considerable sea gaps. However, local development of xeric vegetation at the end of the Tertiary is indicated, for example by the brief appearance of *Acacia* in the pollen record (Mildenhall, 1972).

Pleistocene survival

New Zealand had been a warm archipelago of low relief through most of the Tertiary era, but towards the end, momentum gathered towards the cataclysmic events of the Quaternary, including cooling, orogeny, and in several districts, intense volcanism. Repeatedly during the coldest times, much of the South Island and the highest mountains in the North Island experienced glaciation. Whatever the ultimate origins of the New Zealand alpine flora, most of its speciation has evidently taken place during the Quaternary era. The forest flora, on the other hand, was hard pressed and lost many genera, including *Cocos* and, to judge from the pollen forms, *Bombax*, *Cupania*, *Ephedra* and *Microcachrys*, as well as the "brassii" beeches. The small species content of most forest genera (e.g. two species of *Elaeocarpus* in New Zealand, as compared with 120 in New Guinea according to Good, 1960), doubtless also largely results from Quaternary extinction. Cold-climate pollen assemblages, even from lowland sites as far north as 30°S, indicate grass- and shrub-covered landscapes (Harris, 1963; Moar, 1980). Nevertheless, forest clearly survived within New Zealand, and included species susceptible to cold, although there was no opportunity to migrate northward beyond 34°S, and the maximum lowering of sea level could only partly compensate for the temperature drop of 6°C usually postulated.

Wide gaps in the modern ranges of forest species which appear to predate the postglacial era, together with the return of forest pollen assemblages during apparently brief interstadials, suggest that forest refugia existed even in southern New Zealand, at least during the last, or Otiran, glaciation. The regional diversity of the climate was probably intensified during glacial periods with harsh conditions prevailing inland and near the ice, and relatively mild climates persisting in some coastal areas. At present, *Nothofagus* is disjunct between the northern and southern portions of the South Island, and there is no evidence that it reached the south through postglacial migration. The hypothesis that four species of beech survived the Otiran glaciation in the south is supported by other disjunctions, and concentrations of endemic species. Northern parts of the South Island also probably provided major refugia, with minor ones in other districts. The most important refugium of all would have been north of 38°S, where the most warmth-loving species are restricted at present (e.g. *Beilschmiedia tarairi, Meryta sinclairii, Planchonella novo-zelandica* and many others).

Postglacial recovery and modern distribution

Palynology and carbon-dated wood and charcoal provide growing knowledge of vegetation history during the postglacial period. Glaciers began their lowland major retreat about 14 000 years B.P., and from this time podocarps dominated the pollen spectra in the central parts of the North Island (McGlone and Topping, 1977). Just after 12 000 years B.P. forests spread around Westport in the South Island (Moar and Suggate, 1979), but in other places from Wellington southwards this date marked the replacement of grasses by shrubs, including species of *Coprosma, Dacrydium, Myrsine* and *Phyllocladus*, and it is only after 10 000 years B.P. that tree podocarps or broad-leaved hardwoods show pollen dominance (Moar, 1971).

Pollen percentages for *Nothofagus* are low during the earlier part of the postglacial. Subsequent rises, first recognized in diagrams from Otago and Southland, were attributed to climatic deterioration from a post-glacial optimum (Cranwell and Von Post, 1936), but more recently, the *Nothofagus* rise has been shown to be strongly diachronous. It already constitutes a third of the tree pollen at the base of a profile dated to 10 430 years B.P. near Nelson city (Dodson, 1978) and by about 8000 years B.P. is the dominant pollen there and at the Lake Rotoiti 50 km to the south. In contrast pollen diagrams from north Westland show a rise beginning about 2500 years B.P. (Moar, 1971). The pollen history of *Nothofagus* in the South Island at least is now seen largely as demonstrating a spread to areas deforested during the last glaciation. Presumably, when lowlands, hills and lower mountain slopes became suitable for spread of forest, the small wind-dispersed seed of *Metrosideros* and *Weinmannia*, and the succulent fruit of podocarps were distributed widely and gave rise to the first extensive forests. Beeches with their heavier seeds without special dispersal aids and with somewhat shade-intolerant seedlings would have been slow to spread from the glacial refugia.

Bare surfaces such as landslide scars, and infertile sites with reduced competition, are readily colonized wherever beech seed is available. Streams often provide migration routes for beech, partly because of water transport of seed, but mainly because slips and recent terraces along the banks are available for establishment of seedlings. Especially in the subalpine belt, outlying clumps of beech have become established through wind dispersal (Wardle, 1980b).

The varying opportunities for spread, and the greater competitive ability of beech relative to conifers and other hardwoods at high altitudes, on poor soils and in inland areas with more extreme climates, explain the erratic pattern of forest distribution. Thus, Fiordland, consisting of steep, slip-scarred granitic and gneissic mountains is now almost completely clothed in beech forest. In contrast the more stable and more deeply weathering schistose mountains of central Westland, experiencing a similar climate, support only conifer–broadleaf forest, except in the Landsborough Valley where a migration route from the south allowed beech forest to completely replace conifer–broadleaf within the last 3000 years B.P.

On the Volcanic Plateau in the North Island, most areas affected by volcanic eruption have been occupied by conifer–broadleaf forest. Beech forest exists mainly on sites distant from eruptive centres, as relict, low-altitude stands on pockets of infertile soil which have not been buried and obliterated by loam derived from tephra (Nicholls, 1963). The southern limit of *Agathis australis*, which lies across the northern edge of the Plateau, appears to represent tardy postglacial spread rather than climatic limitations, for its altitudinal limits exceed 700 m while planted trees thrive as far south as Stewart Island.

Recent work has shown that the small hardwood *Ascarina lucida* declined around 5000 years B.P. from relatively high pollen frequencies to its insignificant modern level (McGlone and Moar, 1977). Since this species is demonstrably frost tender, it provides new evidence for some kind of postglacial climatic optimum. Variations in the proportions of conifer pollen further support this view. The change to somewhat harsher regimes, which is also indicated by evidence for glacial advances, may have facilitated the spread of *Nothofagus*.

The Polynesian influence

Direct demands on the forest by the forbears of the Maoris, arriving about 1000 years ago, were modest. The population probably never reached 200 000. Trapping birds and cutting timber were regulated by tapu, and berries were also gathered. Subtler, but more far reaching changes were associated with the introduction of the Polynesian rat and dog and the extinction of the moa, at least some species of which are believed to have been forest dwellers. The main effect, however, was a vast increase in the incidence of fire. By the time Europeans arrived, forest had gone from the drier, eastern side of the South Island and southeast of the North Island. Even in areas of moderately high rainfall, Maori settlements were surrounded by deep hinterlands where fire-induced fern, grass and scrub prevailed. Fires also maintained and greatly extended seral vegetation initiated by volcanic eruption in the central plateau of the North Island.

This period also encompassed climatic variations recorded in the advances and retreats of glaciers. These may have influenced forest ecology, and per-haps also fire risk to vegetation, but it is not easy to distinguish such effects from anthropogenic influences.

European influences

European mission settlements began around 1820, and settlement in earnest after the declaration of British sovereignty in 1840. From these times, timber cutting, agriculture and pastoralism practised by both Europeans and Maoris progressively removed the native forest from lowland terrain. Today, low-altitude forest is restricted to scattered reserves, logged areas not yet cleared and inaccessible steeplands, except for Westland and western Southland, which are still largely forested "frontier" districts where the final assault has not yet taken place.

In parts of the North Island forest has also been widely cleared to pasture on very steep country, often resulting in severe soil erosion. In the South Island the fires of pastoralists, though they have consumed large areas of grassland, scrub and secondary forest, have generally not penetrated virgin forest areas much further than those of their Maori predecessors.

Naturalization of browsing mammals has been extraordinarily thorough, so that nearly all forests are now obviously modified in this way.

TENURE OF FOREST LAND

The 6.25 million ha of indigenous forests are held in various tenures, listed below in approximate order of importance:

State Forest

State Forest totals nearly three million ha, administered by the New Zealand Forest Service, which has developed the following management zoning:

(a) Protection zones on steeplands and at high altitudes include nearly 70% of state forest. Management consists mainly of control of browsing animals by commercial, recreational and state-employed hunters. During the past twenty years, use of exotic species in protection forestry has been investigated, but application is likely to be in areas that do not already support a forest cover.
(b) Production zones, where the timber is made available on a royalty basis. Until recently, loggers generally have been required to extract all merchantable timber, at least of podocarp species, but a recently adopted policy aims at conservation and sustained yield.
(c) Ecological zones are set aside for scientific purposes.
(d) Forest sanctuaries are a stricter version of the preceding zone, and protect especially important areas by parliamentary statute.
(e) Amenity zones are used primarily for protection of scenery.
(f) Historical zones protect Maori and other historical sites.

The following subzones related to recreational use are superimposed on the primary classification: (1) wilderness, which is recognized by statute; (2) natural environment, where tracks, bridges and huts may be provided; (3) recreational development, where intensive public use is catered for; and (4) restricted areas, for reasons of public safety, protection of flora and fauna, etc.

In addition, there are State Forest Parks, which also have statutory recognition. These are substantial areas, which can comprise a range of zones and sub-zones. Generally, emphasis is on right of public access and development of facilities such as tracks and huts. Catchment protection remains an important function, and logging is permitted where deemed appropriate.

National and Maritime Parks

The ten National Parks, together with two Maritime Parks and numerous smaller reserves, comprise 2 620 000 ha or 10% of New Zealand's area; and well over half of this is steepland forest, with most of the remainder lying above the alpine timberline. The Parks are administered by government-appointed boards, serviced by the Department of Lands and Survey, and subject to overall supervision by a National Parks Authority.

Although effectively protected from exploitation by private enterprise, the Parks are vulnerable to decisions by central government. For example, the Mining Act of 1971 allows mining concessions in the parks, although in practice none as yet have been taken up. A government decision in 1959 to raise two large lakes, Manapouri and Te Anau, that would have resulted in drowning a large area of forest in Fiordland National Park, was reversed only after massive protests.

Crown land

Crown land is administered by the Department of Lands and Survey; three categories are relevant:

(1) Crown lease; leased under various tenancy agreements, mainly to farmers. Much forest land is included, although in most cases the tenant does not hold the right to remove timber.

(2) Unoccupied crown land; this is generally inaccessible, mountainous land, largely covered by protection forest. Some lowland forest is included, however, which has conflicting potential for timber production and agricultural development or nature conservation.

(3) Scenic and "allied" reserves. These have been primarily reserved for scenic values, but many have acquired a high biological value, especially in districts where they protect the only remaining lowland forest. "Allied" reserves include those defined for historic and scientific purposes.

Freehold land

Though most freehold land has been cleared of native vegetation, many farmers have retained remnants of forest. Often these are deteriorating fragments used for stock shelter, but some owners have carefully protected their stands for aesthetic reasons. There is provision for designating these as private scenic reserves, or donating them to the State.

Maori land

Maori ownership is complex and on a tribal basis. Considerable tracts of forest are included, but the area is diminishing both through logging and clearing, and through alienation to State or private (not necessarily non-Maori) ownership.

TIMBER PRODUCTION, LOGGING, SILVICULTURE

Historically, native conifers have been the mainstay of the New Zealand timber industry. Today, supplies of kauri (*Agathis australis*) have dwindled to about 2000 $m^3 yr^{-1}$, although there are regenerating stands which should sustain a limited industry in the rather distant future. Sources of *Podocarpus spicatus* and *P. totara* are also nearly exhausted, and permanently so as these species grow on land of high agricultural potential. *Dacrycarpus dacrydioides*, which grows on wet lowlands, is still being heavily logged, even though mature stands now cover scarcely 2000 ha. There exists however a potential for managing regenerating areas as farm woodlots.

Only for *rimu* (*Dacrydium cupressinum*) has a substantial cutting rate been maintained, although between 1960 and 1974 it was halved to 288 000 $m^3 yr^{-1}$ (this figure includes *Podocarpus ferrugineus*, as the two timbers are not distinguished in commerce). By contrast, 1.5 million m^3 of exotic *Pinus* (mainly *P. radiata*) were sawn during 1974 (Anonymous, 1975).

For logged conifer–broadleaf stands, regeneration of the merchantable softwood component is still considered too elusive, and a common practice has been to "convert" to exotic forest, by burning and planting. This practice is now being phased out in State Forest, although it continues on other tenures. Dense *rimu* stands on gley podzol soils in Westland, on the other hand, are considered amenable to selective logging practices, although slow growth and uncertainment recruitment of *rimu* would only allow removal of about a quarter of the timber each 40 to 50 years. Some dense podocarp stands in the central North Island are also zoned for selective logging.

Beilschmiedia tawa is a valued hardwood, and although available areas are rapidly dwindling, some are now included in a sustained yield working circle.

In contrast to the softwoods, the amount of *Nothofagus* timber marketed has been small relative to the potential resource; most lowland beech forests have been merely skimmed for conifers and the best beech logs. There is evidence that much better and more acceptable logs would be yielded by managed beech forest, but a major obstacle has been the large volume of substandard wood that would need to be disposed of in the first cutting cycle. For this reason, in 1971 the Forest Service put forward a draft scheme for basing a wood chips or pulp industry on two areas in the South Island, involving a total area of some 400 000 ha of beech and conifer–beech–hardwood forest. The forests, once cut, would either be regenerated to beech or interplanted with eucalypts or, on some 100 000 ha where regeneration is considered unlikely to succeed, fully converted to exotics. These proposals are still being modified and evaluated.

The Forest Service is now firmly committed to silviculture of native species, even though methods are still somewhat exploratory. Problems posed by recalcitrant regeneration and relatively slow growth rates are compounded by questions of marginal profitability, which in turn are related to the fact that for many years prices of native timber have been set, by legislation, at undesirably low levels. The following notes on silvicultural techniques applicable to the four most important species are drawn from unpublished material of A.E. Beveridge and D.A. Franklin (N.Z. Forest Service).

Agathis australis: Naturally regenerated young stands (which generally have a nurse crop of *Leptospermum* or *Weinmannia*), can be thinned to release about 1000 saplings ha^{-1}, or in older stands, 250 poles ha^{-1}. Planting out of container-grown seedlings also holds promise. It is estimated that to reach 50 cm diameter, trees require 60 to 170 years, according to quality of site and silvicultural treatment; an extra 80 to 100 years may be needed to reach 1 m diameter.

Dacrydium cupressinum: During selection logging, attention should be given to retaining groups of trees, as isolated trees are more subject to mechanical damage to the roots and to wind-throw. Best sites for seedling recruitment are scarified or compacted sites of low fertility, but subsequent growth rates are slow. Seedlings are easily raised in the nursery and can be planted out successfully on to more fertile sites. Given an initial release from competition, they will subsequently continue to grow even when overtopped. Aspects of the silviculture of *D. cupressinum* have recently been discussed by James and Franklin (1978) and Herbert (1980).

Nothofagus menziesii: Best prospects for management are in western Southland. Good seedling establishment before logging should be encouraged, by scarifying to remove undergrowth and to expose mineral soil before a seed year. If regeneration is insufficient, or if it is destroyed during heavy logging (as in chipwood logging), then seed trees will need to be left; possibly also planting with nursery stock will be necessary. Early thinning should be to 2000 to 3000 saplings ha^{-1}, with later thinnings eventually leaving 150 crop trees ha^{-1}. It is estimated that saw logs and veneers could be produced on a rotation of 60 to 80 years.

Nothofagus fusca: Techniques similar to *N. menziesii*, excepting that susceptibility to pin-hole borer *Platypus* means that it is more critical to plan rotations to avoid reservoirs of infection, such as stumps and dead and dying trees, when the new crop is at a susceptible stage. The seedlings demand more light, but growth rates are faster, and early thinning can be much heavier, as the saplings have better apical dominance.

Considering the limited areas of lowland forest that now survive, it is not surprising that there have been strong objections by conservation groups to Forest Service plans for beech, *rimu* and other species (Searle, 1975), and a substantial body of opinion is calling for the earliest possible cessation of logging and clearing in native forests. The next few years will decide whether virgin forests will remain a significant feature of the New Zealand lowland landscape.

SOURCES AND ACKNOWLEDGEMENTS

In preparing this chapter, I am indebted to many New Zealand scientists, particularly in the Forest Research Institute, for information and advice. In addition to specific references in the text, much background information is derived from major publications on New Zealand forests, including those by Cockayne (1928), Hinds and Reid (1957) and McKelvey (1963). Orwin's (1974) bibliography has also been a valuable source.

REFERENCES

Adams, J.A., 1976. Nutrient requirements of four *Nothofagus* species in north Westland, New Zealand, as shown by foliar analysis. *N.Z. J. Bot.*, 14: 211–223.

Anonymous, 1968. Soils of New Zealand. Part 3. *N.Z. Soil Bur. Bull.*, 26: 1–127.

Anonymous, 1975. *Statistics of the Forests and Forest Industries of New Zealand.* N.Z. Forest Service, Wellington, 120 pp.

Atkinson, I.A.E., 1973. Spread of the ship rat (*Rattus r. rattus* L.) in New Zealand. *J. R. Soc. N.Z.*, 3: 457–472.

Atkinson, I.A.E. and Bell, B.D., 1973. Offshore and outlying islands. In: G.R. Williams (Editor), *The Natural History of New Zealand.* A.H. and A.W. Reed, Wellington, pp. 372–392.

Atkinson, I.A.E. and Greenwood, R.M., 1972. Effects of the 1969–70 drought on two remnants of indigenous lowland forest in the Manawatu district. *Proc. N.Z. Ecol. Soc.*, 19: 34–42.

Bagnall, R.G., 1972. The dry weight and calorific value of litter fall in a New Zealand *Nothofagus* forest. *N.Z. J. Bot.*, 10: 27–36.

Barton, I.L., 1972. On the vegetation of the Hunua Ranges, Auckland. *N.Z. J. Bot.*, 10: 8–26.

Baylis, G.T.S., 1959. Effect of vesicular–arbuscular mycorrhizas on growth of *Griselinia littoralis* (Cornaceae). *New Phytol.*, 58: 274–280.

Baylis, G.T.S., 1975. The magnolioid mycorrhiza and mycotrophy in root systems derived from it. In: F.E. Saunders, B. Mosse and P.B. Tinker (Editors), *Endomycorrhizas.* Academic Press, London, pp. 383–389.

Beveridge, A.E., 1964. Dispersal and distribution of seed in central North Island podocarp forests. *Proc. N.Z. Ecol. Soc.*, 11: 48–55.

Beveridge, A.E., 1973. Regeneration of podocarps in a central North Island forest. *N.Z. J. For.*, 18: 23–35.

Bibby, K.M. and Williams, R.W.M., 1953. The effect of various over-winter storage conditions on *Nothofagus menziesii* seed. *N.Z. For. Res. Notes*, 1: 1–8.

Bieleski, R.L., 1959. Factors affecting growth and distribution of kauri (*Agathis australis* Salisb.). *Aust. J. Bot.*, 7: 252–294.

Burke, W.D., 1974. Regeneration of podocarps on Mt. Tarawera, Rotorua. *N.Z. J. Bot.*, 12: 219–226.

Bussell, W.T., 1968. The growth of some New Zealand trees. 2. Effects of photoperiod and temperature. *N.Z. J. Bot.*, 6: 76–85.

Campbell, A.S., 1974. The influence of red beech (*Nothofagus fusca*) on clay mineral genesis. *Trans. 10th Int. Congr. Soil Sci.*, 6: 60–67.

Cameron, R.J., 1954. Mosaic or cyclical regeneration in North Island podocarp forests. *N.Z. J. For.*, 7: 55–65.

Cameron, R.J., 1960. Natural regeneration of podocarps in the forests of the Whirinaki River valley. *N.Z. J. For.*, 8: 337–354.

Cameron, R.J., 1963. A study of the rooting habits of rimu and tawa in pumice soils. *N.Z. J. For.*, 8: 771–785.

Clayton-Greene, K.A., 1977. Structure and origin of *Libocedrus bidwilli* stands in the Waikato district, New Zealand. *N.Z. J. Bot.*, 15: 19–28.

Cockayne, L., 1926. Monograph on the New Zealand beech forests. Part I, the ecology of the forests and the taxonomy of the beeches. *Bull. N.Z. For. Serv.*, 4(1): 1–71.

Cockayne, L., 1928. *The Vegetation of New Zealand.* Engleman, Leipzig, 2nd ed., 456 pp.

Cooper, K.M., 1976. A field survey of mycorrhizas in New Zealand ferns. *N.Z. J. Bot.*, 14: 169–181.

Cranwell, L.M. and Von Post, L., 1936. Post-Pleistocene pollen diagrams from the Southern Hemisphere. I. New Zealand. *Geogr. Ann.*, 3–6: 308–347.

Daniel, M.J., 1975. Preliminary account of litter production in a New Zealand lowland podocarp–rata–broadleaf forest. *N.Z. J. Bot.*, 13: 173–187.

Dodson, J.R., 1978. A vegetation history from north-east Nelson, New Zealand. *N.Z. J. Bot.*, 16: 371–378.

Dugdale, J.S., 1977. Phytophagous insects and their hosts. In: *Symposium on Pest–Plant Relationships. N.Z. Entomol.*, 6: 213–243.

Fitzgerald, A.E. and Wardle, P., 1979. Food of the opossum *Trichosurus vulpecula* (Kerr) in the Waiho Valley, South Westland. *N.Z. J. Zool.*, 6: 339–345.

Fleming, C.A., 1975. The geological history of New Zealand and its biota. In: G. Kuschel (Editor), *Biogeography and Ecology in New Zealand.* W. Junk, The Hague, pp. 1–86.

Forster, R.R. and Wilton, C.L., 1973. The spiders of New Zealand. Part 4. *Otago Mus. Bull.*, 4.

Franklin, D.A., 1968. Biological flora of New Zealand. 3. *Dacrydium cupressinum* Lamb. (Podocarpaceae) Rimu. *N.Z. J. Bot.*, 6: 493–513.

Gilmour, J.W., 1966. The pathology of forest trees in New Zealand. The fungal, bacterial and algal pathogens. *Tech. Pap. For. Res. Inst. N.Z.*, No. 48: 82 pp.

Godley, E.J., 1971. The fruit of *Vitex lucens* (Verbenaceae). *N.Z. J. Bot.*, 9: 561–568.

Godley, E.J., 1979. Flower biology in New Zealand. *N.Z. J. Bot.*, 17: 441–466.

Good, R., 1960. On the geographical relationships of the angiosperm flora of New Guinea. *Bull. Br. Mus. Nat. Hist.*, 2: 203–226.

Greenwood, R.M. and Atkinson, I.A.E., 1977. Evolution of divaricating plants in New Zealand in relation to Moa browsing. *Proc. N.Z. Ecol. Soc.*, 24: 21–23.

Harris, G.P. and Morrison, T.M., 1958. Fixation of nitrogen-15 by excised nodules of *Coriaria arborea* Lindsay. *Nature*, 182: 1812.

Harris, W.F., 1963. Paleo-ecological evidence from pollen and spores. *Proc. N.Z. Ecol. Soc.*, 10: 38–44.

Herbert, J., 1980. Structure and growth of dense podocarp forest at Tihoi, central North Island, and the impact of selective logging. *N.Z. J. For.*, 25: 44–57.

Hinds, H.V. and Reid, J.S., 1957. Forest trees and timbers in New Zealand. *N.Z. For. Serv. Bull.*, 12: 1–211.

Holloway, J.T., 1954. Forests and climate in the South Islands of New Zealand. *Trans. R. Soc. N.Z.*, 82: 329–410.

James, I.L. and Franklin, D.A., 1978. Recruitment, growth and survival of rimu seedlings in selectively logged terrace rimu forest. *N.Z. J. For. Sci.*, 8: 207–212.

June, S.R., 1977. A major range extension for hard beech (*Nothofagus truncata*) in the South Island (note). *Mauri Ora*, 5: 119–121.

June, S.R. and Ogden, J., 1978. Studies of the vegetation of Mt.

Colenso, New Zealand. 4. An assessment of the processes of canopy maintenance and regeneration strategy in a red beech (*Nothofagus fusca*) forest. *N.Z. J. Ecol.*, 1: 7–15.

Killick, N.G., 1976. *New Zealand Official Year Book*. Department of Statistics, Wellington, 1092 pp.

Leamy, M.L. and Fieldes, M., 1976. Soils. In: I. Ward (Editor), *New Zealand Atlas*. Government Printer, Wellington, pp. 122–143.

McColl, H.P., 1974. The fauna of six forest floors on the West Coast of the South Island. A progress report up to September 1973. *Beech Res. News*, 1: 28–31.

McEwan, J.M., 1966. Development of the fruit of *Coprosma lucida*. *N.Z. J. Bot.*, 4: 515–521.

McGlone, M.S. and Moar, N.T., 1977. The *Ascarina* decline and post-glacial climatic change in New Zealand. *N.Z. J. Bot.*, 15: 485–490.

McGlone, M.S. and Topping, W.W., 1977. Aranuian (post-glacial) pollen diagrams from the Tongariro region, North Island, New Zealand. *N.Z. J. Bot.*, 15: 749–766.

McKelvey, P.J., 1963. The synecology of the West Taupo indigenous forest. *Bull. N.Z. For. Ser.*, 14: 1–127.

McKelvey, P.J., 1973. The pattern of the Urewera forests. *N.Z. For. Res. Inst. Tech. Pap.*, 59: 1–48.

MacKenzie, R.M.J. and Gadgil, P.D., 1973. Die-back in tawa. *N.Z. J. For.*, 18: 36–46.

McNabb, R.F.R., 1968. The Boletaceae of New Zealand. *N.Z. J. Bot.*, 6: 137–176.

Mildenhall, D.C., 1972. Fossil pollen of *Acacia* type from New Zealand. *N.Z. J. Bot.*, 10: 485–494.

Miller, R.B., 1963. Plant nutrients in hard beech. *N.Z. J. Sci.*, 6: 365–413.

Milligan, R.H., 1972. A review of beech forest pathology. *N.Z. J. For.*, 17: 202–211.

Mirams, R.V., 1957. Aspects of natural regeneration of the kauri. *Trans. R. Soc. N.Z.*, 84: 661–680.

Moar, N.T., 1971. Contributions to the Quaternary history of the New Zealand flora. 6. Aranuian pollen diagrams from Canterbury, Nelson and North Westland, South Island. *N.Z. J. Bot.*, 9: 80–145.

Moar, N.T., 1980. Late Otiran and early Aranuian grassland in central South Island. *N.Z. J. Ecol.*, 3: 4–12.

Moar, N.T. and Suggate, R.P., 1979. Contributions to the Quaternary history of the New Zealand flora. 8. Interglacial and glacial vegetation in the Westport district, South Island. *N.Z. J. Bot.*, 17: 361–387.

Mokma, D.L., Jackson, M.L., Syers, J.K., Stevens, P.R., 1973. Mineralogy of a chronosequence of soils from greywacke and mica-schist alluvium, Westland, New Zealand. *N.Z. J. Sci.*, 16: 769–797.

Morrison, T.M., 1956. Mycorrhiza of silver beech. *N.Z. J. For.*, 7: 47–60.

Newhook, F.J., 1960. Climate and soil type in relation to *Phytophthora* attack. *Proc. N.Z. Ecol. Soc.*, 7: 14–15.

Nicholls, J.L., 1963. Vulcanicity and indigenous vegetation in the Rotorua district. *Proc. N.Z. Ecol. Soc.*, 10: 58–65.

Nicholls, J.L., 1976. A revised classification of the North Island indigenous forests. *N.Z. J. For.*, 21: 105–132.

Orwin, J., 1974. Annotated bibliography on the ecology of New Zealand indigenous forest and scrub 1929–70. *N.Z. J. Bot.*, 12: 45–113.

Poole, A.L., 1937. A brief ecological survey of the Pukekura State Forest, South Westland. *N.Z. J. For.*, 4: 78–85.

Raven, P.H. and Axelrod, D.I., 1972. Plate tectonics and Australian paleobiogeography. *Science*, 176: 1379–1386.

Robbins, R.G., 1962. The podocarp–broadleaf forests of New Zealand. *Trans. R. Soc. N.Z. (Bot.)*, 1: 33–75.

Rowe, L.K., 1975. Rainfall interception by mountain beech. *N.Z. J. For. Sci.*, 5: 45–61.

Sakai, A. and Wardle, P., 1978. Freezing resistance of New Zealand trees and shrubs. *N.Z. J. Ecol.*, 1: 51–61.

Searle, G.W., 1975. *Rush to Destruction*. Reed, Wellington, 217 pp.

Simpson, G. and Thompson, J.S., 1928. On the occurrence of silver southern-beech in the neighbourhood of Dunedin. *Trans. R. Soc. N.Z.*, 59: 326–342.

Southwood, T.R.E., 1973. The insect/plant relationship—an evolutionary perspective. In: H.F. van Emden (Editor), *Insect/Plant Relationships. Symposium of the Royal Entomological Society of London*, 6: 3–30.

Stevens, P.R. and Walker, T.W., 1970. The chronosequence concept and soil formation. *Q. Rev. Biol.*, 45: 333–350.

Sykes, W.R. and Godley, E.J., 1968. Transoceanic dispersal in *Sophora* and other genera. *Nature*, 218: 495–496.

Veblen, T.T. and Stewart, G.H., 1980. Comparison of forest structure and regeneration on Bench and Stewart Islands, New Zealand. *N.Z. J. Ecol.*, 3: 50–68.

Wardle, J., 1970. The ecology of *Nothofagus solandri*. *N.Z. J. Bot.*, 8: 494–646.

Wardle, P., 1963. The regeneration gap of New Zealand gymnosperms. *N.Z. J. Bot.*, 1: 301–315.

Wardle, P., 1971. An explanation for alpine timberline. *N.Z. J. Bot.*, 9: 371–402.

Wardle, P., 1972. Plant succession on greywacke gravel and scree in the subalpine belt in Canterbury, New Zealand. *N.Z. J. Bot.*, 10: 387–398.

Wardle, P., 1973. New Zealand timberlines. *Arct. Alp. Res.*, 5: A127–A135.

Wardle, P., 1977. The plant communities of Westland National Park. *N.Z. J. Bot.*, 15: 323–398.

Wardle, P., 1978. Regeneration status of some New Zealand conifers, with particular reference to *Libocedrus bidwillii* in Westland National Park. *N.Z. J. Bot.*, 16: 471–477.

Wardle, P., 1980a. Primary succession in Westland National Park and its vicinity, New Zealand. *N.Z. J. Bot.*, 18: 221–232.

Wardle, P., 1980b. Ecology and distribution of silver beech (*Nothofagus menziesii*) in the Paringa district, South Westland, New Zealand, *N.Z. J. Ecol.*, 3: 23–36.

Wardle, P. and Campbell, A.S., 1976. Winter dormancy in seedlings of mountain beech (*Nothofagus solandri* var. *cliffortioides*) near timber line. *N.Z. J. Bot.*, 14: 183–186.

Wardle, P., Mark, A.F. and Baylis, G.T.S., 1970. Vegetation studies on Secretary Island, Fiordland Part 9: Additions to Parts 1, 2, 4 and 6. *N.Z. J. Bot.*, 8: 3–21.

Wells, J.A., 1972. Ecology of *Podocarpus hallii* in Central Otago, New Zealand. *N.Z. J. Bot.*, 10: 399–426.

Wells, J.A. and Mark, A.F., 1966. The altitudinal sequence of climax vegetation on Mt. Anglem, Stewart Island. Part 1. The principal strata. *N.Z. J. Bot.*, 4: 267–282.

Williams, G.R., 1973. Birds. In: G.R. Williams (Editor), *The Natural History of New Zealand*. A.H. and A.W. Reed, Wellington, pp. 304–333.

Chapter 4

TEMPERATE BROAD-LEAVED EVERGREEN FORESTS OF AUSTRALIA[1]

J.D. OVINGTON and L.D. PRYOR

INTRODUCTION

The broad-leaved evergreen forests of temperate Australia are perhaps the most distinctive in the world. They display a remarkable uniformity in that the principal tree genus throughout almost all of them is *Eucalyptus*. In no other country does a single genus play such a dominant role. Eucalypts are dominants over a wide range of environmental conditions. They vary greatly in size and form. Some are the tallest of the hardwoods, trees of *Eucalyptus regnans* have been recorded as being over 100 m tall (Anonymous, 1962). Others are small woodland trees of the peculiar mallee form.

Geographic extent

Because of the aridity of the Australian continent large areas are without a tree cover. The total area of forest in Australia is 42 million ha and of woodland 200 million ha. In general, the forests are found in the better watered parts of the temperate region of Australia. Not more than one half of Australia south of the Tropic of Capricorn is occupied by forest or woodland vegetation (Fig. 4.1).

The natural distribution of eucalypts is largely confined to Australia. About 29 million ha of the forest area is classed as temperate eucalypt forest (Anonymous, 1980) and to describe the forests of temperate Australia is largely to describe the eucalypt forests. A series of maps showing the distribution of individual eucalypt species in Australia has been prepared by Chippendale and Wolf (1981).

Large areas of tall eucalypt woodland and relatively open woodland, as well as mallee communities of shrubby trees, grade imperceptibly into eucalypt forest without any precise boundary. In considering the temperate broad-leaved forests it is thus necessary to take into account the whole of the eucalypt-dominated vegetation of southern Australia.

The greatest area of eucalypt forest occurs in eastern Australia where it forms a more or less continuous coastal band from 150 to 600 km wide from the Tropic of Capricorn in Queensland to South Australia, and extending to the island of Tasmania. Whilst largely limited by environmental conditions to a more or less broad coastal band, the eastern temperate eucalypt forest has a remarkably long north to south extent of some 3000 km.

Apart from the marginal coastal plains and mountains, Australia consists of a flat, eroded plateau of 200 to 500 m elevation. In general there is no distinct climatic upper tree line. However, in the southeast where highland reaches to 2300 m (Mt. Kosciusko) and to 1800 m in Tasmania, an upper climatic tree line occurs. This tree line is approximately 1900 m in southeastern Australia at about latitude 36°S and occurs at a little lower elevation in Tasmania. Eucalypt species vary greatly in susceptibility to snow and frost damage, species such as *E. delegatensis* and *E. niphophila* being more resistant.

There is a second extensive area of eucalypt forest, mainly of *E. marginata* or *E. diversicolor*, in the southwestern part of Western Australia. This is separated from the eastern eucalypt forest by the treeless Nullarbor Plain, a very dry area with predominantly calcareous soils occupied by shrubby and herbaceous species.

[1]Manuscript completed January, 1981.

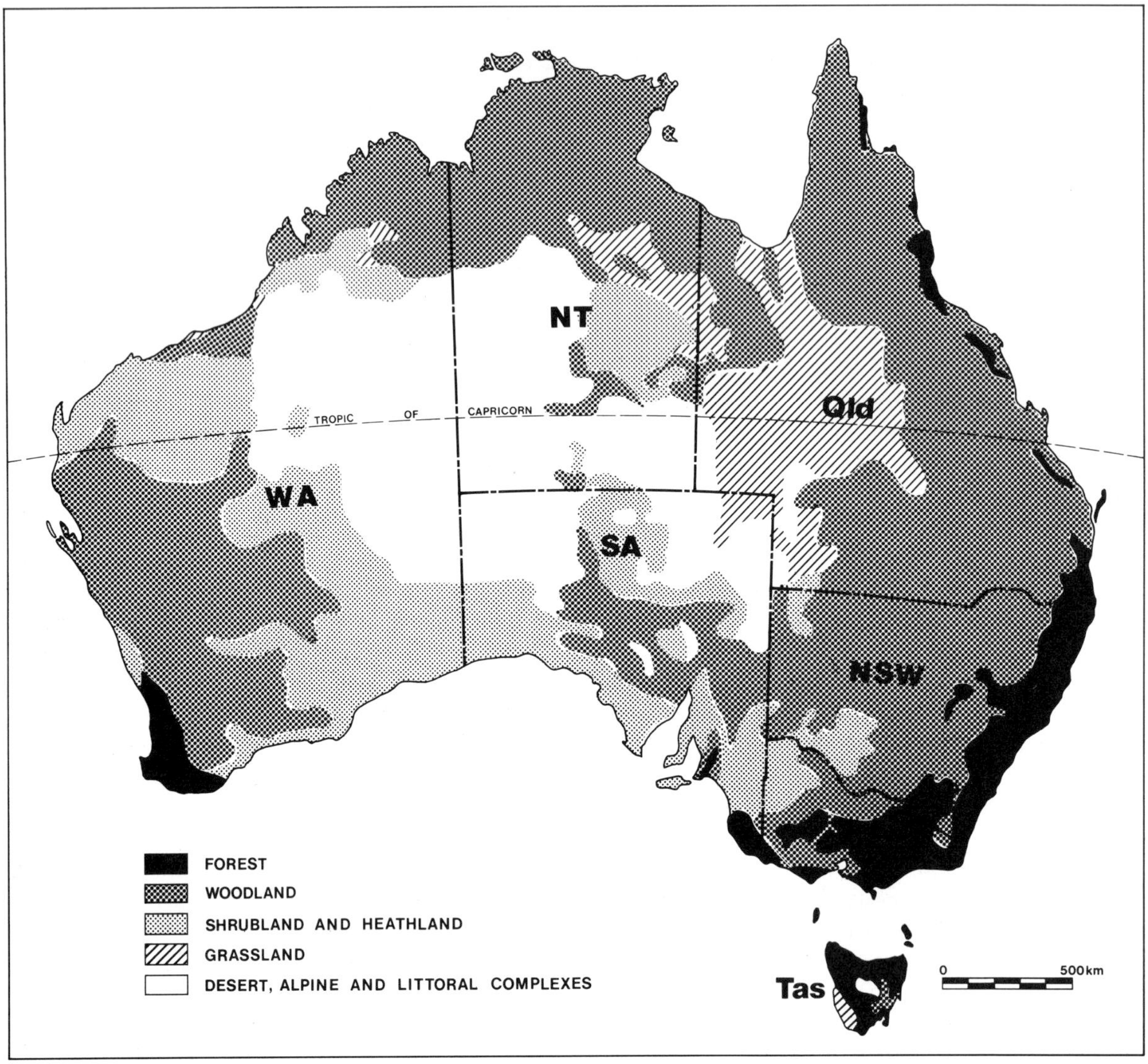

Fig. 4.1. Generalized map of main vegetation formations.

Eucalypt forest is absent from the temperate arid inland zone except for widespread but highly localized stands of *E. camaldulensis*, river red gum. This third area of eucalypt forest is present as thin bands along the inland streams. The stream beds are often dry for most of the year but periodically carry water and are usually deeply filled with sand. The river red gum forest is essentially an outlier occupying a special ecological niche. *E. camaldulensis* has been planted extensively overseas because of its great adaptability, ease of handling in the nursery, potential for rapid growth and the value of its timber.

Two other distinct areas of broad-leaved evergreen forest occur in Australia. One, dominated by trees of *Nothofagus cunninghamii*, is present mainly in the western half of Tasmania and in relatively small areas scattered in south-central Gippsland and in Cape Otway of Victoria. This forest is of particular interest because, in contrast to the eucalypt forests, it has close affinity with the Southern Hemisphere forests of the same tree genus which occur fairly extensively in New Zealand, southern Chile and Argentina.

The other broad-leaved evergreen forest area occurs in the temperate part of eastern Australia as a disjointed coastal strip of rain forest. Essentially Indo-Malayan in character, it contains many of the same genera, and often the same species, as in the forests of tropical Southeast Asia. It is similar in structure to these tropical forests. This type is essentially an extension of tropical forest into the temperate zone resulting from the climatic effects of the warm ocean current which flows southwards down the east coast of Australia. It is not therefore regarded as temperate forest and is not considered further here.

Soil

In Australia the underlying geological formations are mainly rocks of Palaeozoic origin, either metamorphic siliceous sediments or intrusive rocks, often granite or granodiorite. A diversity of soils has developed under arid conditions on these rocks; most are of low plant nutrient status. Eucalypts seem adapted to grow on low nutrient soils, and may have evolved in response to declining soil fertility and a more arid climate. Where basic rocks occur, either as diorite dykes or basalt flows, the soils are more fertile and, provided there is a sufficient soil depth and the climate is wet enough, eucalypt forests thrive.

Beadle (1966) emphasized the role of soil phosphate in determining the distribution of the vegetation of Australia. Attiwill (1972) has suggested that eucalypt forests may be unique among the broad-leaved evergreen forests of the world in their use and conservation of phosphorus. Nevertheless, many eucalypts respond with increased growth rates on more fertile soils. Eucalypt forests as a whole are successful mainly on non-calcareous soil and the great diversity of eucalypt species is found on the light-textured soils of relatively low pH common through much of Australia.

There are a few areas where eucalypt forest has grown successfully on calcareous substrates and where the soil pH remains high. However, most sites with a limestone substrate have an overlying, somewhat acid soil derived over a long period. Tertiary coastal geological formations of limited extent in eastern and western Australia carry eucalypt forest in the appropriate rainfall zones but are of relatively limited extent. One of the most outstanding cases of eucalypt forest on a calcareous soil is that of *E. gomphocephala* in Western Australia, which grows on calcareous sand along the coast. This species is notable for its capacity to tolerate alkaline soils with high pH, often with free lime present.

The broad-leaved temperate forests of Australia have developed during a long evolutionary history in virtual isolation from the rest of the world. They seem to have reached some degree of stability on a relatively unchanging land surface of long duration, much of which displays evidence of earlier peneplanation with the development of lateritic soils.

Fire

There can be no doubt that the broad-leaved evergreen eucalypt forests were subject to wildfire burns in the pre-human phase. In the type of climate where they occur, such forests with their volatile oils and accumulations of bark, leaf and wood debris periodically become highly inflammable. While lightning was no doubt the most common natural source of fire, it is likely that there were other natural causes such as spontaneous combustion, volcanic action or spark ignition from falling rocks. Heavy fuel loads of debris would accumulate in the natural forest during the intervals between fires and periodically would have enabled rapid fire spread and very hot burns.

The forests are well adapted to an environment in which fire is a major ecological factor. In some forests, the natural environment is maintained by occasional heavy burns. This feature is very characteristic of the eucalypt forest adjoining the *Nothofagus* broad-leaved evergreen forests of Tasmania and southern Victoria. Fire is also significant in some of the semi-arid depauperate forests or woodlands. Here good growing seasons occur at rather widely spaced intervals of some decades. Usually during these seasons fuel accumulates on the ground sufficiently to permit a running fire.

Studies of vegetation succession after fire in eucalypt forests have shown that the pattern of regeneration is commonly related to the advent of bushfires. For example, Purdie and Slatyer (1976) reported that vegetative multiplication and seed germination of many species were stimulated by fire. Species vary in their regeneration capacity. Recovery after fires is either by germination of seed residual in the soil and ash or of seed released from trees after a burn; by regrowth from surviving vegetative organs; or by a combination of germination and regrowth.

The resistance of eucalypts to fire in general increases with greater bark thickness. Where broad-leaved eucalypt evergreen forests are subject to regular fires, the species present usually recover quickly after a fire, and seedling and vegetative regrowth rapidly restore a forest cover. In a few of the wetter types damage may be lethal and recovery is by regeneration only from seed.

Most eucalypt species have been associated with fire regimes of some kind throughout their evolutionary history. There is good reason to believe that, if fire could be completely removed from the Australian environment by a policy of complete protection, there would be substantial changes in the distribution of plant species. Some species might even be eliminated.

Bushfires affect timber production, soil conditions, nutrient circulation, wildlife and recreational activities. Within historical times, some have caused loss of human life and serious damage to property. Fire prevention, bushfire control and the use of prescribed burning to reduce the fire hazard are important management activities in Australian eucalypt forests.

Detailed accounts of bushfires in Australia and their effects on the vegetation have been given by Gill (1975) and by Luke and McArthur (1978).

Effects of Aboriginal people

The arrival of Aboriginal people at least 40 000 and perhaps 100 000 years ago began a period of change. The Aboriginal population was a hunting, food-gathering society whose chief impact on the broad-leaved temperate forests was by the use of fire as a hunting tool. The frequency of Aboriginal fires was probably quite different from that of the fires of the pre-human phase. The records of the early European explorers indicate that the smoke of fires was seen frequently over extensive areas; there is little doubt most of these fires were lit by Aboriginal people.

There is growing evidence that Aboriginal use of fire to control the vegetation was more discriminatory than previously believed. The nature of Aboriginal burns varied according to vegetation type and the season and was geared to hunting indigenous animals and collection of useful plants.

The dingo brought to Australia by the Aboriginals probably competed with and preyed upon the indigenous fauna of the forests. Some Australian animals became extinct some time following Aboriginal occupation but the cause is unknown.

Settlement by non-Aboriginal people

Another major event which influenced these forests is the advent of European settlement barely two centuries ago. Besides changing the fire regime, the European settlers brought about major changes in much of the eucalypt forest or woodland by felling for timber but more commonly to provide land for agricultural purposes, cropping and grazing. This sometimes led to the complete removal of eucalypts from extensive areas, but often some scattered trees were retained for shade. It is thought no eucalypt species was entirely eliminated by European settlement but the populations of many tree species were drastically reduced, sometimes with changes in genetic makeup.

Significance of eucalypt forest

Within Australia the eucalypt forests are still the backbone of the timber-using industry although the development of plantations of exotic species, mainly pines, is an increasingly prominent segment of the total wood production. Relatively little sawn eucalypt timber or logs are exported but now there is a growing export of eucalypt woodchips and pulp. The eucalypt forests are important for the conservation of wildlife, the protection of water catchments, prevention of soil erosion, provision of recreation facilities and as a vital component of Australian landscapes.

Outside Australia the use of a number of *Eucalyptus* species, especially of the subgenera *Symphyomyrtus* and *Corymbia*, is having a very substantial effect on the total wood resources of the world because of their capacity for large production of wood in the warmer, drier parts of the world where the natural vegetation is often relatively slow-growing and produces material difficult to use with available technology. The absence of native pest and disease organisms is seen as another factor enabling rapid growth of eucalypts planted overseas. The preservation of the eucalypt gene resource in Australia is regarded as a matter of world importance because of the potential capacity of eucalypts to increase significantly global timber reserves (Eldridge, 1971).

GROSS CHARACTERISTICS

Forest pattern and structure

The most striking characteristic of the broad-leaved evergreen forests of Australia is the fact that the dominant trees are virtually all species of *Eucalyptus*, a large genus of over 450 species (Hall et al., 1963; Blakely, 1965). These are divided into perhaps ten distinct groups, usually regarded as subgenera. Species from these different groups, of which three or four are the most important and numerous, do not interbreed. A feature of eucalypt forest is that commonly there are two or three and occasionally four species growing together as equal co-dominants in the same stand. They retain their genetic separation. In a few forest types only a single tree species is dominant; this is more common in the drier and colder areas of the range of the genus as a whole.

Striking examples of eucalypt forests in which two species make up the tree component are the extensive areas in Western Australia where *E. marginata* and *E. calophylla* occur together apparently in a balanced relationship. This kind of combination is repeated in eastern Australia, as for example in forests of *E. obliqua* and *E. viminalis* in Victoria or in Queensland and New South Wales with *E. pilularis* and *E. microcorys*.

Although the broad-leaved forests have a degree of uniformity because eucalypts are so universally present, a great diversity of very distinct ecosystems occurs within them. Often the pattern of distribution of species is complex, the eucalypt species changing with slight variations in aspect, moisture and the depth, structure and fertility of soils or as a result of fire, logging, and soil disturbance.

Specht et al. (1974) recognized twelve major forest or woodland structural formations (Table 4.1) and gave details of the plant species present in different strata for the various States and Territories. Within the forest or woodland formations they identified 590 major plant alliances out of an Australian total of 1193 for all major plant alliances. The Australian flora includes about 15 000 species of vascular plants. It is not possible here to describe in detail the great variety of plant alliances in the eucalypt forests.

TABLE 4.1

Structural formations (based on Specht et al., 1974)

Life form and height of tallest stratum (m)	Projective foliage cover of tallest stratum			
	dense (70–100%)	mid-dense (30–70%)	sparse (10–30%)	very sparse (< 10%)
Trees[1] > 30	tall closed forest	tall open forest	tall woodland	tall open woodland
Trees 10–30	closed forest	open forest	woodland	open woodland
Trees 5–10	low closed forest	low open forest	low woodland	low open woodland

[1]A tree is defined as a woody plant more than 5 m tall, usually with a single stem.

In forest characteristic of the area of greater rainfall a subordinate layer of medium-sized trees may occur, the species differing from area to area. In some areas the most prominent subordinate trees are the larger *Acacia* species such as *A. melanoxylon* or species of *Casuarina* such as *C. fraseriana* or *C. torulosa*. In one extensive area in the northwest of New South Wales a dense layer of *Callitris huegelii* overtopped by rather widely spaced dominant eucalypts is found over an area some 150 000 ha or more. In southwestern Queensland there is an even larger area of a similar type.

The associated shrub species differ depending on the forest type. For example, in the wet sclerophyll forest, there are tall shrubs with soft leaves and a mesophytic aspect belonging to Australian genera such as the rhamnaceous *Pomaderris* or the tree composites *Olearia* or *Bedfordia*. In the forest characteristic of drier areas, tall shrubs of this kind are missing and there is a rather dense layer of sclerophyll shrubs of distinctive Australian types with genera and species belonging to families like

Epacridaceae, Proteaceae, Rutaceae, and legumes of the tribe of Podalyrieae.

In the relatively open woodlands shrubs are usually infrequent but there is a ground cover principally of grasses and herbs. The estimated areas of different structural formations are given in Table 4.2. Most broad-leaved evergreen forests have a relatively open tree canopy.

Forest types

Because of the large number of eucalypt species and the multiplicity of ecosystems the biological classification of broad-leaved evergreen forests is difficult. In Tables 4.3 and 4.4 details are given of the main alliances identified by Specht et al. (1974) for wet and dry forests respectively, the formations of greatest forestry significance. Most of these alliances are dominated by broad-leaved evergreen tree species (Figs. 4.2–4.9).

The Specht et al. classification is based essentially on the dominant tree species and the forest structure defined in terms of tree height and density of overstorey cover. Within these important broad classes there can be significant variation in the understorey plant species and in the associated fauna.

As part of a wide ranging review of Australian forestry and wood-based industries carried out in 1974 native forest types were grouped for forestry purposes. The eucalypt forests were divided into three productivity classes, class I being the most productive and class III the least productive of the three classes. Details of the main overstorey and understorey species and other information are given in Appendix I for these groups.

Extensive areas in New South Wales, northwest Victoria, South Australia and Western Australia dominated by small eucalypts are described as mallee (see Specht, 1979, 1981). These can scarcely be classified as forest since the dominant eucalypts seldom reach more than 7 m in height. This distinctive life form results from the development of a large lignotuber at ground level or embedded in the soil, from which multiple stems arise bearing the leafy crown. There are many associated rather scattered shrubs in numerous genera, including particularly *Acacia* and *Melaleuca*, in the subordinate layers. There is seldom any herbaceous ground cover except of annuals in occasional favourable seasons.

The *Nothofagus* southern beech or myrtle beech forest of Tasmania and southern Victoria is not extensive and probably consists of less than

TABLE 4.2

Estimated areas (10^3 km^2) of structural formations (based on Specht et al., 1974)

	Queensland	New South Wales and Australian Capital Territory	Victoria	South Australia	Tasmania	Western Australia	Northern Australia	Australia
Tall closed forest	0	0	3.0	0	11.7	0	0	14.7
Tall open forest (wet sclerophyll)	2.1	68.9	27.1	0	21.0	7.3	0	126.4
Open forest (dry sclerophyll)	1.0	92.0	56.3	8.3	11.2	48.8	0	217.6
Tall woodland	5.6	212.0	32.7	15.9	8.3	116.3	0	390.8
Low woodland	125.9	38.0	0	25.3	0	38.4	142.3	369.9
Low open woodland	57.0	100.5	38.7	36.4	0	58.1	112.1	402.8

TABLE 4.3

Important plant alliances of tall open forest (wet sclerophyll forest) (based on Specht et al., 1974)

QUEENSLAND

1. *Araucaria cunninghamii*, *Eucalyptus acmenoides*, *E. andrewsii*, *E. drepanophylla*, *E. grandis*, *E. maculata*, *E. microcorys*, *E. pellita*, *E. pilularis*, *E. propinqua*, *E. saligna*, *E. torelliana*, *Syncarpia glomulifera*, *Tristania conferta*

NEW SOUTH WALES, including AUSTRALIAN CAPITAL TERRITORY

1. *Eucalyptus delegatensis–E. dalrympleana*, containing as well:
   E. dalrympleana subsp. *dalrympleana*; *E. dives*; *E. pauciflora* subsp. *pauciflora* (montane race); *E. radiata* subsp. *radiata*; *E. radiata* subsp. *robertsonii*; *E. viminalis*

2. *Eucalyptus fastigata–E. viminalis*, containing as well:
   E. andrewsii subsp. *campanulata*; *E. bridgesiana* subsp. *bridgesiana*; *E. cypellocarpa*; *E. dalrympleana* subsp. *dalrympleana*; *E. deanei*; *E. dives*; *E. elata*; *E. fraxinoides*; *E. globoidea*; *E. laevopinea*; *E. macrorhyncha*; *E. maidenii*; *E. mannifera* subsp. *maculosa*; *E. nitens*; *E. obliqua*; *E. pauciflora* subsp. *pauciflora* (montane race); *E. radiata* subsp. *radiata*; *E. radiata* subsp. *robertsonii*; *E. rubida*; *E. sieberi*; *E. smithii*; *E. st-johnii*

3. *Eucalyptus saligna–E. resinifera–E. pilularis*, containing as well:
   Angophora spp.; *E. gummifera*; *E. acmenoides*; *E. botryoides*; *E. deanei*; *E. globoidea*; *E. grandis*; *E. microcorys*; *E. quadrangulata*; *Syncarpia glomulifera*; *Tristania conferta*; *T. suaveolens*

4. *Eucalyptus acmenoides–E. propinqua–E. microcorys*, containing as well:
   E. eugenioides; *E. paniculata*; *E. pilularis*; *E. punctata*; *E. resinifera*; *E. umbra* subsp. *carnea*; ± *Tristania conferta*

5. *Eucalyptus maculata–E. paniculata*

VICTORIA

1. *Eucalyptus delegatensis–E. dalrympleana*
   a. *E. delegatensis–E. dalrympleana*
   b. *E. delegatensis–E. viminalis*
   c. *E. nitens–E. fastigata*
2. *E. regnans*
3. *E. sieberi–E. globoidea*
4. *E. globoidea–E. botryoides*
5. *E. obliqua–E. cypellocarpa–E. st-johnii*
6. *E. obliqua–E. radiata*

TASMANIA

1. *Eucalyptus delegatensis* ± *E. dalrympleana*
2. *E. regnans*
3. *E. obliqua*–(*E. viminalis*)
4. *E. robertsonii–E. dalrympleana*
5. *E. globulus*–(*E. viminalis*)
6. *E. nitida*–(*E. ovata*)

WESTERN AUSTRALIA

1. *Eucalyptus diversicolor*
2. *E. marginata–E. calophylla*
3. *E. gomphocephala*

250 000 ha. It has a close ecosystem affinity with forests dominated by species of the same genus in New Zealand and also in Chile and Argentina. The dominant tree, and, in many areas, the only tree present is *Nothofagus cunninghamii*, which forms a dense canopy under which there are relatively few associated species. Where there are associated trees the most common are *Atherosperma moschatum* (southern sassafras), *Acacia melanoxylon*, *Eucryphia lucida* and *Phyllocladus asplenifolius*. Subordinate and minor tree species are *Anodopetalum biglandulosum* and *Leptospermum* spp. Other understorey species are *Anopterus glandulosus*, *Cenarrhenes nitida*, *Leptospermum lanigerum* and *Tasmannia lanceolata*, whilst the tree fern *Dicksonia antarctica* and ferns of *Blechnum* spp. are common, as are lichens and mosses.

Nothofagus forests occur in high rainfall areas sometimes with more than 2500 mm yr^{-1} and many rainy days during the year. The long history of development of this ecosystem is associated with extremely low nutrient levels in the soils and with

TABLE 4.4

Important plant alliances of open forest (dry sclerophyll forest) (based on Specht et al., 1974)

QUEENSLAND

1. *Eucalyptus tetrodonta–E. miniata–Eucalyptus* sp.
2. *E. polycarpa–E. alba–E. tessellaris*
3. *E. brownii*
4. *E. populnea*
5. *E. crebra–E. drepanophylla–E. melanophloia*
6. Eucalypt open-forest alliance, containing:

E. acmenoides	*E. phaeotricha*
E. baileyana	*E. pilularis*
E. citriodora	*E. planchoniana*
E. cloeziana	*E. propinqua*
E. crebra	*E. resinifera*
E. curtisii	*E. saligna*
E. eugenioides	*E. seeana*
E. gummifera	*E. siderophloia* (syn. *E. fibrosa*)
E. intermedia	
E. maculata	*E. sideroxylon*
E. major	*E. signata*
E. melanophloia	*E. tereticornis*
E. melliodora	*E. tessellaris*
E. microcorys	*E. trachyphloia*
E. moluccana	*E. umbra*

7. *Tristania laurina–Casuarina cunninghamiana*
8. Eucalypt open-forest, containing:

E. acmenoides	*E. phaeotricha*
E. alba	*E. polycarpa*
E. citriodora	*E. resinifera*
E. cloeziana	*E. siderophloia* (syn. *E. fibrosa*)
E. crebra	
E. drepanophylla	*E. tereticornis*
E. intermedia	*E. tessellaris*
E. melanophloia	*E. trachyphloia*
E. moluccana	*E. umbra*

9. *Callitris columellaris*
10. *Melaleuca quinquenervia*
11. a. *Acacia harpophylla*
    b. *A. harpophylla–Geijera parviflora*
    c. *A. harpophylla–Brachychiton rupestre*
    d. *A. harpophylla–Casuarina cristata*
    e. *A. harpophylla–Terminalia oblongata*
12. *Acacia argyrodendron*
13. "Sandstone open-forest", containing:

Angophora costata	*E. peltata*
Callitris columellaris	*E. polycarpa*
Eucalyptus cloeziana	*E. tenuipes*
E. crebra	*E. watsoniana*
E. maculata	

NEW SOUTH WALES, including AUSTRALIAN CAPITAL TERRITORY

1. *Eucalyptus delegatensis–E. dalrympleana*
2. *E. fastigata–E. viminalis*
3. *E. saligna–E. pilularis–E. resinifera*
4. *E. acmenoides–E. propinqua–E. microcorys*
5. Sandstone complex, containing:

Angophora costata	*E. piperita*
Casuarina spp.	*Syncarpia glomulifera* + rain forest elements
E. pilularis	

VICTORIA

1. *Eucalyptus delegatensis–E. dalrympleana*
    a. *E. delegatensis–E. dalrympleana*
    b. *E. delegatensis*
2. *E. sieberi–E. globoidea*
3. *E. globoidea–E. botryoides*
4. *E. obliqua–E. cypellocarpa–E. st-johnii*
5. *E. obliqua–E. radiata*
6. *E. baxteri–E. nitida*
7. *E. macrorhyncha–E. dives–E. polyanthemos*
    a. *E. macrorhyncha–E. polyanthemos–E. dives* (plus *E. obliqua, E. baxteri, E. goniocalyx*)
    b. *E. macrorhyncha–E. polyanthemos–E. dives* (plus *E. sideroxylon*)
8. *E. viminalis–E. ovata*
9. *E. radiata–E. rubida*
10. *E. pauciflora*
11. *E. sideroxylon–E. leucoxylon–E. woollsiana* subsp. *microcarpa*
12. *E. albens–Callitris columellaris* subsp. *glauca*
13. *E. gummifera–Angophora floribunda*

SOUTH AUSTRALIA

1. *Eucalyptus rubida*
2. *E. obliqua–E. baxteri*
3. *E. baxteri* (± *E. fasciculosa*)
4. *E. obliqua–E. goniocalyx* (± *E. fasciculosa*)
5. *E. cladocalyx–E. goniocalyx*
6. *E. macrorhyncha*

TASMANIA

1. *Athrotaxis cupressoides*
2. *Eucalyptus delegatensis* (± *E. dalrympleana*, or: *E. coccifera, E. gunnii, E. archeri, E. urnigera, E. subcrenulata, E. pauciflora, E. rodwayi*)
3. *E. coccifera* (± *E. gunnii, E. urnigera, E. subcrenulata, E. pauciflora*)
4. *E. gunnii*
5. *E. archeri*
6. *E. urnigera*
7. *E. subcrenulata*
8. *E. pauciflora*
9. *E. rodwayi*
10. *E. obliqua* ± *E. amygdalina, E. viminalis, E. nitida*
11. *E. sieberi* ± *E. amygdalina–E. viminalis*
12. *E. amygdalina–E. viminalis*
13. *E. pulchella–E. viminalis* ± *E. globulus*
14. *E. nitida–E. ovata*

TABLE 4.4 (*continued*)

15. *E. globulus* ± *Callitris tasmanica*	4. *E. gomphocephala*
16. *E. tenuiramis–E. risdonii–E. viminalis*	5. *E. loxophleba*
17. *E. morrisbyi–E. tenuiramis–E. amygdalina*	6. *E. redunca* var. *elata*
18. *E. st-johnii–E. ovata–E. viminalis*	7. *E. cornuta*
	8. *Agonis flexuosa*
WESTERN AUSTRALIA	9. *E. dundasii*
	10. *E. diversicolor*
1. *Eucalyptus marginata–E. calophylla*	11. *E. tetrodonta–E. miniata*
2. *E. rudis–Melaleuca* spp.	
3. *E. marginata–E. calophylla*	

Fig. 4.2. New South Wales, *Eucalyptus pilularis* (blackbutt). Photo Forestry Commission N.S.W.

Fig. 4.3. New South Wales, *Eucalyptus grandis* (flooded gum). Photo Forestry Commission N.S.W.

an accumulation of organic matter in a duff layer, often not very thick, over siliceous mineral soil.

Of particular interest is the relationship between the *Nothofagus* forest and the wetter phases of the eucalypt forest. *Nothofagus* and some of the associated species in the absence of fire will invade the adjoining eucalypt forest, first as an understorey species. If there is no disturbance by fire the eucalypts die and the forest progressively becomes *Nothofagus* forest. An oscillating balance seems to have been maintained between these two kinds of forest depending on fire, which apparently occurred infrequently. Studies of the age-classes of the trees involved indicate in the pre-European period a fire frequency of only once in 50 to 100 years.

Ecophysiology of tree species

In many parts of the world a single dominant may range over large geographic distances,

Fig. 4.4. New South Wales, *Eucalyptus maculata* (spotted gum). Photo Forestry Commission N.S.W.

Fig. 4.5. New South Wales, *Eucalyptus camaldulensis* (red gum). Photo Forestry Commission N.S.W.

frequently in more or less continuous stands. In contrast, the dominant eucalypts are relatively restricted geographically and, if they have an extensive total range, are nevertheless confined to highly specific microhabitats within that range. Following a long history of adaptation eucalypt species appear to have become highly adapted to the particular sites in which they thrive. Their actual distribution is a fine mosaic of relatively small stands separated one from the other according to slight changes in habitat conditions determined by land form, minor topography and geological substrate.

Within a *Eucalyptus* species there may be genetic variations between different populations. Provenances may differ in their growth rate, stem straightness, frost resistance, drought resistance and reaction to different soil conditions (Eldridge, 1971). Ladiges and Ashton (1977) found that seedlings of *E. viminalis* raised from seed gathered from populations growing on acid or alkaline soils differed in their tolerance of calcareous soils and their susceptibility to lime chlorosis. Eldridge (1972) reported significant genetic variation in the growth of *E. regnans* over a 4.8-km transect from 370 to 900 m altitude on Mount Erica, Victoria. Higher altitude populations produced trees with slower growth and straighter stems. Genetic variation within a species is seen as providing an opportunity to improve the growth of plantations of eucalypts by careful selection for desirable characteristics.

Throughout the eucalypt forests, the two predominant habitat factors controlling the species composition and growth of the eucalypt forests are water availability broadly determined by regional climate, together with microhabitat conditions, especially as related to topography. The soil-nutritional level, which by world standards is frequently very small especially for phosphorus, also plays a prominent role in the distribution pattern of the species.

The interplay of these factors and the great

Fig. 4.6. Victoria, *Eucalyptus sieberi* (silvertop ash). Photo Forests Commission Vic.

Fig. 4.7. Victoria, *Eucalyptus regnans* (mountain ash). Photo Forests Commission Vic.

number of eucalypt species available give an unusual character to eucalypt forests. Because the dominant eucalypt species are very sensitive to habitat conditions, the changes from site to site of dominant species occur more frequently and rapidly than in many Northern Hemisphere forests, where the number of dominant species in the temperate zone is very much fewer and the site requirements of each species much less exacting than for Australian species.

The various species of eucalypts which make up the forest, each on their own site, are highly efficient in maintaining their population boundary. Species between which there is no breeding barrier seldom occur together as co-dominants in a single stand unless there has been some mixing as a result of disturbance by people. Even where occasional interspecific hybrids are found, the boundaries do not break down to give any genetic blending.

Most eucalypt forests regenerate through the creation of small openings in the canopy as old over-mature trees die or become moribund or are blown down through environmental accidents such as wind storms. Regeneration follows in these small openings, and uneven-aged forests with a finely graded sequence of age-classes result. However, there are exceptions. As described previously, at the wetter limit of eucalypt forest there is a dynamic ecological situation which in the absence of fire leads to the replacement of eucalypts by *Nothofagus* broad-leaved forest. The maintenance and regeneration of species such as *Eucalyptus regnans* results from intense and extensive fire. Stands of this species, often some hundreds of hectares in extent, are the product of rather widely spaced—perhaps 50 to 100 years—regeneration periods and are even-aged from that time. The same perhaps may be largely true of some of the semi-arid forest types such as mallee.

Some seed is produced each year by trees of most eucalypt species but the amount produced varies greatly from tree to tree and from year to

Fig. 4.8. Victoria, *Eucalyptus sideroxylon* (red ironbark). Photo Forests Commission Vic.

year (Cremer et al., 1978). Seed distribution is mainly by the wind but dispersal is rarely greater than the height of the tree. Adequate light, plant nutrients and water are essential for seedling establishment. A hot fire may be a prerequisite for successful regeneration. The rapid growth which follows intense fire is known as the "ash bed" effect and probably results from the increased availability of nutrients, a reduction in competition by other plants and changes in the microflora. Some eucalypts appear to be allelopathic in that they suppress the regeneration of their progeny, probably by chemical inhibitors.

Fauna

The eucalypt forests provide a wide diversity of habitats and are a very important refuge for Australian wildlife. Just as for plants, there is a high level of endemism in Australian animals. Some animal species display remarkable specialization, for example the food of the koala (*Phascolarctos cinereus*) is largely confined to leaves of a few species of *Eucalyptus*. Some species are restricted to one or several forest types whilst others are very tolerant being distributed widely and not restricted to forest habitats. Frith (1973) has pointed out the difficulty of classifying wildlife habitat areas because a classification based on vegetation does not necessarily have much to do with the distribution of vertebrate fauna. He draws attention not only to differences between the mammals and birds of sclerophyll forest and woodland but also to differences between subdivisions of forest or woodland.

Detailed inventories of the mammals inhabiting the tall, open *Eucalyptus* forests of southeastern

Fig. 4.9. Tasmania, *Nothofagus cunninghamii* (myrtle rainforest). Photo Forestry Commission Tas.

and southwestern Australia and Tasmania have been prepared by Tyndale-Biscoe and Calaby (1975). They conclude that the *Eucalyptus* forests of southeastern Australia and of Tasmania support a rich and varied fauna of mammals and birds and together form the single most important refuge for wildlife in Australia. For tall open forest Tyndale-Biscoe and Calaby identified 14 species of mammals for which the habitat is optimum in southeastern Australia, 5 for Tasmania and 2 for Western Australia. Most mammal inhabitants of these forests proved to be marginal species—33 in southeastern Australia, 17 in Tasmania and 16 in southeastern Australia. They explained the greater number of species in the mainland eastern forests as resulting from separation, with the forest mammal fauna originating in mainland eastern Australia and following the Pleistocene becoming isolated from Tasmania and the southwest by Bass Strait and the Nullarbor Plain respectively.

Because of the importance of eucalypt forests for the conservation of wildlife, some interest has centred on the effects of forestry practices on wildlife (McIlroy, 1978). There is evidence of faunal differences between natural and managed forests. Forestry practices vary considerably between the States and Territories but those of greatest significance for wildlife are stand improvement operations and selective logging, clearfelling, establishment of plantations, management for the supply of wood-chips and pulp and prescribed burning. Woinarski (1979) found that more bird species were present in natural eucalypt forest than in an adjacent eucalypt plantation. Driscoll (1977) investigated the change in the avifauna as native vegetation was replaced by plantations of *Pinus radiata* and found that the pine plantations had fewer bird species than wet or dry sclerophyll eucalypt forest and that the kinds of birds present changed.

Newsome et al. (1975) have pointed out that the fauna existing in dry sclerophyll forests is a fire-climax assemblage and that, whilst fire may limit faunal diversity, timely fires may assist the survival of certain species. Christensen and Kimber (1975) found that, in the southwest forests, fire temporarily reduced the number of small mammals, but there were relatively small changes in the populations of birds and macropods. Leadbeater's possum (*Gymnobelideus leadbeateri*), previously very rare, is reported as flourishing in regenerated areas after the 1939 fires in Victoria (Ride, 1970). The preservation of old trees with suitable hollows as homesites is probably important for the survival of such arboreal species of mammals. Partial clearing of forest may increase the numbers of ground-living wildlife such as kangaroos, wallabies and wombats.

Open eucalypt forest and woodland provide grazing for sheep, and cattle and the commercial production of honey from eucalypts is a multimillion dollar industry. Introduced mammals, particularly rabbits, horses, pigs and goats, sometimes do considerable damage in eucalypt forest.

Jenkins and Bartell (1980) record 32 species of reptiles in dry sclerophyll forest but only 9 species in wet sclerophyll forest in the Australian high country. They attribute this difference to the more luxuriant understorey of the wet sclerophyll forest, which reduces solar radiation reaching the forest floor so that the habitat is less suitable for reptile species which rely on sun-basking to gain body heat.

As a result of co-evolution there is an enormous variety of insects associated with eucalypts. Some appear beneficial whilst others on occasion cause wide-scale damage.

Pests and diseases

Australia is fortunate in that as yet no introduced animal species has become a major threat to the native forest resource. The effects of rabbits and feral animals on tree regeneration and the clearing of forest for pasture for domestic stock have been mentioned previously.

In undisturbed forest there are relatively few native grazing animals because of the paucity of food. Where the forest is disturbed, either naturally or by forest operations, animal numbers may increase in and around clearings. Here regeneration may be limited by the browsing of kangaroos, wallabies, possums and rats (Gilbert, 1961). Populations of koalas can become so abundant that substantial numbers of trees of the desirable food species are defoliated by browsing and may die. The koala population has been drastically reduced by disease and by killing for their pelts by Europeans, nevertheless the effects of koala browsing can be observed in the tree crowns from time to time and may create local problems when the protection of koalas results in greatly increased numbers.

An enormous variety of insects is associated with eucalypts (Carne and Taylor, 1978), but remarkably few are regarded as pests. Apparently the diversity of insect life, predators, parasites and diseases provides a biological check preventing individual species becoming so numerous as to reach pest proportions. The extent of the loss of wood production in native forests by insect attack is unknown.

In eucalypt forests partial or complete defoliation resulting in a decrease in the photosynthetic capacity of the trees occurs at intervals with outbreaks of the stick insects *Didymuria violescens* and *Podacanthus wilkinsoni*; the leaf beetle *Chrysophthara bimaculata*; the christmas beetle *Anoplognathus chloropyrus*; the moths *Perthida glyphoga*, *Uraba lugens* and *Doratifera* spp.; the lerp insects *Cardiaspina* spp., *Glycaspis* spp. and *Hylinaspis* spp.; and the saw flies, *Perga* spp. Sap-sucking bugs (*Amorbus* and *Gelonus* spp.) and weevils may damage the apical shoots of eucalypts to cause their death and multiple leaders (brooming). Insects of many species affect seed production.

Other insects feed on the wood and inner bark. Greaves et al. (1967) have shown that significant financial loss can occur by a reduction in log quality due to the formation of a central hollow pipe in the tree bole by termites. They found that termite infestations begin in damaged tissues such as fire scars and that termite damage can be reduced by fire protection and the control of termites in old trees. Degrading of eucalypt timber is also caused by a variety of pinhole borers and other insects (Neumann and Marks, 1976).

Several species of the family Loranthaceae (mistletoes) are obligate parasites on eucalypt crowns and by reducing photosynthesis of the trees cause a reduction in basal area increment of the host trees. The management of native stands is believed to have increased the incidence of mistletoe infestations by opening up the tree canopy (Penfold and Willis, 1961).

Information on the diseases of eucalypt leaves, branches and stems, roots and stem butts has been summarized by Heather and Griffin (1978). Some pathogens are host-specific whilst others are less selective, but the overall impact of disease on eucalypt forest is unknown.

Over the last decade attention has been directed to forest destruction, especially of *Eucalyptus marginata* jarrah forest in Western Australia by the soil fungus *Phytophthora cinnamomi* (Newhook and Podger, 1972; Pratt and Heather 1973; see also Malajczuk and Lamont, 1981). The disease has reached epidemic proportions in the *Eucalyptus marginata* forests of Western Australia causing serious financial loss. In contrast some eastern eucalypts co-exist with the disease organism, but possibly with some loss of wood production. Christensen (1973) has emphasized that many plant species of the native jarrah forest are affected. Susceptible species of *Banksia* and *Xanthorrhoea* suffer considerable mortality and this in turn has adverse repercussions on the variety of forest fauna associated with them. Forest hygiene measures designed to prevent the spread of the fungus by reducing the transfer of soil on vehicles have been introduced to contain the fungus.

Investigations of other kinds of dieback in native eucalypt forest have failed to identify insects or diseases as the prime cause. The decline of some eucalypt species in some areas of rural Australia is causing increasing concern (Kile et al., 1980). Ellis et al. (1980) have suggested that a possible cause of the dieback of *Eucalyptus delegatensis* in Tasmania is due to soil sickness with limited breakdown of phytotoxins under the existing conditions.

Forestry practices which are seen as reducing the outbreaks of pests and diseases include the maintenance of vigorous tree growth, preservation of diversity, prevention of wildfire and mechanical damage and the removal of infested trees. Wide-scale application of chemicals to control pests and diseases has not been used. There is concern that large uniform plantations of eucalypts may constitute a serious pest risk and because of the financial investment may require increasing the use of chemical control measures.

FOREST FUNCTIONAL PROCESSES

Succession

More detailed study is needed of succession in eucalypt forest but is difficult because of the complex pattern of species distribution reflecting very sensitive delineation of eucalypt species. Whilst differences in micro-topography, aspect and soil may be the basis for the pattern of species distribution in eucalypt forest, superimposed upon these are the effects of disturbances caused by wildfires, logging and grazing.

The natural processes of stand regeneration and development vary according to the species present and their ability to tolerate the reduced levels of light within the forest and competition for water and nutrients in the root-growing space. Succession is also influenced by the extent to which soil conditions have changed under the previous eucalypt stand, the extent of soil disturbance and its effects on seed bed conditions and the effects of pests and diseases (Florence, 1969).

Ecosystem dynamics

Relatively few studies have been made of the production and dissipation of organic matter, energy flow and the circulation of nutrients and water in eucalypt forest ecosystems. The available data are so scanty and eucalypt ecosystems are so diverse that it is unwise to make generalizations. More comprehensive studies of the dynamics of eucalypt ecosystems could contribute to a better understanding of the environmental basis of species pattern and assist in developing management procedures appropriate to eucalypt forest (Ovington, 1968).

Organic production, accumulation and decomposition greatly influence nutrient circulation. In eucalypt forests large amounts of litter may accumulate and the nutrients it contains are unavailable to the trees until released by decomposition or by fire. O'Connell et al. (1978) found that in forests of *E. marginata* the amount of litter on the ground and its nutrient content varied greatly. They considered that periodic burning by releasing nutrients in the litter may be particularly important in the cycling of nitrogen and phosphorus which are in short supply in these forests. Limited information is available of the role of soil fauna in litter breakdown (Howard, 1975; Plowman, 1979).

Harwood and Jackson (1975) showed that considerable losses of phosphorus, potassium, calcium and magnesium to the atmosphere occur when a mixed forest of *Eucalyptus regnans*, *Nothofagus cunninghamii*, *Atherosperma moschatum*, *Acacia melanoxylon* and *A. dealbata* is burnt. Loss of the four nutrients as a percentage of that in the litter before the burn was 10, 9, 7 and 17 respectively. Such losses represent a significant drain on the ecosystem particularly as the soils are generally poor in nutrients.

Lee and Correll (1978) found that, in a forest of *Eucalyptus obliqua*, the litter accumulated on the forest floor was 980 g m^{-2} representing 4.2 years of the mean litter fall. The energy content of the annual litter fall was 4900 kJ m^{-2}. They summarized the information on litter fall for eight forest types and pointed out it ranges from 210 to 995 g m^{-2} yr^{-1}, with litter accumulation on the forest floor of up to 2180 g m^{-2}. The maximum amounts of nutrients in the annual litter fall, as mg m^{-2}, were: nitrogen 5760, phosphorus 190, potassium 830, calcium 4880 and magnesium 832.

Attiwill (1966, 1968 and 1972) has investigated the production and turnover of organic matter and the cycling of nutrients in a forest of *E. obliqua* (Table 4.5). Flinn et al. (1979) found that, for forest catchments dominated by *E. radiata*, nutrient input in the precipitation exceeded that removed in stream flow from the catchments.

In general these studies of nutrient circulation emphasize that much of the nutrient capital within eucalypt ecosystems is in the plant organic matter. Because of the infertility of Australian soils, the retention of as much as possible of this nutrient capital within the ecosystem by sensitive management is very important. From estimates based on biomass and nutrient concentrations for six species

TABLE 4.5

Biomass and nutrient circulation in *Eucalyptus obliqua* forest (based on Attiwill, 1966, 1972)

	Dry weight (t ha^{-1})	Nutrients (kg ha^{-1})		
		P	Ca	K
Tree crowns	29	9	93	64
Stem bark	44	6	189	117
Stem wood	228	12	28	51
Litter	18	5	64	16
Understorey	1	4	26	24
Dead on trees	4	–	–	–
Total	334	36	400	272
Annual return from biomass to soil	4	1	26	17

of eucalypts, Wise and Pitman (1981) have shown that nutrient removals under a short rotation (ten-year) plantation system would be much greater than those under a native forest sawlog regime but much less than removals experienced with agricultural crops.

Although water is an important product of forest land, few comprehensive studies are available of the hydrologic cycles of different types of eucalypt forest to provide a guide to management.

Forest activities associated with hardwood logging may adversely affect water quality by increasing turbidity (Cornish, 1980). Shea et al. (1975) found that *E. marginata* forest transpires at a high rate during the dry, hot summer months and forested catchments yield less than 10% of the rainfall. Disturbance of the forest cover affected water quality and resulted in the discharge of highly saline water. They conclude that on non-saline areas there is a large potential to increase water yield by manipulation of the forest vegetation. Tunstall and Walker (1975) found that when the trees were killed in a *E. populnea* community there was an increase in soil salinity but little change in water storage in the soil.

Smith (1974) compared throughfall, stemflow and interception in a mature plantation of *Pinus radiata* and in a mature native forest of *Eucalyptus rossii*, *E. maculosa* and *E. dives*. Interception in the pine plantation was 18.7% of the precipitation and only 10.6% in the native eucalypt forest. More recently Feller (1981) has shown that the interception of precipitation is greater in *Pinus radiata* plantations than in adjacent *Eucalyptus obliqua* forest and has suggested that replacement of dry sclerophyll eucalypt forest with pines may lead to declines in water yield unless the pines are heavily thinned.

Flinn et al. (1979) found that, in small catchments with a forest cover dominated by *E. radiata*, streamflow from the catchments represented only 7 to 14% of the total rainfall. Bren and Turner (1980) after studying the hydrology of a small catchment of native forest, predominantly *E. dives* and *E. viminalis*, concluded that manipulation of eucalypt forest is a feasible means of increasing water yield, although the extent of the change in output cannot be predicted. They found that forests in areas of shallow ground water close to a stream have a much greater water usage than the same type of forest located upslope, so that special care is needed in managing eucalypt forest in the riparian zone.

USES OF EUCALYPT FOREST

Wood production

Detailed records of the volume of wood in the growing stock and of the annual production of wood for some forest types are available from measurements taken in sample plots. Unfortunately information is not available for all species of potential importance. Borough et al. (1978) gave yield statistics for species currently important and for species of potential importance and some of these are summarized in Table 4.6.

Curtin (1970) pointed out that, whilst the annual increment of wood exceeds 14 $m^3 ha^{-1}$ in some eucalypt forests, the average annual harvest rarely exceeds 0.5 $m^3 ha^{-1}$—much less than that of eucalypt plantations overseas. Curtin emphasizes that Australian eucalypt forests have experienced a continuous period of exploitation for up to 100 years so that the abundance and quality of the growing stock are far below what is necessary for high sustained wood production. He advocates management and silvicultural policies to restrict the yield, so that the present growing-stock resource is not further depleted through exploitation. It is suggested that suitable silvicultural treatments to improve regeneration will eventually result in the development of a more vigorous growing stock in the long-term, with consequent enhanced productivities.

The finance available for the silvicultural treatment of eucalypt forest stands is limited. Dargavel (1980) has pointed out that decisions about the future of Australian forests and financial investment are a matter of public decision, most eucalypt forests being in public ownership.

The standards of wood utilization vary considerably. In many forests only a small proportion of the trees reach merchantable size for sawlogs. The integration of utilization operations for sawlogs, chip board, plywood, woodchips and pulp would increase the overall utilization.

In Table 4.7 data are given of the removal of wood from Australian broad-leaved forests, and it

TABLE 4.6

Stand measurements for some eucalypt species (based on Borough et al., 1978)

Species	Status of stand	Age (yr)	Height of dominant trees (m)	Number of stems ha^{-1}	Stem volume (m^3 ha^{-1})	Mean stem increment (m^3 ha^{-1} yr^{-1})
E. delegatensis	unthinned natural	60	42	292	763	13
E. diversicolor	unthinned natural	101	50	92	575	6
E. globulus	unthinned plantation	19	13	939	136	7
E. grandis	thinned natural	40	–	77	189[1]	5
E. grandis	thinned plantation	36	47	267	582	16
E. nitens	natural	12	17	1025	176	15
E. obliqua	thinned natural	60	–	22	134	2
E. pilularis	natural uneven-aged	70	–	103	373	5
E. pilularis	natural even-aged	49	–	652	443[1]	9
E. regnans	natural unthinned	60	56	195	839[1]	14
E. sieberi	natural untinned	60	–	778	615	10
E. maculata	natural thinned	80	–	99	200[1]	3

[1]Merchantable volume.

TABLE 4.7

Removal of wood from Australian broad-leaved forests in the year 1978/79 (10^6 m^3)

Produce	Removal
Sawlogs	4.61
Veneer logs	0.16
Sleeper logs	0.60
Pulpwood for hardboard	0.11
Pulpwood for particleboard	0.03
Pulpwood for paper and paperboard	4.20
Fencing	0.10
Mining poles	0.21
Hewn wood	0.09
Poles and piles	0.17
Other	0.01
Total	10.29

is evident that the two most important products are sawlogs and pulpwood. Some minor forest products are harvested—in 1978 and 1979 26 617 t of charcoal, 80 000 kg of *Eucalyptus* oils and 201 t of gums and resins.

The harvesting of sawlogs and pulpwood is highly mechanized and depends upon the creation of an adequate road system, which is also essential for fire control. In general most broad-leaved evergreen forests now have a reasonable system of roads.

At the Forestry and Wood-based Industries Development Conference held in Canberra in 1974 it was concluded that the wood-based industries have an enormous potential to increase their contribution to the social and economic life of the nation but that this will depend upon a successful interaction with the public in the areas of resource and environmental pollution. It was felt the annual domestic consumption of timber products would increase greatly—to 31 million m^3 of wood in the year 2010 with additional wood harvested for export. Some people regard this forecast as optimistic. However, much of this wood is expected to be obtained from softwood plantations rather than native eucalypt forest. Consequently, there would not be a corresponding increase in the utilization of native hardwoods.

Other uses

Since most of the eucalypt forests are publicly owned, the Australian community is requiring that they be managed to satisfy all reasonable needs and desires of the community. Multiple use allows the meeting of different community requirements and enables the people to gain maximum benefit.

Multiple use management involves making decisions on the relative priority and importance for the various possible uses. In addition to production of wood and other forest products, other uses of forest land are regarded as important, especially water supply and recreation.

Australia's future development is more likely to be limited by the availability of water than by any other natural resource constraint. Consequently forest management on water catchments may be critical. This management must be such as to ensure good water yields and by preventing soil erosion provide good quality water whilst reducing siltation in natural waterways and water reservoirs. In Western Australia there is concern that the clearance of the eucalypt forest for bauxite may well increase water salinity and threaten the water supply to settlements.

The forest services have developed programmes to provide recreational facilities in native forest areas. It is difficult to estimate the future demand for open-space recreation in Australia. The increasing costs of travel seem likely to increase the pressures to increase recreation in the eucalypt forests near centres of population. This will occur both in national parks and forests managed primarily for wood production. The use of forested areas for outdoor recreation is likely to continue to increase at a faster rate than population growth as more time is available for leisure activities.

The eucalypt forests are also seen as being important as wildlife refuges, as scientific reference points for ecological research, for grazing of domestic animals (in some areas grazing is used to reduce the understorey and lessen the risk of wildfire) and for honey production. They contain a variety of animals and plants which in the future may be of benefit to people. They are used for education, scientific study, fitness campaigns and various sporting activities. The eucalypt forests also have aesthetic appeal and are increasingly seen as being associated with the national identity.

Because of the urgent need to find a substitute for petroleum the use of forests for methanol production is under consideration. Preliminary studies indicate that the production of methanol from wood would be more expensive than the conversion of natural gas or coal (Wise, 1980).

It seems likely that in future Australian eucalypt forests will be managed as multiple use forests. The extent to which the different uses are compatible will depend on individual circumstances but emphasis will be given to water supply, recreation and wood production. The problems of reconciling different uses, or of providing for individual uses by forest zoning, will require increasing attention by forest managers.

MANAGEMENT

Forest legislation and administration

Of the 42.4 million ha of forest in Australia, 26.8 million ha are in public ownership and 15.6 million ha in private ownership. Of the 42 707 ha of eucalypt plantations, 15 612 ha are privately owned and the rest are publicly owned.

On the formation of the Commonwealth of Australia in 1901, responsibility for forestry remained under the control of the individual states. Each state maintains its own forest service and has sole control and management rights over forests within its boundaries.

All of the governments, State and Federal, have well-developed forest legislation from which the various government services derive their legal status. Parallel legislation provides similar status for the national park services charged with forest management in national parks.

The functions of the state forest services include the reservation of land, the protection and control of forests, the marketing and utilization of forest products and the establishment and maintenance of coniferous plantations. Some state governments provide financial assistance to private land owners, primarily to encourage the planting of conifers.

About half of the forest area vested in the state governments is crown land. In the 1970s there has been a progressive transfer of crown land to the national park system and to the state forest services. The area of native eucalypt forest may be decreasing due to clearing for agriculture, communications and urban developments.

Much of the privately owned natural broadleaved evergreen forests is in agricultural areas. At present, and especially in Tasmania, the value of native forests is being appreciated increasingly by private land owners and increasingly forests in agricultural areas are being managed more intensively for timber production. At the same time the value of small woodlots for wildlife, soil protection and climate amelioration is being recognized.

Management systems and objectives

Whilst some individual eucalypt stands may have growth rates in excess of 14 $m^3\ ha^{-1}\ yr^{-1}$, the average yield is about a tenth of this (Curtin, 1970). The relatively poor productivity of native eucalypt forest results from several causes. Much of the most productive forest has been cleared to provide agricultural land. Past logging has removed the better trees of species with commercially valuable wood and the remaining trees have often been damaged by fire and other agencies so that there is a high percentage of defective wood.

The early management practices were designed to improve forest production by eliminating useless trees by ringbarking and so releasing from competition the undergrowth of useful tree species. Most fast-growing eucalypts are intolerant of shade and competition, especially *Eucalyptus delegatensis*, *E. diversicolor*, *E. grandis*, *E. nitens*, *E. regnans* and *E. saligna*. Whilst some improvement occurred as a result of the early management, the results were generally disappointing. Suppressed trees did not always respond and regeneration, especially in the wetter forests, was inadequate. It is now recognized that appreciable site disturbance is required to create a suitable seed bed (Florence and Shepherd, 1975).

In order to meet the current need for wood which varies greatly for different parts of Australia, the management of the eucalypt forest varies according to local conditions and requirements. Under extensive forest use, management may involve exploitative logging of old growth forest, heavy cutting with minimal improvement treatment and selective logging to improve uneven-aged stands. Elsewhere forest use is intensive and clear cutting may be practised to develop fully stocked even-aged stands. Alternatively there may be detailed selective felling with improvement and regeneration operations carefully planned and supervised. Under extensive systems the production of wood as $m^3\ ha^{-1}\ yr^{-1}$ ranges from 1.5 to 6 and under intensive systems from 2 to 25.

As areas of softwood plantations come into production progressively over the next two decades, the demands placed on the eucalypt forests to supply sawlogs and pulpwood may decrease. It is envisaged that increasingly the wood requirement from hardwood forests will come from more intensive management of a relatively small part of the forest resource close to markets and possibly from eucalypt plantations. In some of the forests which are logged at present, in future timber production may be a minor objective compared with forest conservation, wildlife management, water supply and recreation.

Eucalypts are one of the most important trees planted in the middle and lower latitudes. About 4 million ha of eucalypt plantations have been established throughout the world, particularly in South America and Africa. The area of eucalypt plantations in Australia is relatively small, in 1979 amounting to only 42 407 ha compared with 681 530 ha of softwood plantations. Most of these eucalypt plantations are in New South Wales, Victoria and Western Australia. Intensively managed plantations can be very productive and the area of eucalypt plantation may expand to meet the needs of paper production and other uses. Some planting of eucalypts has been undertaken by large private forestry corporations, and the area of these plantations is broadly equal to that of the state forest services.

The following eucalypt species are important for planting for wood production in Australia (Turnbull and Pryor, 1978): *E. delegatensis*, *E. diversicolor*, *E. globulus*, *E. grandis*, *E. nitens*, *E. obliqua*, *E. pilularis*, *E. regnans*, *E. saligna*, *E. sieberi*.

There is evidence that eucalypts can be intensively managed on about ten-year rotations either as coppice crops or by re-establishment planting (Carter, 1974). This would require increased mechanization, and short rotation crops are seen primarily as a source of material for pulpwood. Mean annual increments of 21 $m^3\ ha^{-1}$ of wood are anticipated.

ECONOMIC STATUS

Employment

The forestry and wood-based industries are involved in the development, management and processing of one of Australia's major renewable natural resources. They are an important component of the Australian rural economy and contribute to rural employment.

Information gathered in a 1971 census indicated that of all persons employed in the primary production industries, about 3% (12 328 persons) were employed in production forestry (Wesney and Freestone, 1973). The employment of some of these would be related to tropical forests or softwood plantations and probably 2% of the total employment derives from eucalypt forests. In 1978 and 1979, 17 594 people were employed in forestry, 47 942 were industrial workers concerned with wood and wood products (timber, plywood, containers, woodchips) and 25 980 in industries related to paper and paper products. This gives a total employment in forestry and forest industries of 91 516 (Anonymous, 1980).

Economic considerations

Ferguson and Dargavel (1978) examined the economic returns of some eucalypt forests with intensive management regimes for the production of sawlogs and pulpwood. They concluded there is scope for more intensive management of eucalypt forests; but this depends on markets, sites, costs and prices.

In 1978 and 1979 the values of sales of wood in Australia were $A 1820 million[1] and pulp and paper products $A 1505 million. In the same year the value of exports of forest produce was $A 141 million and of imports $A 770 million (Anonymous, 1980).

Future trends seem clear. Australian timber needs will be met increasingly by conifer plantations, mainly of *Pinus radiata*, planted in the broad-leaved temperate hardwood zone. Special funds have been provided to encourage the planting of softwoods, and on March 31, 1979 there were almost 500 000 ha of coniferous plantations compared with less than 27 000 ha of eucalypt plantations. In 1978 and 1979 23 760 ha of softwoods were planted and only 853 ha of eucalypts. The establishment of some of these plantations may involve clearance of eucalypt forest but others will be planted on abandoned farms which may have had a eucalypt forest cover before clearing for agriculture.

Future utilization

There is still a considerable volume of uncommitted wood in the natural eucalypt forest, especially of pulpwood grade, but the competing demands of recreation, conservation and amenity are removing this progressively as a potential resource for the wood-processing industries. This trend can be expected to continue.

Although eucalypt plantations are only a small proportion of all plantation forests, the area of eucalypt plantations is likely to increase. These eucalypt plantations would contribute to sawlog and pulpwood production.

Because of the natural advantages of European, North and South American and African producers and the high freight charges, it seems that the export market for eucalypt wood is restricted. The greatest potential markets are Japan and the countries of Southeast Asia.

CONSERVATION STATUS

Changing attitudes to Australian forests

Forestry played an important part in the economic development of Australia following the proclamation of British sovereignty in 1788. For example, in the oldest colony—New South Wales—the extraction of the highly prized red cedar *Toona australis* for export to Europe resulted in the building of roads and ports. This facilitated the opening up of the eucalypt forest and subsequently many saw mills and settlements were established. The early settlers regarded the eucalypt forest either as an obstacle to be removed to make way for pasture and agriculture or as a source of timber. The need for timber for export and for building materials was paramount and there was little interest in forest conservation.

As the forests were cleared and the value of timber increased, concern to protect the forests grew. As early as 1875 legislation was enacted in South Australia to conserve the remaining forests and by 1920 forest legislation had been enacted in all states.

[1] In 1981 $A 1.00 = $U.S. 1.15.

State forest services were established to look after the forests so as to maintain an inexpensive supply of wood. Progressively, professional foresters were appointed, and they recognized the need to combine providing wood for the timber industry with forest conservation. Their objective was to ensure the survival of the forest in perpetuity whilst providing for a sustained supply of wood for industry.

Within the state forests logging was normally by a selection system, the trees removed consisting of single or small groups of trees scattered over forest blocks. In general only the better trees of the species of greatest value were harvested. Although the nature of the forests changed, large tracts of eucalypt forest were left and some of this was largely unexploited.

The foresters recognized the multiple roles of forests, for example for fauna and flora protection, as water catchments, for recreation and amenity, for the production of a variety of minor products such as honey, tannin and eucalypt oil and in some areas for grazing. Multiple use management was limited by financial constraints. Nevertheless, the State Forest Services provided for some uses other than wood production. For example, the Queensland Service was responsible for national parks and set high standards in their management. As a result of increasing demands on land and the need for some intensive land use, there has been increasing pressure for multiple use management of public owned forest.

Woodchips

Within the last quarter of a century, there have been several proposals to use eucalypt forests as the basis for the export of woodchips or woodpulp to Japan for the paper industry. Some of these proposals have been implemented and large-scale woodchip operations established. Wood for the woodchip and pulp industries is obtained largely by clearing the forests in large blocks and these operations have become a major conservation issue (Anonymous, 1977). There has been considerable opposition to the woodchip–pulpwood operations (Routley and Routley, 1973). Environment impact statements have been prepared and there has been heated public debate.

The proponents for woodchip operations see them as a measure of providing rural employment, of obtaining financial resources necessary to improve forests and forest management and as a means of providing roads and other facilities for a variety of uses, for example recreation as well as wood production. It is suggested that forest cutting for woodchip production is comparable in effect to natural wildfires.

Besides questioning the economics of woodchip operations, conservationists have criticized them in terms of alleged adverse environmental effects—soil erosion, siltation, increased salinity, eutrophication of natural waters, the spread of pests and diseases and the loss of native species of plants and animals and of forest ecosystems. Some woodchip operations are in areas of high recreation value and have been criticized for their impact on landscapes and for the loss of amenity. Another argument used against woodchipping is that there is insufficient knowledge of the long-term environmental effects (Anonymous, undated).

The economic and environmental aspects of the export hardwood woodchip industry was studied by a working group set up by the Australian Ministers for the Environment and Conservation and of Agriculture (Anonymous, 1975). The working group estimated the net value of three existing woodchip projects as over $A 28 million on the basis of a 100-year planning period and at a discount rate of 10%. The logging operations necessary to meet the then existing contracts for woodchips in 1975 would need to be carried out on approx. 600 000 ha. There would also be some provision of timber for sawmills from this area.

Protected areas

Much publicly owned forest land is not dedicated primarily to timber production. Large areas of eucalypt forests of different kinds are included in national parks, nature reserves, water catchments and protected areas. The removal of timber from such areas is usually banned and the forests remain largely natural.

The management plans for state forests afford considerable protection to the forest reserves. Even where timber production is the prime objective, forest management tends to be conservative and aimed at keeping a forest dominated by native species. In addition representative areas of different forest types or areas of special interest are set aside by the State forest services from normal forest management, with logging and other forms of destructive management banned.

The area of eucalypt forest protected in these different ways is unknown but is considerable. Specht et al. (1974) estimate that 61% of closed forest alliances and 76% of open forest–woodland alliances in Australia are represented in existing protected areas.

TABLE 4.8

Eucalyptus species of highly restricted occurrence, and endangered or potentially endangered

E. acies	*E. kitsoniana*
E. alpina	*E. kruseana*
E. approximans spp. *approximans*	*E. lane-poolei*
E. aquilina	*E. lansdowneana* spp. *lansdowneana*[1]
E. archeri	*E. largeana*
E. argophloia	*E. luehmanniana*
E. badjensis	*E. megacornuta*
E. baeuerlenii	*E. michaeliana*
E. bakeri	*E. mitchelliana*
E. barberi	*E. morrisbyi*
E. beardiana[2]	*E. neglecta*
E. benthamii	*E. nigra*[1]
E. brockwayi	*E. olsenii*
E. burdettiana	*E. paliformis*
E. burgessiana	*E. parvifolia*
E. caesia	*E. pendens*
E. calcicola	*E. pulverulenta*
E. camfieldii	*E. pumila*
E. cneorifolia	*E. remota*
E. conglomerata	*E. rhodantha*
E. cordata	*E. risdonii*
E. coronata	*E. rummeryi*
E. crenulata	*E. saxatilis*
E. cupularis[1]	*E. scoparia*
E. curtisii	*E. sepulcralis*
E. dendromorpha	*E. squamosa*
E. desmondensis	*E. steedmanii*
E. ficifolia	*E. stenostoma*
E. fitzgeraldii[1]	*E. stoatei*
E. froggatii	*E. stowardii*[1]
E. georgi[2]	*E. sturgissiana*
E. gregsoniana	*E. tetrapleura*
E. halophila[2]	*E. triflora*
E. imlayensis[2]	*E. woodwardii*
E. insularis	*E. xanthonema*[1]
E. johnsoniana[2]	*E. yarraensis*
E. kartozoffiana	

[1] Distribution inadequately known and in *E. nigra* taxonomic boundaries are not clear.

[2] Species newly described and distribution inadequately known.

Some eucalypt species, perhaps half of all species, are in danger of being reduced to critical population levels (Pryor, 1981). The risk is both from the direct impact of land use developments and the more insidious effect of inhibition of regeneration. The species most threatened are those of restricted distribution; sometimes the only remaining stands are on road verges or private land (Table 4.8).

Land use and national policies

The coincidence that the main eucalypt forests are in the coastal zone of greater rainfall where most Australians live means that the land use pressures on the remaining forest land will continue to increase. Since forests are assuming increasing significance as a natural environment important to many people they are seen as a recreational facility and a fundamental part of the national heritage.

At the first meeting of the Australian Forestry Council in 1964 it was agreed that Australia should aim for self-sufficiency in forest products by the year 2000. This was to be achieved largely by the establishment of softwood plantations. Hanson (1980) has suggested that there is a need to consider establishing forests in Australia to compete in the overseas markets for forest products and to form the basis of future employment. However, national land use policies with respect to the eucalypt forest are influenced by many considerations including international agreements relating to trade in a diversity of materials and products.

APPENDIX I: FOREST TYPES (based on Anonymous, 1974)

EUCALYPT PRODUCTIVITY CLASS I: Estimated total area is 1.6×10^6 ha

(a) Forest type name: Moist Eucalypt
Location: Queensland and New South Wales

Floristic composition

(1) Major tree species
Eucalyptus pilularis (blackbutt)
E. grandis (flooded gum)
E. saligna (Sydney blue gum)
E. microcorys (tallowwood)
E. resinifera (red mahogany)
Tristania conferta (brush box)
Syncarpia glomulifera (turpentine)

(2) Minor tree species
Eucalyptus campanulata (New England blackbutt)
E. acmenoides (white mahogany)
E. eugenioides (thin-leaved stringybark)
E. drepanophylla (grey ironbark)
E. botryoides (southern mahogany)
E. deanei (round-leaved gum)
Syncarpia hillii (satinay)
Eucalyptus cloeziana (Gympie messmate)

(3) Understorey species
Arecaceae (palms)
Casuarina torulosa (forest oak)

Mature stand height: 40–60 m
Altitudinal range: 0–1000 m
Rainfall: 1000–1800 mm yr^{-1}
General comments: A wet sclerophyll forest with a dense understorey of mesophytic shrubs.

(b) Forest type name: Ash
Location: New South Wales

Floristic composition

(1) Major tree species
E. delegatensis (alpine ash)

(2) Minor tree species
Eucalyptus viminalis (manna gum)
E. dalrympleana (mountain gum)
E. cypellocarpa (mountain grey gum)
E. radiata (narrow-leaved peppermint)
E. robertsonii (narrow-leaved peppermint)

Mature stand height: 50–60 m
Altitudinal range: 100–1700 m
Rainfall: 1000 mm yr^{-1}
General comments: An alpine tall wet sclerophyll forest.

(c) Forest type name: Ash
Location: Victoria

Floristic composition

(1) Major tree species
Eucalyptus regnans (mountain ash)
E. delegatensis (alpine ash)
E. nitens (shining gum)

Mature stand height: 75–90 m
Altitudinal range: 200–1400 m
Rainfall: in excess of 1000 mm yr^{-1}
General comments: Major tree species tend to occur in pure stands. Requires cool, moist, sheltered conditions.

(d) Forest type name: Ash

Location: Tasmania

Floristic composition

(1) Major tree species
Eucalyptus obliqua (messmate)
E. delegatensis (alpine ash)
E. regnans (mountain ash)
E. sieberi (silvertop ash)
E. globulus (Tasmanian blue gum)
E. viminalis (manna gum)
E. dalrympleana (mountain gum)

(2) Minor tree species
Nothofagus cunninghamii (myrtle beech)
Atherosperma moschatum (southern sassafras)
Acacia dealbata (silver wattle)
A. melanoxylon (blackwood)
Beyeria viscosa
Olearia argophylla (musk)
Pittosporum bicolor (pittosporum)
Pomaderris apetala
Eucryphia lucida (leatherwood)
Leptospermum lanigerum (woolly tea-tree)

(3) Understorey species
Anopterus glandulosus
Bedfordia salicina
Cassinia aculeata
Goodenia ovata
Gahnia psittacorum
Lepidosperma spp.
Prostanthera lasianthos
Blechnum spp.
Dicksonia antarctica
Histiopteris incisa

Mature stand height: 40–90 m
Altitudinal range: 0–900 m
Rainfall: 850–1650 mm yr^{-1}

(e) Forest type name: Karri
Location: Western Australia

Floristic composition

(1) Major tree species
Eucalyptus diversicolor (karri)
E. calophylla (marri)

(2) Minor tree species
Eucalyptus marginata (jarrah)
E. jacksonii (red tingle)
E. guilfoylei (yellow tingle)
E. brevistylis (Rate's tingle)
E. patens (W.A. blackbutt)
E. megacarpa (bullich)

(3) Understorey species
Casuarina decussata (karri sheoak)
Agonis flexuosa (W.A. peppermint)
Banksia littoralis

B. verticillata
Trymalium spathulatum
Bossiaea aquifolia
Acacia pentadenia

Mature stand height: 50–80 m
Altitudinal range: 0–300 m
Rainfall: in excess 1000 mm yr^{-1}
General comments: Commonly occurs on acidic red sandy loam soils of low nutritive value.

EUCALYPT PRODUCTIVITY CLASS II: Estimated total area is 3.2×10^6 ha

(a) Forest type name: Silvertop
Location: Victoria and New South Wales

Floristic composition

(1) Major tree species
Eucalyptus sieberi (silvertop ash)

(2) Minor tree species
Eucalyptus obliqua (messmate)
E. cypellocarpa (mountain grey gum)
E. muellerana (yellow stringybark)
E. radiata (narrow-leaved peppermint)
E. eugenioides (thin-leaved stringybark)
E. globoidea (white stringybark)
E. gummifera (red bloodwood)

Mature stand height: 30–45 m
Altitudinal range: 0–1000 m
Rainfall: 650–1250 mm yr^{-1}
General comments: *E. sieberi* is the dominant species with a wide range of other species associated. It is generally dry sclerophyll associated with soils of moderate to low fertility.

(b) Forest type name: Stringybark
Location: Northern New South Wales

Floristic composition

(1) Major tree species
Eucalyptus andrewsii (New England blackbutt)
E. campanulata (New England blackbutt)
E. planchoniana (needlebark stringybark)
E. agglomerata (blue-leaved stringybark)
E. globoidea (white stringybark)
E. laevopinea (silvertop stringybark)

(2) Minor tree species
Angophora costata (smooth-barked apple)
Eucalyptus punctata (grey gum)
E. caliginosa (broad-leaved stringybark)
E. youmanii (Youman's stringybark)
E. cameronii (diehard stringybark)

Mature stand height: 20–40 m
Altitudinal range: 600–1300 m
Rainfall: 750–1300 mm yr^{-1}
General comments: Dry sclerophyll forest with low understorey.

(c) Forest type name: Stringybark
Location: Southern New South Wales and Victoria

Floristic composition

(1) Major tree species
Eucalyptus muellerana (yellow stringybark)
E. eugenioides (thin-leaved stringybark)
E. globoidea (white stringybark)
E. baxteri (brown stringybark)
E. obliqua (messmate)
E. viminalis (manna gum)
E. radiata (narrow-leaved peppermint)
E. robertsonii (narrow-leaved peppermint)

Mature stand height: 30–45 m
Altitudinal range: 0–900 m
Rainfall: 600–1300 mm yr^{-1}
General comments: Mixed forest containing a wide range of eucalypts. Dry sclerophyll in nature, generally occupying moderately fertile sites.

(d) Forest type name: Mixed Eucalypt
Location: Tasmania

Floristic composition

(1) Major tree species
Eucalyptus delegatensis (alpine ash)
E. obliqua (messmate)
E. sieberi (silvertop ash)
E. globulus (Tasmanian blue gum)
E. viminalis (manna gum)
E. dalrympleana (mountain gum)
E. pulchella (white peppermint)
E. amygdalina (black peppermint)
E. simmondsii (Smithton peppermint)
E. tasmanica (silver peppermint)
E. johnstonii (Tasmanian yellow gum)
E. ovata (swamp gum)

(2) Minor tree species
Acacia dealbata (silver wattle)
A. mearnsii (black wattle)
Hakea acicularis
Casuarina suberosa (black sheoak)
Exocarpus cupressiformis (native cherry)

(3) Understorey species
Anopterus glandulosus
Bedfordia salicina
Cassinia aculeata
Goodenia ovata
Lepidosperma spp.
Prostanthera lasianthos
Blechnum spp.

Dicksonia antarctica
Histiopteris incisa
Plus a wide range of shrub species vernacularly known as "shark prickles"

Mature stand height: 15–40 m
Altitudinal range: 0–900 m
Rainfall: 650–1500 mm yr^{-1}

(e) Forest type name: Jarrah
Location: Western Australia

Floristic composition

(1) Major tree species
Eucalyptus marginata (jarrah)
E. calophylla (marri)

(2) Minor tree species
Eucalyptus patens (W.A. blackbutt)
E. megacarpa (bullich)
E. wandoo (wandoo)
E. gomphocephala (tuart)
E. rudis (W.A. flooded gum)

(3) Understorey species
Casuarina fraseriana
Banksia grandis
Persoonia longifolia
Xanthorrhoea preissii
Kingia australis
Macrozamia reidlei

Mature stand height: 20–30 m
Altitudinal range: 0–300 m
Rainfall: 650–1300 mm yr^{-1}
General comments: Commonly occurs on the lateritic gravels of the Darling Range, but attains a greater size in the red loam soils of the deeply dissected river valleys.

(f) Forest type name: Spotted Gum
Location: Queensland and New South Wales

Floristic composition

(1) Major tree species
Eucalyptus maculata (spotted gum)
E. citriodora (lemon-scented gum)

(2) Minor tree species
Eucalyptus muellerana (yellow stringybark)
E. longifolia (woollybutt)
E. saligna (Sydney blue gum)
Tristania conferta (brush box)
Syncarpia glomulifera (turpentine)
E. cloeziana (Gympie messmate)
E. crebra (narrow-leaved red ironbark)
E. decorticans (gum-top ironbark)
E. fibrosa (blue-leaved ironbark)
E. tereticornis (forest red gum)
E. acmenoides (white mahogany)

(3) Understorey species
Macrozamia spp. (burrawangs)
Acacia spp. (wattles)

Mature stand height: 20–50 m
Altitudinal range: 0–700 m
Rainfall: 600–1250 mm yr^{-1}
General comments: The type is largely dominated by spotted gum with a large range of species as associates. It varies from dry to wet sclerophyll and is generally found on sandy or well-drained soils or poor rocky sites.

EUCALYPT PRODUCTIVITY CLASS III: Estimated total area is 8.3×10^6 ha

(a) Forest type name: Dry Eucalypt
Location: Queensland and New South Wales

Floristic composition

(1) Major tree species
Eucalyptus paniculata (grey ironbark)
E. microcarpa (grey box)
E. propinqua (grey gum)
E. melliodora (yellow box)
E. albens (white box)
E. tereticornis (forest red gum)
E. crebra (narrow-leaved red ironbark)
E. moluccana (grey box)

(2) Minor tree species
Eucalyptus racemosa, *E. haemastoma*, *E. rossii* (scribbly gum)
E. cloeziana (Gympie messmate)
E. fibrosa (blue-leaved ironbark)
E. phaeotricha (Queensland white stringybark)
E. acmenoides (white mahogany)
E. gummifera (red bloodwood)

(3) Understorey species
Acacia spp. (wattles)

Mature stand height: 20–30 m
Altitudinal range: 0–900 m
Rainfall: 500–1250 mm yr^{-1}
General comments: An extremely varied miscellaneous type including a wide range of ecological types. Although comprised largely of dry inland forests, paperbark, mangrove and casuarina forests of low productivity have been included.

(b) Forest type name: Ironbark, Box
Location: Victoria

Floristic composition

(1) Major tree species
Eucalyptus sideroxylon (red ironbark)
E. microcarpa (grey box)

E. leucoxylon (yellow gum)

Mature stand height: 25–30 m
Altitudinal range: 100–500 m
Rainfall: 400–500 mm yr^{-1}
General comments: Open forest on poor soils in areas of low rainfall and hot dry summers.

(c) Forest type name: Mallet
Location: Western Australia

Floristic composition

(1) Major tree species
Eucalyptus astringens (mallet)

(2) Minor tree species
Eucalyptus gardneri (blue mallet)
E. falcata (white mallet)
E. spathulata (swamp mallet)
E. occidentalis (flat-topped yate)
E. accedens (powderbark wandoo)
E. salmonophloia (salmon gum)
E. wandoo (wandoo)
E. longicornis (red morrell)
E. loxophleba (York gum)

(3) Understorey species
Acacia acuminata (raspberry jam)
Casuarina hugeliana (rock sheoak)
Dryandra spp.

Mature stand height: 20 m
Altitudinal range: 100–300 m
Rainfall: less than 600 mm yr^{-1}
General comments: On laterite breakaways and screes below.

(d) Forest type name: Low Grade Eucalypt
Location: Queensland

Floristic composition

(1) Major tree species
As no individual tree species dominate in this extremely variable type no differentiation into major and minor tree species is made.
Eucalyptus racemosa (scribbly gum)
E. gummifera (red bloodwood)
E. trachyphloia (brown bloodwood)
E. tereticornis (forest red gum)
E. alba (poplar gum)
E. dichromophloia (gum-topped bloodwood)
E. ochrophloia (yapunyah)
E. coolibah (coolibah)
Angophora costata (smooth-barked apple)
Eucalyptus propinqua (grey gum)
E. moluccana (grey box)
E. crebra (narrow-leaved red ironbark)
E. tessellaris (carbeen)
E. leptophleba (Molloy red box)
E. cambageana (Coowarra box)
E. bloxsomei (yellow jacket)
Lysicarpus angustifolius (brown hazelwood)
Acacia doratoxylon (brown lancewood)

Mature stand height: 20–30 m
Altitudinal range: 0–900 m
Rainfall: 600–1250 mm yr^{-1}
General comments: An extremely variable type covering a large range of site types.

(e) Forest type name: Red Gum
Location: Victoria and New South Wales

Floristic composition

(1) Major tree species
Eucalyptus camaldulensis (river red gum)

Mature stand height: 20–40 m
Altitudinal range: 100–600 m
Rainfall: 500 mm yr^{-1}
General comments: Open forest invariably occupying low riverside site subject to periodic flooding. Particularly well developed on the flood plain of the Murray River.

REFERENCES

Anonymous, undated. *The Great Forest Sell Out. The Case Against the Woodchip Export Industry*. Australian Conservation Foundation, Melbourne, Vic., 12 pp.

Anonymous, 1962. *Forest Trees of Australia.* Forestry and Timber Bureau, Canberra, A.C.T., 230 pp.

Anonymous, 1974. *Forwood Report of Panel 2. Forest Resources.* Australian Government Publishing Service, Canberra, A.C.T., 56 pp.

Anonymous, 1975. *Economic and Environmental Aspects of the Export Woodchip Industry.* Australian Government Publishing Service, Canberra, A.C.T., Vol. 1, 84 pp., Vol. 2, 119 pp.

Anonymous, 1977. *Woodchips and the Environment.* Report from the Standing Committee on Science and the Environment. Australian Government Publishing Service, Canberra, A.C.T., 454 pp.

Anonymous, 1980. *Australian Forest Resources 1979.* Australian Government Publishing Service, Canberra, A.C.T., 51 pp.

Attiwill, P.M., 1966. The chemical composition of rainwater in relation to cycling of nutrients in mature eucalypt forest. *Plant Soil*, 24: 390–406.

Attiwill, P.M., 1968. The loss of elements from decomposing litter. *Ecology*, 49: 142–147.

Attiwill, P.M., 1972. On the cycling of elements in mature *Eucalyptus obliqua* forest. In: R. Boardman, *Australian Forest Tree Conference*. Forestry and Timber Bureau, Canberra, A.C.T., pp. 39–46.

Beadle, N.C.W., 1966. Soil phosphate and its role in moulding segments of the Australian flora and vegetation, with special reference to xeromorphy and sclerophylly. *Ecology*, 47: 992–1007.

Blakely, W.F., 1965. *A Key to the Eucalypts*. Forestry and Timber Bureau, Canberra, A.C.T., 359 pp.

Borough, C.J., Incoll, W.D., May, J.R. and Bird, T., 1978. Yield statistics. In: W.E. Hillis and A.G. Brown (Editors), *Eucalypts for Wood Production*. Griffin Press, Adelaide, S.A., pp. 201–225.

Bren, L.J. and Turner, A.K., 1980. Hydrologic output of small forested catchment: implications for management. *Aust. For.*, 43: 118–123.

Carne, P.B. and Taylor, K.L., 1978. Insect pests. In: W.E. Hillis and A.G. Brown (Editors), *Eucalypts for Wood Production*. Griffin Press, Adelaide, S.A., pp. 155–168.

Carter, W.G., 1974. Growing and harvesting eucalypts on short rotations for pulping. *Aust. For.*, 36: 215–225.

Chippendale, G.M. and Wolf, L., 1981. *The Natural Distribution of* Eucalyptus *in Australia*. Special Publication, Australian National Parks and Wildlife Service, Canberra, A.C.T.

Christensen, P., 1973. Some ecological aspects of jarrah dieback. *For. Focus*, 10: 11–16.

Christensen, P.E. and Kimber, P.C., 1975. Effects of prescribed burning on the flora and fauna of south-west Australian forests. *Proc. Ecol. Soc. Aust.*, 9: 85–106.

Cornish, P.M., 1980. Water quality studies in New South Wales State Forests. 1. A north coast eucalypt forest near Lismore. *Aust. For.*, 43: 105–110.

Cremer, K.W., Cromer, R.N. and Florence, R.G., 1978. Stand establishment. In: W.E. Hillis and A.G. Brown (Editors), *Eucalypts for Wood Production*. Griffin Press, Adelaide, S.A., pp. 81–135.

Curtin, R.A., 1970. Increasing the productivity of *Eucalyptus* forests in New South Wales. *Aust. For.*, 34: 97–106.

Dargavel, J.B., 1980. The political detention of an Australian forestry perspective. *Aust. For.*, 43: 5–15.

Driscoll, P.V., 1977. Comparison of bird counts from pine forests and indigenous vegetation. *Aust. Wildl. Res.*, 4: 281–288.

Eldridge, K.G., 1971. Genetically improved eucalypt seed for Australian pulpwood forests. *Appita*, 25: 35–50.

Eldridge, K.G., 1972. Genetic variation in growth of *Eucalyptus regnans*. *Bull. For. Timber Bur., Canberra*, No. 46: 72 pp.

Ellis, R.C., Mount, A.B. and Mattay, J.P., 1980. Recovery of *Eucalyptus delegatensis* from high altitude dieback after felling and burning the understorey. *Aust. For.*, 43: 29–35.

Feller, M.C., 1981. Water balances in *Eucalyptus regnans*, *E. obliqua* and *Pinus radiata* forests in Victoria. *Aust. For.*, 44: 153–161.

Ferguson, I.S. and Dargavel, J.B., 1978. Economic considerations. In: W.E. Hillis and A.G. Brown (Editors), *Eucalypts for Wood Production*. Griffin Press, Adelaide, S.A., pp. 363–374.

Flinn, D.W., Bren, L.J. and Hopmans, P., 1979. Soluble nutrient inputs from rain and outputs in stream water from small forested catchments. *Aust. For.*, 42: 39–49.

Florence, R.G., 1969. The application of ecology to forest management with particular reference to eucalypt forests. *Proc. Ecol. Soc. Aust.*, 4: 82–100.

Florence, R.G. and Shepherd, R.K., 1975. The role of eucalypt forests in wood production. *Aust. For.*, 38: 100–116.

Frith, H.J., 1973. *Wildlife Conservation*. Angus and Robertson, Sydney, N.S.W., 414 pp.

Gilbert, J.M., 1961. The effects of browsing by native animals on the establishment of seedlings of *Eucalyptus regnans* in the Florentine Valley, Tasmania. *Aust. For.*, 25: 116–121.

Gill, A.M., 1975. Fire and the Australian flora: a review. *Aust. For.*, 38: 4–25.

Greaves, T., Armstrong, G.J., McInnes, R.S. and Dowse, J.E., 1967. Timber losses caused by termites, decay and fire in two coastal forests in New South Wales. *C.S.I.R.O., Canberra, Div. Entomol., Techn. Pap.* No. 7: 4–18.

Hall, N., Johnston, R.D. and Marryatt, R., 1963. The natural occurrence of eucalypts. *For. Res. Inst., Canberra, Leafl.*, No. 65: 12 pp.

Hanson, A.G., 1980. Should Australia plan to export forest products? *Aust. For.*, 43: 70–74.

Harwood, C.E. and Jackson, W.D., 1975. Atmospheric losses of four plant nutrients during a forest fire. *Aust. For.*, 38: 92–99.

Heather, W.A. and Griffin, D.M., 1978. The potential for epidemic disease. In: W.E. Hillis and A.G. Brown (Editors), *Eucalypts for Wood Production*. Griffin Press, Adelaide, S.A., pp. 201–225.

Howard, T., 1975. Litter fauna in *Nothofagus cunninghamii* forests. *Proc. R. Soc. Vic.*, 87: 207–213.

Jenkins, R.W.G. and Bartell, R., 1980. *Reptiles of the Australian High Country*. Inkata Press, Melbourne, Vic., 278 pp.

Kile, G.A., Greig, P.J. and Edgar, J.G., 1980. *Tree Decline in Rural Victoria*. Institute of Foresters of Australia, Inc., Melbourne, Vic., 20 pp.

Ladiges, P.Y. and Ashton, D.H., 1977. A comparison of some populations of *Eucalyptus viminalis* Labill. growing on calcareous and acid soils in Victoria, Australia. *Aust. J. Ecol.*, 2: 161–178.

Lee, K.E. and Correll, R.L., 1978. Litter fall and its relationship to nutrient cycling in a South Australian dry sclerophyll forest. *Aust. J. Ecol.*, 3: 243–252.

Luke, R.H. and McArthur, A.G., 1978. *Bushfires in Australia*. Australian Government Publishing Service, Canberra, A.C.T., 359 pp.

McIlroy, J.C., 1978. The effects of forestry practices on wildlife in Australia: a review. *Aust. For.*, 41: 78–94.

Malajczuk, N. and Lamont, B.B., 1981. Specialized roots of symbiotic origin in heathlands. In: R.L. Specht (Editor), *Heathlands and Related Shrublands, B. Analytical Studies*. Ecosystems of the World, 9B. Elsevier, Amsterdam, pp. 165–182.

Neumann, F.G. and Marks, G.C., 1976. A synopsis of imported pests and diseases in Australian forests and forest nurseries. *Aust. For.*, 39: 83–102.

Newhook, F.J. and Podger, F.D., 1972. The role of *Phytophthora cinnamomi* in Australian and New Zealand forests. *Annu. Rev. Phytopathol.*, 10: 299–326.

Newsome, A.E., McIlroy, J. and Catling, P., 1975. The effects of an extensive wildfire on populations of twenty ground vertebrates in south-east Australia. *Proc. Ecol. Soc. Aust.*, 9: 107–123.

O'Connell, A.M., Grove, T.S. and Dimmock, G.M., 1978. Nutrients in the litter on jarrah forest soils. *Aust. J. Ecol.*, 3: 253–260.

Ovington, J.D., 1968. Considerations of forest productivity. *Aust. For.*, 32: 191–204.

Penfold, A.R. and Willis, J.L., 1961. *The Eucalypts.* World Crop Series. Interscience, New York, N.Y., 550 pp.

Plowman, K.P., 1979. Litter and soil fauna of two Australian subtropical forests. *Aust. J. Ecol.*, 4: 87–104.

Pratt, B.H. and Heather, W.A., 1973. Recovery of potentially pathogenic *Phytopthora* and *Pythium* spp. from native vegetation in Australia. *Aust. J. Biol. Sci.*, 26: 575–582.

Pryor, L.D., 1981. *Australian Endangered Species: Eucalypts.* Special Publication, Australian National Parks and Wildlife Service, Canberra, A.C.T.

Purdie, R.W. and Slatyer, R.O., 1976. Vegetation succession after fire in sclerophyll woodland communities in south-eastern Australia. *Aust. J. Ecol.*, 1: 223–236.

Ride, W.D.L., 1970. *A Guide to the Native Mammals of Australia.* Oxford University Press, Melbourne, Vic., 250 pp.

Routley, R. and Routley, V., 1973. *The Fight for the Forests.* Research School of Social Sciences, Australian National University, Canberra, A.C.T., 290 pp.

Shea, S.R., Hatch, A.B., Havel, J.J. and Ritson, P., 1975. The effect of changes in forest structure and composition on water quality and yield from the northern jarrah forest. *Proc. Ecol. Soc. Aust.*, 9: 58–73.

Smith, K.M., 1974. Throughfall, stemflow and intercept on in pine and eucalypt forest. *Aust. For.*, 36: 190–197.

Specht, R.L., 1979. The sclerophyllous (heath) vegetation of Australia: the eastern and central states. In: R.L. Specht (Editor), *Heathlands and Related Shrublands, A. Descriptive Studies.* Ecosystems of the World, 9A. Elsevier, Amsterdam, pp. 125–210.

Specht, R.L., 1981. Mallee ecosystems in southern Australia. In: F. di Castri, D.W. Goodall and R.L. Specht (Editors), *Mediterranean-Type Shrublands.* Ecosystems of the World, 11. Elsevier, Amsterdam, pp. 203–231.

Specht, R.L., Roe, M.E. and Boughton, V.H., 1974. Conservation of major plant communities in Australia and Papua New Guinea. *Aust. J. Bot., Suppl.*, 7: 667 pp.

Tunstall, B.R. and Walker, J., 1975. The effect of woodland disturbance on soil water. *Proc. Ecol. Soc. Aust.*, 9: 49–57.

Turnbull, J.W. and Pryor, L.D., 1978. Choice of species and seed sources. In: W.E. Hillis and A.G. Brown (Editors), *Eucalypts for Wood Production.* Griffin Press, Adelaide, S.A., pp. 6–65.

Tyndale-Biscoe, C.H. and Calaby, J.H., 1975. Eucalypt forests as refuge for wildlife. *Aust. For.*, 38: 117–133.

Wesney, D. and Freestone, R., 1973. The significance of forestry in the national economy. *Occas. Pap., Bur. Agric. Econ., Canberra*, No. 19: 36 pp.

Wise, P.K., 1980. Forests for methanol production. *Aust. For.*, 43: 44–51.

Wise, P.K. and Pitman, M.G., 1981. Nutrient removal and replacement associated with short rotation eucalypt plantations. *Aust. For.*, 44: 142–152.

Woinarski, J.C.Z., 1979. Birds of a *Eucalyptus* plantation and adjacent natural forest. *Aust. For.*, 42: 243–247.

Chapter 5

TEMPERATE BROAD-LEAVED EVERGREEN FORESTS OF SOUTHEASTERN NORTH AMERICA

D.F. OLSON, Jr.

INTRODUCTION

In the southern United States, commercial forests occupy approximately 81 million ha of land, and are extremely important in the regional economy, as well as the national economy of the United States. The temperate broad-leaved evergreen forest types are minor components of this vast timbered region, assuming only limited importance locally for timber and other miscellaneous products, wildlife habitat, and forest recreation.

There are no large contiguous areas of broad-leaved evergreen forests in the region, but rather the types, and the individual species exhibiting this life form, comprise small, scattered stands from Virginia to Texas, and southward to central Florida. In much of the southern mixed hardwood forests (Quarterman and Keever, 1962), broad-leaved evergreens grow in mixture with deciduous hardwoods and conifers. Monk (1965) observed that in north-central Florida this hardwood forest type may contain from 8 to 35 tree species and vary from 7 to 100% in degree of evergreenness.

GROSS CHARACTERISTICS

In southeastern North America, relatively few species of temperate broad-leaved evergreens reach tree size. The principal species, alphabetically by generic name, are:

loblolly-bay	*Gordonia lasianthus*
evergreen hollies	*Ilex*, principally *I. opaca*
southern magnolia	*Magnolia grandiflora*
sweetbay	*Magnolia virginiana*
southern bayberry	*Myrica cerifera*
redbay	*Persea borbonia*
myrtle oak	*Quercus myrtifolia*
live oak	*Quercus virginiana*
sand live oak	*Quercus virginiana* var. *maritima*
scrub live oak	*Quercus virginiana* var. *fusiformis*
cabbage palmetto	*Sabal palmetto*

Another oak, laurel oak (*Quercus laurifolia*) bears leaves which are persistent until early spring. Because of this extremely long leaf retention, laurel oak reacts much more like an evergreen oak than a deciduous oak. All of these trees are dicotyledonous except cabbage palmetto, one of the few monocotyledonous trees in North America.

Temperate broad-leaved evergreen trees assume major importance in only four forest types, as listed by the Society of American Foresters (Eyre, 1980):

Type No.	Type name
74	Cabbage palmetto
88	Willow oak–water oak–diamond leaf–laurel oak
89	Live oak
104	Sweetbay–swamp tupelo–redbay

Küchler (1964) described three major and three minor vegetation types in which temperate broad-leaved evergreens are components, as follows:

Type No.	Type name
111	Oak–hickory–pine forest
112	Southern mixed forest
113	Southern floodplain forest
86	Juniper–oak savanna
90	Live oak–sea oats
114	Pocosin

The major types of Küchler (111, 112, 113) correspond generally to the southern mixed hardwood forest described by Quarterman and Keever (1962) as climax in the southeastern coastal plain of the United States.

Along the Gulf and Atlantic Coasts, forests dominated by live oak (*Quercus virginiana*) occupy many offshore banks and narrow fringes along the coast, immediately behind the foredunes. These forests form a salt spray climax along the North Carolina coast (Wells, 1939; Bourdeau and Oosting, 1959), and a subclimax of the mixed hardwood type in other portions of the range. In northeastern Florida live oak forests may be found on inland as well as coastal sites (Laessle and Monk, 1961). Live oak is found on inland sites in Alabama, Mississippi, and Louisiana as well. The variety *fusiformis* is found in southeastern Oklahoma, central Texas, and mountains of northeastern Mexico (Woods, 1965). The variety *maritima* is a principal taxon in the scrub communities of north-central Florida (Laessle, 1958).

In the southern portion of Florida, there are many other minor broad-leaved evergreen types, but they occur in a subtropical climatic zone rather than a temperate climatic zone, and are not discussed. Principal among these forest types is the mangrove. Other minor types in south Florida include everglade hammocks, bayheads, and cypress savanna, all of which contain sweetbay and redbay. Mahogany (*Swietenia mahagoni*), *Melaleuca*, and several other members of the Myrtaceae also assume local importance in south Florida. These wetland forests are discussed at length elsewhere in this series (Hofstetter, 1983; Lugo et al., in prep.).

FOREST FUNCTIONAL PROCESSES

Because of the relatively minor role of broad-leaved evergreens in southern forests, there has been little research conducted on forest functional processes in these forest types.

The distribution pattern of the broad-leaved evergreen trees in the southern mixed hardwood forest in north-central Florida is somewhat related to the chemical and physical properties of the soil (Monk, 1965, 1966a). There is a strong tendency for soils low in calcium, potassium, phosphorus, and moisture equivalent to support a community dominated by evergreen trees, both conifers and angiosperms. It was suggested that an evergreen community on dry, infertile sites may have selective advantage over a deciduous community by the gradual return of nutrients to the soil or the establishment of a more efficient nutrient cycle through: (1) leaching of nutrients from leaves; (2) year-round leaf-fall; and (3) possible slower decay and subsequent loss of nutrients from evergreen leaves.

In a study of hardwood swamps, including evergreen hardwood swamps known locally as bayheads, Monk (1966b) discovered that, in general, the bayhead habitat is more sterile, more acid, and not flooded as deeply as the mixed swamp habitat. The latter contains mainly deciduous species, while the bayheads are dominated by *Gordonia*, *Magnolia*, *Myrica* and *Persea*.

PAST, PRESENT AND POTENTIAL USE

Of the broad-leaved evergreens in southern United States, only southern magnolia and sweetbay, known collectively as magnolia, are much used for timber. Because of the scattered occurrence of these species, forest inventories generally combine them with other hardwoods. Consequently, accurate estimates of timber volume are not available. In the total southern forest, they are a minor component, comprising only a fraction of a percent of the whole.

Magnolia is used for lumber, veneer, and plywood (Maisenhelder, 1970). The sapwood of magnolia is creamy white, and the heartwood is light to dark brown. The wood is even-textured and moderately heavy (568 kg m^{-3} at 15% moisture content). Lumber from magnolia is used principally for furniture and boxes. Veneer goes into containers and packages, or plywood, which is then used for such items as cabinet work and doors.

Because of their handsome foliage and showy flowers, magnolias are highly prized as ornamentals.

American holly (*Ilex opaca*) is used sparingly for lumber (Krinard, 1973). It has been described as the whitest wood known, and this characteristic makes American holly distinct among North American woods. It is used principally for specialty products, such as ornamental inlays, turnery and carving.

The principal use of American holly is not as wood, but as foliage and fruit for Christmas wreaths and decorations. The destructive harvesting used in gathering this seasonal product has nearly eliminated American holly from natural stands in certain sections of the country.

Southern bayberry (*Myrica cerifera*) is, like the holly, a small broad-leaved tree with persistent leaves that has a special use. The waxy coating of its fruits is refined and used for making candles and scented soaps (Harrar and Harrar, 1946).

Live oak possesses outstanding durability and strength characteristics (Cooper and Watt, 1973). In the early history of the United States, until steel ships replaced wooden ones, live oak was an important source of ship timber because of this extreme toughness. In the modern era, live oak timber has little commercial use. With its shiny evergreen leaves and wide-spreading branches, live oak is a favorite landscaping tree, and is often picturesquely draped with Spanish moss (*Tillandsia usneoides*). Like many other oaks, live oak acorns supply an important source of food for wildlife. The acorns are large and sweet, and especially attractive to wild turkey.

REFERENCES

Bourdeau, P.F. and Oosting, H.J., 1959. The maritime live oak forest in North Carolina. *Ecology*, 40: 148–152.

Cooper, G.A. and Watt, R.F., 1973. Oak—An American wood. *U.S.D.A. For. Serv., Am. Woods, Pap. FS-247*: 9 pp.

Eyre, F.H., 1980. *Forest Cover Types of the United States and Canada*. Society of American Foresters, Washington, D.C., 148 pp.

Harrar, E.S. and Harrar, J.G., 1946. *Guide to Southern Trees*. McGraw-Hill, New York, N.Y., 712 pp.

Hofstetter, R.H., 1983. Wetlands in the United States. In: A.J.P. Gore (Editor), *Mires: Swamp, Bog, Fen and Moor, B. Regional Studies*. Ecosystems of the World, 4B. Elsevier, Amsterdam, pp. 201–244.

Krinard, R.M., 1973. American holly—An American wood. *U.S.D.A. For. Serv., Am. Woods, Pap. FS-242*: 5 pp.

Küchler, A.M., 1964. Potential natural vegetation of the conterminous United States. *Am. Geogr., Soc., Spec. Publ.*, No. 36.

Laessle, A.M., 1958. The origin and successional relationship of sandhill vegetation and sand-pine scrub. *Ecol. Monogr.*, 28: 361–387.

Laessle, A.M. and Monk, C.D., 1961. Some live oak forests of northeastern Florida. *Q.J. Fla. Acad. Sci.*, 24: 39–55.

Lugo, A.T., Brinson, M.K. and Brown, S.L. (Editors), in press. *Wetland Forests*. Ecosystems of the World, 15. Elsevier, Amsterdam.

Maisenhelder, L.C., 1970. Magnolia—An American wood. *U.S.D.A. For. Serv., Am. Woods, Pap.* FS-245: 7 pp.

Monk, C.D., 1965. Southern mixed hardwood forest of north-central Florida. *Ecol. Monogr.*, 35: 335–354.

Monk, C.D., 1966a. An ecological significance of evergreenness. *Ecology*, 47: 504–505.

Monk, C.D., 1966b. An ecological study of hardwood swamps in north-central Florida. *Ecology*, 47: 649–654.

Quarterman, E., and Keever, C., 1962. Southern mixed hardwood forest: climax in the southeastern coastal plain, U.S.A. *Ecol. Monogr.*, 32: 167–185.

Wells, B.W., 1939. A new forest climax: the salt spray climax of Smith Island, N.C. *Bull. Torrey Bot. Club.*, 66: 629–634.

Woods, F.W., 1965. Live oak (*Quercus virginiana* Mill.). In: *Silvics of Forest Trees of the United States*. Agricultural Handbook, 271. U.S. Dep. Agric., Washington, D.C., pp. 584–587.

Chapter 6

TEMPERATE BROAD-LEAVED EVERGREEN FORESTS OF THE MEDITERRANEAN REGION AND MIDDLE EAST[1]

E.F. DEBAZAC

INTRODUCTION

Geographical extent

On a 1:5 000 000 scale vegetation map as published by UNESCO–FAO (1969), the broad-leaved evergreen forest appears as an almost continuous but relatively narrow ring around the Mediterranean Sea. However, the limits of that zone may be more broadly defined. For example, on the Iberian Peninsula and North Africa, this type of forest also extends along the Atlantic coast. It can be found even on the western slopes of the Himalayan Mountains (Afghanistan and Pakistan), an area that—although far away, and in many other respects distinctly different—belongs to the same floral region. The small scale of 1:5 000 000 imposes obvious limitations to accuracy and the geographical distribution of the vegetation consequently appears mainly related to general climatic factors of the environment. This fact becomes evident when comparing the vegetation map with the bioclimatic map of the same area, also published by UNESCO–FAO (1963).

Within the Mediterranean Floral Region, the temperate broad-leaved evergreen forest, which consists mostly of medium-sized trees, rarely over 18 m tall, contains components belonging to an old regional floral group descended from the late Tertiary era. It has remained unchanged through the Quaternary era, despite climatic variations which caused drastic variations elsewhere. This forest type is characterized by a dominance of evergreen species, but often a few deciduous and coniferous trees are mixed in with it.

One feature of the vegetative cover, the density of the canopy, depends on climatic and soil factors. For example, very dense evergreen oak forest of *Quercus ilex* exists where ecological conditions are close to those of the deciduous *Q. pubescens* forest, which is generally found at a higher elevation or more northern latitude. In contrast, the shrubby vegetation of *Pistacia lentiscus*, found in North Africa and often mixed with a conifer, *Tetraclinis articulata*, along its southern border, has an open cover, even on relatively well-protected sites.

The structure and floral composition of the present forest are primarily the result of intensive past and present human influence. All of the region has been inhabited by people for at least several thousand years. The development of successive flourishing civilizations around the Mediterranean Sea—Phoenician, Greek, Roman and Arabic to mention only the historical period—brought severe overuse of the natural forest vegetation. As a result, very few examples of truly climax forest exist at present. Fortunately, the process of man-made degradation is fairly well known and a model of climax community composition can be re-created with an acceptable degree of accuracy by analyzing the existing vegetation.

A simple way to define the limits of the Mediterranean temperate broad-leaved evergreen forest is by a short description of the two main types of deciduous forests to the north and south where moisture conditions are more or less favourable.

To the north of the Mediterranean Region, the broad-leaved evergreen forest zone forms an almost continuous strip along the coast, occasionally interrupted by valleys with deep alluvial soils and

[1]Manuscript completed November, 1980.

consequently different vegetations. From the coast, the evergreen forest extends to various altitudes where a more humid climate encourages the development of deciduous broad-leaved forest dominated by oak species. On the Iberian Peninsula, as well as in southern France and Italy, the most widely distributed deciduous oak is *Quercus pubescens*. The forest constitutes the transition to central European broad-leaved forest dominated by *Quercus sessiliflora* and *Fagus silvatica*. In the northeastern part of the Mediterranean zone and in the Middle East, the variety of deciduous oaks is greater and includes species such as *Quercus aegilops*, *Q. cerris*, *Q. conferta*, *Q. infectoria*, *Q. libani*, and, in the extreme east, *Q. persica*. All these species are elements of the deciduous forest belt surrounding the evergreen zone.

To the south of the Mediterranean Sea, the evergreen broad-leaved forests also extend from the sea level to various altitudes. In the upper parts of the coastal North African mountains, climatic conditions suit deciduous oaks, such as *Quercus fagina* s.l., or the conifers *Cedrus* and *Abies*. Thus, the vegetation distribution pattern is similar to that already mentioned for southern Europe, but with a more developed evergreen zone at lower elevations and deciduous or conifer forests at higher elevations. Further south, beyond the mountainous zone where the climate is becoming warmer and drier, Mediterranean evergreen vegetation enters into competition with other types of arid deciduous open forests. The dominant tree species vary according to specific geographical conditions: *Argania sideroxylon* in south Morocco, *Acacia* and deciduous *Pistacia* in North Africa, and in the Middle East *Acacia gummifera*, *A. raddiana*, *Pistacia atlantica*, *P. khinjuk*, *P. mutica*. Alternatively, local features of topographic climate history may have favoured a direct change to grassland of subdesertic character, dominated by the grasses *Aristida* and *Stipa*.

Within an area extending 6000 km longitudinally and 1500 km latitudinally, two species—*Pistacia lentiscus* (Terebinthaceae) and *Quercus ilex*, an evergreen oak—are present throughout despite the considerable heterogeneity (Fig. 6.1).

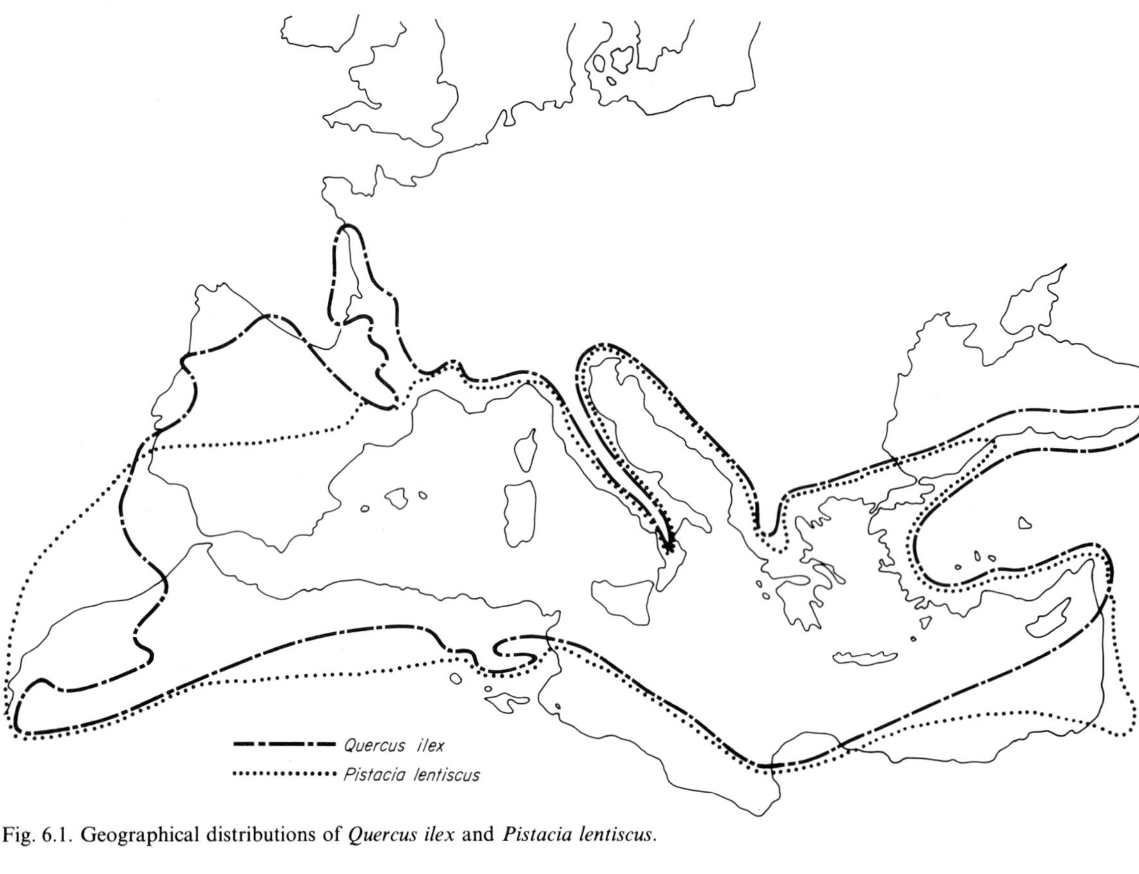

Fig. 6.1. Geographical distributions of *Quercus ilex* and *Pistacia lentiscus*.

Both species can be used as valuable ecological indicators. *Pistacia lentiscus* is characteristic of the warmer and drier part of the evergreen area, close to the Mediterranean Sea. *Quercus ilex* grows in more humid conditions and extends to the extreme limits of the evergreen forest zone, where it grows in competition with deciduous trees.

The geographical and ecological distributions of both species reflect biological features such as cold hardiness and low-temperature resistance, which have been studied experimentally (Larcher and Mair, 1969; Larcher, 1970). Recent scientific work on the climatic classifications of perennial plants, using sophisticated methods (Brisse and Grandjouan, 1974), confirm the validity of the above ecologic indicators. Of course, there are many variations at regional levels; for example, in southern France, *Pistacia lentiscus* is restricted to altitudes below 250 m. In the same area, stands of *Quercus ilex* can be found up to 500 to 600 m. On the Atlas slopes of Morocco, *Pistacia lentiscus* extends from sea level up to 1200 m and *Quercus ilex* will grow up to 2000 m.

In addition to *Pistacia lentiscus* and *Quercus ilex*, other evergreen trees and shrubs, which are less widely distributed, can be classified according to climatic relationships. For instance, with increasing hardiness and resistance to low winter temperatures and preference for more favourable moisture regimes, the species sequence is: *Ceratonia siliqua*, *Pistacia lentiscus*, *Olea oleaster*, *Quercus ilex* and *Phillyrea media*. This is confirmed by their relative geographical distribution (Mooney, 1969; Debazac and Mavrommatis, 1971) and has proved very useful for a quick evaluation of the ecological situation of any pure or mixed stands of evergreen vegetation.

Climatic and topographic conditions

The entire zone of evergreen broad-leaved forests is characterized by climates of strong seasonal variations. The rainfall is nearly always concentrated in the cold season, from autumn to spring, with one or two peak falls. The dry season, which lasts from one to eight months, coincides with the period of longer days.

Under such conditions, average annual climatic parameters can only be of limited value. Nevertheless, there have been detailed investigations to determine any correlations between climate and natural or introduced vegetation. Many classifications of climate types and indices have been proposed in order to determine the relationships between climate and vegetation., Most use various combinations of two parameters commonly available, namely the annual and monthly values of temperature and rainfall.

For example, the pluviothermic index (Q) (Fig. 6.2) proposed by Emberger (1932, 1952, 1955) takes into account the average annual rainfall in mm (P), the monthly mean maximum temperature of the warmest month (M) and the monthly mean minimum temperature of the coldest month (m) both in °C. The formula given is:

$$Q = \frac{P}{2\left[\frac{(M+m)}{2}(M-m)\right]} \times 100$$

In the Mediterranean Region, this index ranges from 20 to over 250. A bioclimate is considered arid with $20 < Q < 30$, semi-arid with $30 < Q < 50$, subhumid with $50 < Q < 90$ and humid with Q over 90. In addition, subdivisions have been introduced with reference to m: a bioclimate is considered warm if m exceeds 7°C, temperate if m ranges from 3 to 7°C, chilly if between 0 and 3°C, cold if between -5 and 0°C and very cold if m is below -5°C.

Emberger's approach has proved very valuable in assessing the relationships between climate and the main species or vegetation units. These relationships appear more clearly when using the parameters Q and m in an orthogonal diagram. For example, the occurrence depending on climate of the two species mentioned previously, *Quercus ilex* and *Pistacia lentiscus*, is very well defined. *Quercus ilex* is only found in cold semi-arid to temperate humid bioclimates and *Pistacia lentiscus* in temperate semi-arid to warm humid bioclimates.

The xerothermic coefficient (x), proposed by Bagnouls and Gaussen (1952), represents the annual average number of biologically dry days per year and is another way of classifying mediterranean bioclimates. A climate is considered mediterranean when x falls between 0 and 200. There are four main subdivisions: submediterranean ($0 < x < 40$), mesomediterranean ($40 < x < 100$), thermomediterranean ($100 < x < 150$) and xeromediterranean ($x > 150$). The concept has been used for mapping. Another comparison between the vegeta-

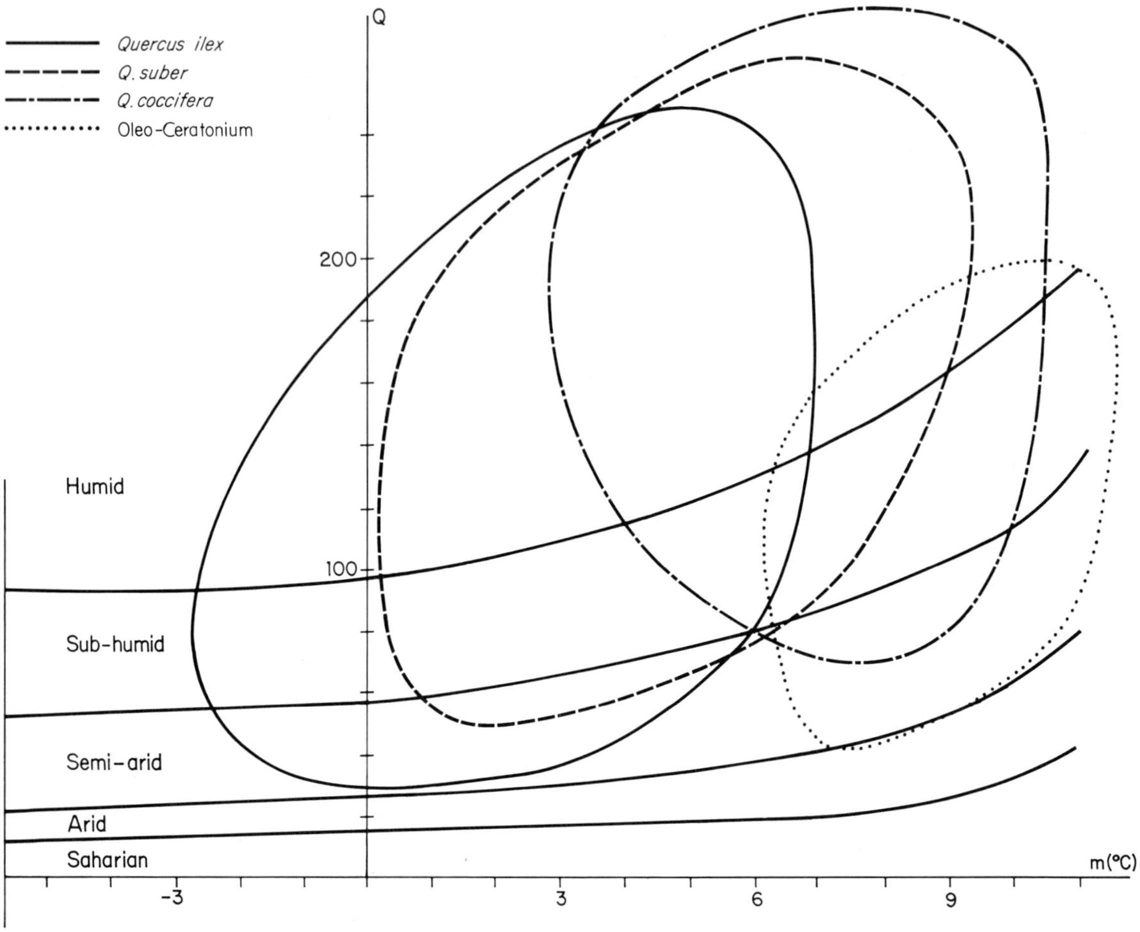

Fig. 6.2. Relationships between presence of *Quercus ilex*, *Q. suber*, *Q. coccifera* and Oleo-Ceratonion and the pluviometric index (Q) of Emberger.

tion map and the bioclimatic map (UNESCO–FAO, 1963, 1969) shows that part of the evergreen forest vegetation area falls into the zone of the thermomediterranean bioclimate, that is, where x is between 100 and 150. This applies to a coastal strip from Lisbon in the west to Alicante in the east (southern half of the Iberian Peninsula), to the south of Italy, to the coastal strip around the Greek peninsula and western and southern Turkey, and also in the Middle East to a strip between Latakia and Tel-Aviv (Jaffa) and extending without interruption eastward along the northern fringe of the Tigris watershed. There is even a definite correlation between thermomediterranean climate and the occurrence of evergreen vegetation in an area as isolated as the Crimea. It is also a feature of the western Mediterranean islands (Balearics, Corsica, Sardinia) and those of the eastern part (Aegean Islands, Crete, Cyprus). In North Africa, a similar zone stretches over a continuous coastal area from Morocco to Tunisia, but further eastward, between Tunis and the Nile Valley, it only occurs in two isolated areas, one south of Tripoli and the other east of Benghāzi.

However, evergreen broad-leaved forests also occur in the mesomediterranean bioclimate of either a long dry season ($75 < x < 100$) or a short one ($40 < x < 75$). Examples are found in the southern half of the Iberian Peninsula, in a coastal strip in northeastern Spain, southern France, Italy, the southern part of the Adriatic coast of the Balkans, and in the mountainous areas of North Africa (Rif, Algero-Tunisian Tell). The division between evergreen and deciduous temperate forests falls within this mesomediterranean zone.

Bioclimatic definitions based only on climatic

data are not sufficiently accurate to take into account local changes in elevation and aspect. These local changes may easily become more important than general climatic factors and are readily expressed by changes in the native vegetation.

The water balance between the soil/plant system and atmosphere is also very important, a significant ecological factor being the duration of the dry period. This balance can be quantitatively expressed by evapotranspiration, a concept first introduced by Thornthwaite (1948). Unfortunately the data necessary to quantify potential evapotranspiration by different methods (Turc, 1961) are only available for relatively few localities and mostly for cultivated areas. In addition, it is difficult to evaluate the amount of water which can be stored in forest soils and becomes readily available to trees and shrubs. This is the reason why it has not been possible to establish a strict quantitative relationship between the type of vegetation and the dryness of a site as expressed by an evapotranspiration diagram.

Computers now provide more sophisticated methods of analysis. A recent attempt (Brisse and Grandjouan, 1974) using simultaneously monthly minimum temperature, mean monthly rainfall and monthly evapotranspiration shows that Mediterranean conditions for evergreen forests are a winter minimum temperature of above 2°C, summer minimum temperature of above 15°C, and summer rainfalls of less than 150 mm. Within these conditions, detailed analysis highlights the previously mentioned distinctive differences between the *Pistacia lentiscus* and the *Quercus ilex* vegetation types.

Geological and soil conditions

All available field data show a close relationship between the general soil type and the prevalence of evergreen broad-leaved forests. This must not be interpreted as indicating an exact correlation between the occurrence of evergreen forests and the special soil evolution processes of the present. It seems that, despite inevitable geomorphological processes, pedogenesis in the Mediterranean Region presently covered by evergreen vegetation has been less affected by the generally strong climatic changes of the Quaternary era than neighbouring areas in central and northern Europe. Pedogenesis has been continuous without serious disturbances, at least up to the time when human influence became an important factor.

The rather high mean temperature and the alternation of dry and humid seasons have accelerated the weathering of the parent materials and the disintegration of organic matter. The principal feature of these soils is therefore the development of a thick oxic B horizon. The colour of such soils is generally reddish due to unhydrated iron oxides, more or less linked to clay components. This colour does not necessarily reflect the present pedogenesis but may also be indicative of near-tropical conditions during earlier periods (Duchaufour, 1970). This type of soil is often referred to as *terra rossa* or "red mediterranean soil".

Since numerous geological and lithological structures occur within the region, this mediterranean soil is likely to have developed from a variety of parent materials. In fact, calcareous parent materials are more frequent where the process of oxidation is facilitated by a calcium and magnesium soil complex. Red mediterranean soils also occur on metamorphic shales with a high content of alterable minerals, but rarely on granite.

Where the duration of the dry season is shorter, either because of general climatic differences or because of local topography, and where, as a result, there is competition between evergreen and deciduous forests, the soil is brown not red. This colour change may occur only in the upper part of the profile (A horizon), or most of the B horizon as well.

Under drier climatic conditions, where the evergreen forest is competing with an open steppe vegetation, other soil types have developed, showing different organic matter distributions, leaching of argillic elements and, sometimes, the formation of a subsurface crust.

According to the soil classification adopted by FAO–UNESCO (1970) for international use, the red mediterranean soil is classified either as ferrasol (usually rhodic ferrasol with a red to dusky oxic B horizon) or a humic ferrasol when the B horizon has a relatively high content of organic matter. The French classification describes them as fersiallitic (Duchaufour, 1970). According to the latest U.S. classification (Anonymous, 1975) they could belong to the group of Inceptisols (Xerochrepts), mainly to the Alfisols (in particular Rhodoxeralf and Haploxeralf) and partly to the Ultisols (if formed on non calcareous subsoils such as sandstones, Rhodoxerults).

On sloping land, the disturbance of the dense protective ground cover was responsible initially for the removal by erosion of the top soil layers. For example, rendzina soils of the region are the result of such erosion processes and often are remnants of a red mediterranean soil. Continuous erosion eventually removed all soil and subsoil. Where the exposed parent material is a calcareous rock, some sparse woody vegetation may have continued to persist. On shales and clay parent rock barren badlands are the end-result.

Floral features

Within the bioclimatic area defined previously, the evergreen broad-leaved forest forms the general climax vegetation except where adverse edaphic factors, such as sodium accumulation, have prevented its development (Molinier, 1954). The flora of the Mediterranean Region is considered relatively rich in species by comparison with the central European one. This is not true for the flora of Mediterranean woodlands because most of the species grow in open areas, in grasslands or on rocky sites. The number of dominant woody species in evergreen forests is small and their high competitiveness almost prevents the development of other vegetation elements in a close-canopy forest. Floral families of trees, shrubs and other woody species are limited in number and each family is represented in the regional flora by a few and sometimes only one species.

The main trees and shrubs listed below are arranged in three family groups taking into account the relationship between the flora of the Mediterranean Region and other regional floras.

(1) Families represented in the temperate zone by both evergreen and deciduous species

Fagaceae

Quercus ilex (Mediterranean evergreen oak) is the main tree species. It has a circum-Mediterranean distribution, but is much more abundant in the west.

Quercus suber (cork oak) is restricted to the Western Mediterranean (Portugal, Spain, France, Italy, Corsica, Sardinia, Sicily and North Africa) and exclusively to non-calcareous soils (Fig. 6.3).

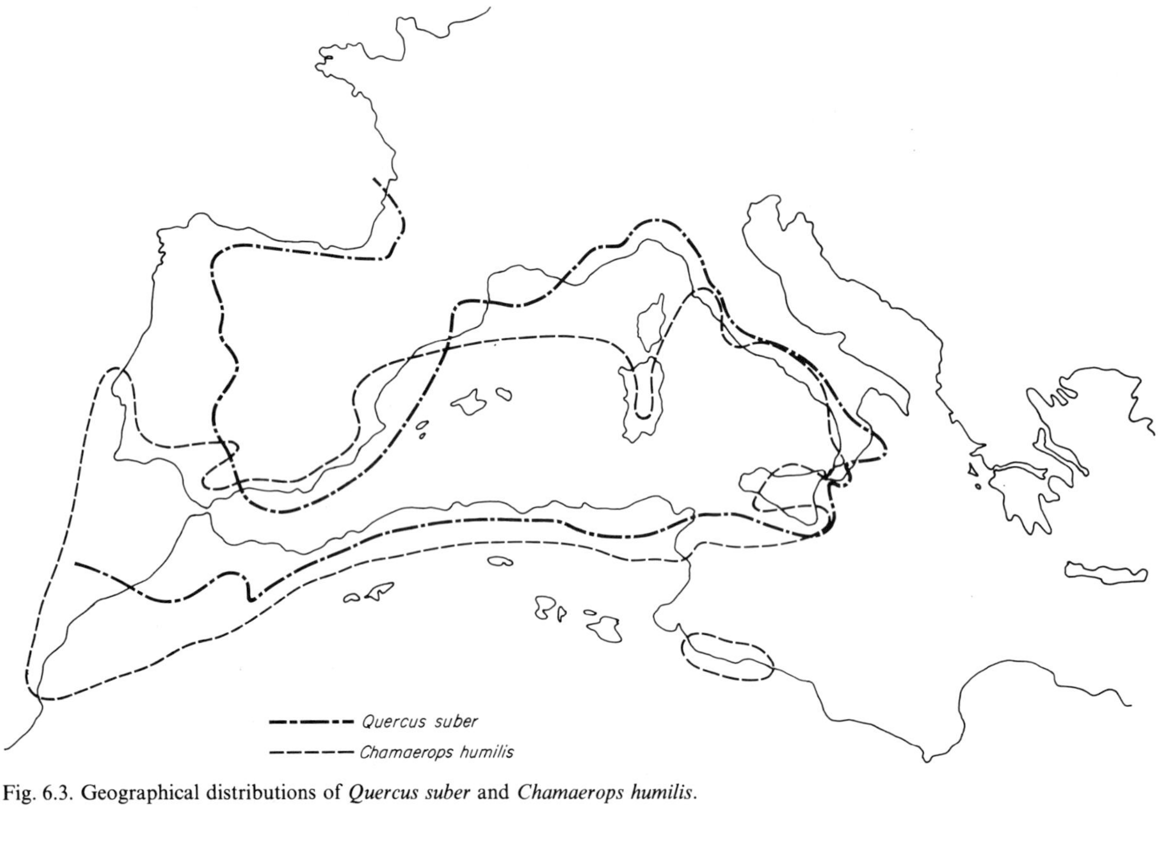

Fig. 6.3. Geographical distributions of *Quercus suber* and *Chamaerops humilis*.

Quercus coccifera, another evergreen oak, also occurs exclusively in the Western Mediterranean. It is replaced in the Eastern Mediterranean (Turkey, Middle East) by a closely related species, *Quercus calliprinos*. Both do not grow over 10 m tall. They have a powerful reproductive capacity by layering, and become highly competitive after forest fires.

Quercus alnifolia is an endemic evergreen oak occurring only in Cyprus.

Rosaceae

Rosa sempervirens is a climbing species and the only evergreen woody species of this family belonging to the forest flora. Other deciduous small trees and shrubs of this family, such as *Crataegus* and *Prunus* become common when the forest vegetation is degraded.

Fabaceae

Ceratonia siliqua (carob tree) is the only evergreen tree of this family, and belongs to a subfamily, Caesalpinioideae, which has its principal representation in tropical regions. Because of the value of its fruit it is often specially protected, and sometimes cultivated, when other trees and shrubs are removed to make room for cultivation.

Shrubs of many other genera, such as *Adenocarpus*, *Calicotome*, *Coronilla*, *Colutea*, *Cytisus*, *Genista*, also play important roles in degraded forest lands because of their high capacity of dissemination. All are deciduous.

Rhamnaceae

Rhamnus alaternus, a medium-size evergreen shrub, grows in mixed stands together with *Quercus ilex* and *Pistacia lentiscus*. Other deciduous shrubby species of the same genus, such as *Rhamnus infectoria* and *R. oleoïdes*, are also floral elements of evergreen forest, but again are more frequent in degraded situations.

Caprifoliaceae

Lonicera implexa is a climbing evergreen species occurring specifically in the *Pistacia lentiscus* vegetation type.

Viburnum tinus is a shrub of the understorey in typical *Quercus ilex* forest of the northwestern part of the region.

Apocynaceae

Nerium oleander is well known for its abundant flowers and is frequently used as an ornamental species. It occurs in the *Pistacia lentiscus* zone, where it is a feature of river and creek banks and depends on permanent water supply.

Liliaceae

Asparagus acutifolius, a wild asparagus, and *Smilax aspera* are also elements of a group of climbing species belonging to the forest flora. *Ruscus aculeatus*, a small evergreen species, occurs frequently in the lowest forest stratum in dense *Quercus ilex* stands, but its distribution is not restricted to the Mediterranean Region.

Only a few bulbous plants naturally grow in woodland although they are numerous in open and grassland areas.

Ericaceae

Arbutus unedo is a small tree, more frequently a shrub, with a circum-Mediterranean distribution whilst a second species of the same genus, *A. andrachne*, is limited to the east. Several species of *Erica* play an important role in different types of degraded forest vegetation: on non-calcareous soils, *E. arborea* and *E. scoparia* are characteristic and dominant shrubs of the "maquis" vegetation; *E. multiflora* often becomes abundant in advanced stages of degradation and on marl soils. *E. verticillata* is limited to a narrow strip along the eastern shores of the Mediterranean and is associated with shrubs of the *Pistacia lentiscus* zone.

Cistaceae

Evergreen forest communities include many shrubs belonging to genera such as *Cistus*, *Halimium* and *Helianthemum*. However, these are only important in degraded situations and especially after fires when they frequently become dominant species.

Lamiaceae

Many shrubs, known either for their flavour or smell, are natives of the evergreen Mediterranean vegetation, and species belonging to genera such as *Lavandula*, *Salvia*, *Saturea* and *Thymus* are well represented components of degraded woodlands.

Buxaceae

Buxus sempervirens, an evergreen shrub, although not limited to the Mediterranean area, is worthy of mention because of its typical association with *Quercus ilex* on dolomitic soils.

(2) Families of principal geographic distribution in warm-temperate and subtropical zones

Anacardiaceae

As emphasized earlier, *Pistacia lentiscus* is a species of particular interest because of its abundance and indicative value. Other deciduous species of the genus *Pistacia*, like *P. atlantica*, *P. palaestina* and *P. terebinthus*, may occur mixed with different types of evergreen vegetation and are also valuable geographical and ecological indicators.

Several deciduous species belonging to the genus *Rhus* are found in the Mediterranean evergreen forest zone. Some of them are indicative of ecological conditions, as for example *Rhus pentaphylla* which occurs in the most arid conditions in the type of vegetation already characterized by the occurrence of *Pistacia lentiscus* in Morocco.

Oleaceae

Olea europaea, the wild parent of the cultivated olive tree, has been given such importance as an indicator that it was often used to define a climatic zone around the Mediterranean. However, *Pistacia lentiscus* is even more suited for that purpose since its distribution has not been changed so much by changing cultivation patterns.

The genus *Olea* is also represented by other species in Ethiopia and on the Arabian peninsula by *O. chrysophylla* and in India by *O. cuspidata*. This is an example of the affinity between the Mediterranean and tropical flora.

Evergreen species of the genus *Phillyrea*, mostly *P. angustifolia*, *P. latifolia* and *P. media*, are common shrubs of the Mediterranean evergreen vegetation.

(3) **Families with principally tropical distribution**

Lauraceae

This family, which includes numerous species in the tropical area but none in the temperate zone, is represented in the Mediterranean Region by only one tree, *Laurus nobilis*, a native of rare humid sites and sometimes cultivated.

Myrtaceae

This family, well known in the Southern Hemisphere for its very numerous species, especially *Eucalyptus*, is again only represented by one species, *Myrtus communis*, which occurs in the *Pistacia lentiscus* zone.

Arecaceae (Palmae)

Chamaerops humilis is the only representative of this family in the Mediterranean Region although of great importance in tropical areas. It is a palm tree which rarely is allowed to develop fully because its leaves are frequently harvested. It is distributed throughout the western Mediterranean Region (Fig. 6.3), especially in North Africa and to a lesser extent along the Spanish coast and in southern Italy. It is a valuable indicator of the warmer conditions of the *Pistacia lentiscus* zone in a subhumid climate.

A few species of gymnosperms also are natural components of the broad-leaved evergreen forests. *Juniperus phoenicea* (red juniper), with generally a broad geographical distribution, has two curiously distinct patterns of distribution, one in the coastal areas of the *Pistacia lentiscus* zone, the other one in mountainous arid areas. *Juniperus oxycedrus* also is widespread and commonly associated with *Quercus ilex*.

Natural stands of *Cupressus sempervirens*, a typical tree of the Mediterranean landscape which is frequently planted for ornamental purposes and as wind breaks, are scattered throughout the Eastern Mediterranean area. They are found at the upper part of the dry *Pistacia lentiscus* zone, especially on hard calcareous soils in Libya, Crete and the Aegean islands, southwest Turkey, Cyprus and, more eastward, in Iran. A closely related species, *Cupressus atlantica*, is endemic to the Moroccan Atlas and another species to the Hoggar mountains in southern Algeria.

Tetraclinis articulata, a North African cedar, is the only species of a subfamily (Callitrinae) of which all other species are found in the Southern Hemisphere. Thus, from a phytogeographical point of view, it is similar to *Myrtus communis* within the Myrtaceae.

Historical elements

Because of the mildness of climate and the relative proximity to the sea, the evergreen broad-leaved forest of the Mediterranean Region has been a privileged environment for the development of human civilization. The influence of people, perhaps from as early as the Neolithic era, has been profound.

The following summary of the main consequences of human intervention helps us to understand the present condition of the broad-leaved evergreen forests.

A major part of the forest cover has been removed to allow cultivation not only along deep-soil plains and gently undulating foothills but also in mountain areas of steep slopes and shallow soils. Increasing population pressure led to the clearing of even more marginal land so that the remaining broad-leaved evergreen forests are relicts of the original cover. Through continuous overuse and overgrazing and frequent fires, forests gradually became degraded and sometimes converted to bush or grassland. However, these processes indirectly have contributed to an extension of the limits of evergreen forest at the expense of adjacent deciduous forest on a relatively small scale. The principal cause has been a general degradation of the vegetative cover, which increased aridity. This favoured evergreen species such as *Quercus ilex* which were then better able to compete with deciduous species like *Q. pubescens* and gradually replaced them. At the same time, within the evergreen forest zone, species better adapted to arid conditions, as for instance *Quercus coccifera*, replaced those requiring more temperate conditions, like *Quercus ilex*.

Three major types of degraded forest vegetation can be distinguished (see Di Castri et al., 1981).

(1) The "garrigue" vegetation type is found on calcareous or marl soils and is dominated by the evergreen oaks *Quercus coccifera* and *Q. calliprinos*, which have replaced *Q. ilex* or even deciduous oaks.

(2) The "maquis" vegetation type is found on non-calcareous soils. The climax species, like the evergreen oaks *Quercus ilex* and *Q. suber*, have almost disappeared, and the dominant species are now either *Arbutus unedo* and *Erica arborea* (high "maquis") or Cistaceae (low "maquis").

(3) Pine forests play a major role in the present forest scene, due to their powerful seed dissemination and quick early growth after forest fires. Mediterranean pines are of five species. *Pinus halepensis* (Aleppo pine) occurs in the western part of the region as far as Greece, and a related species, *P. brutia*, in the eastern part (Greece, Turkey, Cyprus and the Middle East). They cover extensive areas principally in the *Quercus coccifera–Q. calli-*

prinos zone but also in the *Q. ilex* and even in the deciduous oak zone. They form an upper stratum over a shrub vegetation. *Pinus pinaster* (cluster pine) has spread from its native areas in the Western Mediterranean mountains (Spain, southern France, Italy, Corsica and North Africa) to non-calcareous soils elsewhere. It is frequently found mixed with maquis vegetation. *P. pinea* (umbrella pine) is different, for apparently its areas of origin are somewhat isolated, poor sites near the coast. Because of its pleasing shape and edible seeds, it was frequently planted for ornamental purposes and seed production, especially in Spain and Lebanon.

The geographically different subspecies of *P. nigra* (Austrian, Dalmatian and Corsican pine), from their typically scattered mountainous sites, have also invaded degraded woodlands and are competing with broad-leaved trees and other pines (*P. brutia* and *P. pinaster*) in colonizing forest lands frequently ravaged by fires.

Consequently, most pine forests of the Mediterranean Region must be considered secondary forests and a result of human influence on the environment. They have replaced part of the broad-leaved evergreen forests and, since their products are presently very marketable, can be considered valuable stands when properly managed.

Rehabilitation of forest vegetation related to the improvement and intensification of land use in the plains occurred in the last part of the nineteenth century, and went hand in hand with rural migration, industrialization and urbanization. This trend was first noticed in southern France, somewhat later in Italy and Spain and more recently in Greece. With few exceptions, the socio-economic conditions in the Middle East and North Africa have not yet allowed a similar development in these areas.

The natural rehabilitation of degraded woodlands as well as the reafforestation of abandoned agricultural lands pose many problems. The growth rate of native evergreen trees is low, and lack of economic viability and clear objectives provide few incentives for their re-establishment. Present-day afforestation is mainly based on pine planting, despite the high susceptibility of the plantations to fires. Fire still remains a main obstacle to the restoration of a stable forest vegetation.

Current research is aimed at finding suitable broad-leaved trees adapted to the ecological conditions of the Mediterranean Region and at the same time having considerable wood production potential. Good results have been obtained with species of *Eucalyptus*, especially *E. camaldulensis* and *E. globulus*, planted in areas where there is little frost hazard and the soil capability is rather high. Mainly since the Second World War, industrial plantations of *Eucalyptus* have been established successfully in Spain, Morocco and Italy, on lands previously covered by native evergreen forest. Already *Eucalyptus* wood is being used extensively, especially for the production of paper pulp.

GROSS CHARACTERISTICS

Tree and understorey species composition

The size of the forest area, the history of the flora and the numerous combinations of ecological factors are obvious reasons for the wide variability in composition of the Mediterranean evergreen forest. Furthermore, ecological factors are subject to gradual change so that the vegetation cover should realistically be considered as a continuum.

Plant-sociologic methodology has proved to be a successful way to analyze and describe vegetation. Much research work has been devoted to the Mediterranean Region, which, in this regard, has almost become a pilot research area. A commonly used method of vegetation analysis, description and classification is that of the Zürich–Montpellier school. This analysis is based mainly on floral features but also takes account of ecological factors. As a first step, the vegetation is analyzed by sampling. All the species of a sample area, together with the relevant information concerning their abundance and their structural place, are listed. By comparing different floral lists for a local area, it is possible to group samples according to affinity. Such groupings can be arranged at different levels of synthesis. At a local level, a group of similar samples is used as a basic unit. This is called "plant association", which is characterized by a specific species composition. At a regional level, plant associations of definite affinity are grouped into "alliances" and on a still larger scale, "alliances" are grouped into "orders". This special terminology is similar to that adopted in taxonomy.

In the previous concept, all the evergreen forest plant associations of the Mediterranean region are grouped into only one "order", *Quercetalia ilicis* (Braun-Blanquet, 1952). Within this all-embracing vegetation unit, only two "alliances" have been defined, Quercion ilicis and Oleo-Ceratonion. These closely correspond with the two units characterized by dominance of *Quercus ilex* and *Pistacia lentiscus*, as described previously from a phytogeographical viewpoint. Both units are well defined by floral features and can be used as guidelines to describe forest composition in degraded situations.

(1) *Quercus ilex* zone

In the *Quercus ilex* forest zone, *Q. ilex* is the dominant tree. However, around the western part of the Mediterranean (Portugal, Spain, southern France, Italy, Corsica, Sardinia and North Africa), another evergreen oak, *Q. suber* (cork oak), is found on large areas of non-calcareous soils. It grows either in pure stands or mixed with *Q. ilex*. No other evergreen tree can compete effectively with these oaks which, when in the upper storey, grow to a height of up to 18 m. Two other evergreen species which can reach tree size, *Arbutus unedo* and *Phillyrea media*, are usually dominated by these oaks. Some evergreen shrubs, such as *Rhamnus alaternus* and *Viburnum tinus*, remain in the understorey. Other deciduous medium-size shrubs, like *Crataegus monogyna* and *Pistacia terebinthus*, are found in the associations.

Apart from the dominance of evergreen trees and shrubs, a few permanent evergreen and deciduous climbing plants are also a characteristic feature of the evergreen Mediterranean vegetation. Such species are *Asparagus acutifolius*, *Clematis flammula*, *Lonicera* sp., *Rosa sempervirens*, *Rubia peregrina* and *Smilax aspera*. They play, in some respects, a similar role to that of the lianas in tropical forests.

When forests are properly managed, either for wood production by long-rotation coppicing or for the harvest of cork, their natural stucture often is not greatly changed. However, in some Mediterranean countries, such as Spain and Portugal, woodlands are kept very open and are almost like parklands. They are used simultaneously for the production of firewood, cork, acorns and forage. These woodlands, called *montado*, consequently have a completely different structure.

Once forests have been degraded through overuse and fire, their structure and composition are greatly changed. Numerous ecological factors in various combinations are responsible for a wide range of degraded forest situations but only a few examples can be given here.

In the western part of the Mediterranean Region a high proportion of the native *Quercus ilex* forests on calcareous soils has been replaced by garrigue-type shrublands, where *Q. coccifera* has become the dominant species. The canopy may be only 1 m high and more or less open, leaving room for a number of small shrubs, most frequently belonging to the Lamiaceae (*Phlomis*, *Rosmarinus*, *Satureja*, *Thymus*). In the eastern Mediterranean Region, the *Quercus calliprinos* woodlands are of the same type, but it is not always clear if this vegetation is a substitute for, or the original, vegetation type. Following a long period of overuse and abuse when nearly all woody species have disappeared, the last stage of degradation is grassland, dominated by species like *Brachypodium ramosum* and usually on rendzina soils (Biswell, 1964). On non-calcareous soils of the *Quercus ilex* zone, degradation has transformed forests, usually cork oak forests, into maquis dominated by such ericaceous shrubs as species of *Arbutus* and *Erica*. The ground cover may be so dense as to be impenetrable, but, at an advanced stage of degradation, it is more likely to be open with an increased proportion of other shrubby species, mostly Cistaceae (*Cistus*, *Halimium*, *Helianthemum*).

The degradation of the forest vegetation into garrigues or maquis types has frequently been associated with the previously described extension of pine areas.

(2) *Pistacia lentiscus* zone

In the *Pistacia lentiscus* zone, the original forest cover may have been closed canopy stands of evergreen trees. However, well-preserved examples of such forests are rare. The two evergreen oaks, *Quercus coccifera* in the Western Mediterranean and *Q. calliprinos* in the east are important but not dominant elements. Frequently four other evergreen species, distributed throughout the region, are also present: *Ceratonia siliqua* (carob tree), *Olea oleaster* (wild olive tree), *Myrtus communis* and *Pistacia lentiscus*. Each of these has slightly different ecological requirements and their proportional representation depends on the local conditions. The presence of other species is related to special conditions: *Nerium oleander*, as mentioned previously, is found wherever plants have access to permanent water. The occurrence of three conifers, *Cupressus sempervirens*, *Juniperus phoenicea* and *Tetraclinis articulata*, depends on the ecological conditions and floral history.

Associated mainly with different soil conditions, other types of degraded forest vegetation, similar to those described for the *Quercus ilex* zone can be found throughout the Mediterranean Region. These include many different garrigue or maquis types, shrub or grassland. A thorny deciduous shrubland is worth special mention because of its wide distribution throughout the Eastern Mediterranean Region (Greece, Turkey and the Middle East). The dominant species is usually *Poterium spinosum* and this type is found on the most depleted areas.

Ecophysiology of dominant species

The physiological features of the evergreen trees and shrubs of the Mediterranean Region are not basically different from those of temperate deciduous forest species.

The main annual period of growth is early spring from about March to May, with a possible secondary growth phase in June. In fact, a species is only considered evergreen when leaves of two successive years coexist during a period. Most evergreen species retain their leaves for about two years. However, *Quercus suber* may lose part of its foliage before new leaf growth occurs. Differences may exist in the annual leaf production; for example, *Quercus ilex* has a bi-annual rhythm by dropping a greater mass of leaves in alternate years (Rapp, 1969).

Flowering generally occurs in early spring, except for a few species such as *Arbutus unedo*, *Ceratonia siliqua* and *Smilax aspera*, which flower in autumn. The fruit usually ripens in the fall of the same year. *Quercus coccifera* is exceptional in that its acorns take two years to mature.

Most of the dominant species have no very strict requirements with regard to soil-chemical conditions. However, *Quercus suber* and some *Erica* spp. do not tolerate a soil pH value above 7.

The fundamental ecophysiological differences between temperate evergreen and deciduous trees are drought resistance and frost hardiness. Both features are reflected in their relative geographical distribution. More specific relationships between physiologic features and climatic factors have been established experimentally. For example, within the genus *Pistacia*, *P. lentiscus* uses less water than the deciduous *P. terebinthus*, frequently associated with *Quercus ilex* (Mooney, 1969; Ilijanic and Graucacin, 1972). Similarly, *Quercus ilex* has a better drought resistance than the deciduous *Q. pubescens*, which grows in a transitional zone between the Mediterranean and central European regions. Drought resistance, which is certainly partly related to various anatomical features (Kupferschmidt-Werssbluet and Fahn, 1965), definitely appears a principal limiting factor to plant distribution in Mediterranean bioclimates.

Low temperatures also have specific and complex selective influences. The presence of a plant at a particular site indicates that every part of that plant—leaves, cambium, buds and flowers—is able at any stage of its development to tolerate the lowest temperature. Significant observations are made at every exceptional climatic event. In addition to these observations in natural conditions, experiments on frost hardiness have enabled a classification of Mediterranean evergreen trees and shrubs based on frost damage of leaves. This damage occurs at the following temperatures: *Ceratonia siliqua* −7°C; *Myrtus communis* and *Nerium oleander* −8°C; *Quercus suber*, *Q. coccifera* and *Rhamnus alaternus* −11°C; *Quercus ilex* −15°C; *Phillyrea media* −16°C (Larcher and Mair, 1969; Loissant and Rapp, 1971).

The occurrence and frequency of low winter temperatures are important limitations to the introduction into the Mediterranean Region of species with a tropical affinity.

Forest structure

The few remaining examples of vegetation which are considered to be near-climax communities have structures similar to tall forest systems. Competition between the few species able to form a canopy is so strong as to prevent other species becoming established.

The dominant trees of Mediterranean broad-leaved evergreen forest, such as the evergreen oaks *Quercus coccifera*, *Q. ilex*, *Q. suber* and *Q. calliprinos*, *Ceratonia siliqua* and *Olea europaea*, reproduce from seeds. All evergreen trees and shrubs produce an abundant regrowth after cutting. Most of the present evergreen forests therefore have a coppice structure with trees and shrubs of either uniform or varying age, depending on the coppicing system applied. Since formal management is rarely practised, uneven or irregularly structured stands are frequent. The forest structure can be also expressed by the distribution of the plant biomass. Few data are available on this subject but some values are given later.

Faunal composition

Descriptive lists of insects living in the soil or on dead or live plants are available for many different forest types (Herve, 1959; Halperrin, 1963). However, quantitative data with evaluation of zoo-biomass, which could indicate the role of insects in the secondary productivity of forest vegetation, are not available (Silva et al., 1968).

The situation is similar for soil invertebrates and terrestrial vertebrates (amphibians, reptiles, birds, mammals).

The Mediterranean broad-leaved evergreen forest fauna differs in some important aspects from that of the temperate deciduous forest. For example, the fodder base is inadequate for large mammals like deer and roe deer, but wild pigs are common. With regard to birds, the geographical distributions of some non-migratory species like the red-legged partridge (*Alectoris rufa*), are rather closely linked with Mediterranean conditions, but again, quantitative data are not available.

Pests and diseases

The relevant literature (Binyamini and Avizohar-Hershenzon, 1973) gives a long list of fungi living on Mediterranean trees, but very few are considered damaging parasites (Malençon and Marion, 1951). This suggests that the native flora has built up a considerable resistance to diseases through a long history of natural selection. In contrast, pines and introduced deciduous broad-leaved trees like poplars are far more susceptible to pests and diseases.

Insect attacks of epizootic proportions may cause temporary defoliation but rarely threaten the survival of species (Anonymous, 1966). Insects such as *Lymantria dispar*, *Malacosoma neustria* and *Tortrix viridina* which are widely distributed throughout the Mediterranean Region may cause widespread defoliation of evergreen oak forest in the area from Portugal to Greece. *Euproctis chrysorrhea* and *Thaumatopoea processionea* may cause similar damage in areas from southern France to Greece and from Italy to Turkey.

Leaves are not the only part of trees attacked by insects. Larvae of *Coroebus bifasciatus* (Coleoptera) are damaging parasites on young twigs of *Quercus ilex*. *Cerambyx cerdo* lives on stems and *Balaninus elephas* is a parasite on acorns.

In some cases, damage may result in increment loss of wood or fruit production. Insect damage may affect forest operations—for example when defoliation by *Lymantria* makes cork harvesting impossible.

FOREST FUNCTIONAL PROCESSES

Organic production and decomposition

Information on the production of organic matter by Mediterranean evergreen forest is far more limited than for temperate deciduous and coniferous forests (Eckardt, 1967). Until a few years ago, almost nothing was known of organic production, but valuable knowledge is now being gained through the work of IBP and the programme on Man and the Biosphere (M.A.B.), sponsored by UNESCO. The first data are only just becoming available.

The following examples are the results of the well-advanced integrated project in southern France. The evergreen forest vegetation being studied consists of stands of *Quercus ilex* and garrigue of *Q. coccifera* (Rapp, 1969; Loissant and Rapp, 1971).

The above-ground biomass of a 150-year-old *Q. ilex* stand was estimated as 264 t ha^{-1} made up as follows: 230 t for boles and timber (over 7 cm diameter on bark); 25 t of small wood; 2 t of 1- and

2-year-old shoots; 4.5 t of leaves of the current year and 2.5 t of leaves of the previous year. The underground biomass was estimated at 40 to 50 t ha^{-1}. The total mineral content of the stand was evaluated as 5.7 t (about 4% for the leaves): calcium accounts for 68%; nitrogen for 13%; potassium for 11% and phosphorus for 4%. From these figures, the part of the annual organic production accumulated in standing wood elements was estimated as 1.7 t ha^{-1} yr^{-1}, whilst from 3.8 to 7.8 t ha^{-1} yr^{-1} was estimated as the annual litter fall. These data on organic productivity are slightly less than the annual productivity already found in studies of Mediterranean xerophytic forests. Bazilevich et al. (1971) gave for Mediterranean vegetation an aggregate annual increment ranging from 10.1 to 15 t of dry matter ha^{-1}. The leaf biomass found for the *Quercus ilex* stand is close to the values of green assimilating organs (6.1–12.5 t of dry matter ha^{-1}; reported by Bazilevich et al., 1971). It is also estimated that the decomposition of half of the annual litter fall takes about 28 months. The studies give values for the recycling of minerals amounting to a total annual turnover of 120 to 270 kg ha^{-1} yr^{-1} [calcium 50%, nitrogen 30%, potassium 13%, magnesium 4%, phosphorus 2% and micronutrients (iron, manganese and copper) 1%].

The above-ground biomass in a *Quercus coccifera* experimental plot located in the same region was determined as 23.5 t ha^{-1} (including 4 t for leaves) for a 17-year-old stand and 37.1 t ha^{-1} (including 5.6 t for leaves) for a stand more than 30 years old. The annual litter fall was 2.3 and 2.6 t ha^{-1} respectively. These values, like the primary productivity evaluated at 3.4 t ha^{-1} yr^{-1} for the 17-year-old stand (standing woody material 1.1 t ha^{-1} yr^{-1}), are much less than the corresponding values for *Quercus ilex*.

Data on the mineral input through precipitation also became available (Rapp and Romane, 1968; Rapp, 1973). These showed that, of the total mineral input from precipitation (155 kg ha^{-1} yr^{-1}), more than half (98 kg) came from airborne dust particles deposited on leaf surfaces and washed down by precipitation passing through the canopy. With regard to the relative proportion of the principal elements, there is an important qualitative difference between the mineral input by precipitation and that from litter fall. Sodium is much more abundant in precipitation. The amount of potassium in precipitation reaching the forest floor can only be explained by the addition of potassium by leaching from the leaves. A nutrient deficit was observed in the mineral cycle but the difference between the total input to the soil (litter fall, precipitation) and uptake may be logically explained by the standing crop increment of minerals.

Energy flow

Prior to the development of sophisticated measuring apparatus and obtaining a better comprehension of the general concepts of the functioning of ecosystems, the measurement of energy flow was by indirect means, which at best gave incomplete results. Studies of the influence of forest cover on air and soil temperatures are examples of this indirect approach. Pavari (1937) and FAO (1962) reported that in winter the mean air temperature under the canopy of a dense evergreen oak stand was slightly less (1°C) than the outside temperature. Conversely, in the same plot, during the vegetative season (from February to September), the air temperature under the forest canopy was greater than that outside. The forest cover influences the air temperature above the forest to a height of 500 m. The soil temperature was less under the forest canopy throughout the year, unlike deciduous forests where a similar difference occurs only during summer. The measurement of evapotranspiration provides another indirect way to study energy flow. The present trend is to adopt a more comprehensive approach in order to establish relationships between radiation, interception, gas exchanges and organic matter production (Eckardt et al., 1973). More experimental data are needed to achieve an understanding of the complexity of energy transfer in Mediterranean forest ecosystems, despite the recent increase in research (Galoux, 1971).

Water circulation

The influence of an evergreen forest cover on the water cycle has been the subject of many discussions over a long period of time because the availability of water profoundly affects any human activity in the Mediterranean Region. This is in accord with the modern ecosystem approach which is concerned with the influence of biological factors on the water cycle and the influence of hydrology on ecosystem functioning. As part of the M.A.B. programme, quantitative data on the role of the broad-leaved evergreen forest cover in the water cycle are being collected and analyzed.

The forest cover is a transfer zone between the atmosphere and the soil, for both precipitation and evapotranspiration. The following components of the hydrological cycle are recognized: throughfall, which reaches the soil through the canopy with no interception; stemflow, which is intercepted by the trees and runs along the branches and the trunks down to the soil; and finally interception, which is the difference between the total amount of rainfall above the canopy and that reaching the soil either directly or indirectly.

In Italy, interception was estimated as 35% of the total rainfall for a 12-year-old *Quercus ilex* coppice, 3 m tall (Pavari, 1937; and FAO, 1962). Measurements made in a 150-year-old *Quercus ilex* stand in France (Rapp and Romane, 1968) gave the following annual means: throughfall 65%; stemflow 4%; and interception 30%. The ratio between throughfall, stemflow and interception can vary considerably according to the intensity and duration of each precipitation period. Rainfall of less than 2 mm, for example, produces no throughfall at all. When this minimum is exceeded, the relationship between the total amount of rainfall above the canopy and the throughfall is linear as is the relationship between the total amount of rainfall and stemflow. Water reaching the soil may either infiltrate or run off the soil surface depending on the infiltration capacity of the soil and rainfall intensity. Under the same *Quercus ilex* stand in southern France, the soil infiltration capacity was estimated at 2.5 mm min^{-1}, making run-off unlikely. In Morocco, under a dense cover of a mountain stand of *Quercus suber*, no run-off was observed (Malencon and Marion, 1951).

The influence of an evergreen forest cover on the soil/atmosphere phase of the water cycle may be estimated from evaporation rates. From measurements of major components of the solar radiation balance of scrub forest of *Quercus calliprinos* and *Pistacia lentiscus* and pasture, Stanhill et al. (1973) concluded that differences in the evapotranspiration rates of the two vegetations are the cause rather than the result of differences in their radiation balances. Water yield measurements are also a means of comparing the influences of a forest with those of other vegetation covers (Scachori and Michaeli, 1965).

During a full year, soil water reserves are replenished during the autumn to spring period by the percolation of rain water into the soil and subsoil. The soil water reserve is depleted during the subsequent rainless period from late spring to autumn. Since the infiltration–percolation period is shorter than the period of high evapotranspiration, soil leaching processes are of very limited importance. On the whole, the water storage of a forest soil under an evergreen cover seems greater than for the soil of open grassland (Gindel, 1964).

Global dynamics

With regard to the functional processes, the available knowledge is much too limited to draw a general picture of their global implications.

Much time and effort have been devoted to a purely descriptive approach to Mediterranean evergreen forests. Now, in order to orientate forest policy, there is an obvious need for a quantitative and integrated approach to the functioning of broad-leaved evergreen forests as ecosystems.

PAST, PRESENT AND POTENTIAL USE; ECONOMIC STATUS

From the beginning of historical time, there has been in the land-use pattern of the Mediterranean Region a clear difference between cultivated lands, usually privately owned, and woodlands, frequently publicly owned and utilized by the community in accord with local rules. In the early local economy, woodlands provided vital products unobtainable from cultivated lands. These products included timber, firewood, fodder for livestock and food (fruit, acorns). At the present time, the economic importance of Mediterranean forest products depends on the level of industrial development in individual countries.

Despite their potential biological productivities, which are equivalent to those of temperate deciduous or coniferous forest, Mediterranean broad-leaved evergreen forest areas play different roles to those played by temperate forest. First of all, native Mediterranean trees produce structural timber of no marketable value in a modern open-market economy. Construction timber production from evergreen oaks for example is of little importance, even at the village level. Nearly all the volume of timber utilized in the populated areas of the region is imported both from temperate and tropical zones. Local wood is still a raw material for handicrafts, but this is of very limited importance. Even when wood can be chipped to be used in the production of particle board, technical and economic difficulties of utilizing the wood of broad-leaved evergreen trees can rarely be solved. In addition, the native woods are unsuitable for plywood or veneer.

The main forest product is firewood which may be used directly or converted into charcoal, but the market for this and the production per hectare is rather limited. As an example, the average volume of firewood obtained from good stands of *Quercus ilex* managed under a coppice system with rotations from ten to twenty years does not exceed 2 m^3 ha^{-1} yr^{-1}. Values of forest productivity are based on local estimates, and volume and yield tables have rarely been established (Castellani, 1970). Forest products include a high proportion of small round wood and firewood but, because of large transport costs, this cannot compete with other sources of available energy. In such conditions, firewood consumption has gradually decreased in industrialized countries and, in many cases, has almost disappeared. New management systems have been applied in order to accommodate to this situation. For example, longer rotations, up to thirty years, have been introduced to increase the average assimilating biomass and the proportion of large round wood in coppice with standards systems. Unfortunately this has not solved the problem. However, firewood still remains the prime source of fuel where there is little or no industrial development. In such areas the overuse of woodlands, which may even involve the extraction of tree stumps, still exists and is causing a progressive depletion of the biomass and consequently a decrease in productivity.

Cork, which began to be used on an industrial basis in the eighteenth century, remains a forest product of international importance, despite competition with man-made materials of similar physical properties. Cork is harvested from the same tree every nine or twelve years, depending on growth rate. Harvest plans applied to cork oak stands, either natural or planted, are similar to those used in coppice systems. A harvest of 100 kg ha^{-1} yr^{-1} is considered a good yield of cork. Cork oak (*Quercus suber*) is the only native evergreen broad-leaved Mediterranean tree which has been deliberately planted. Even where cork is the main product, cork oak forests are managed on a multiple use basis, other uses are firewood production, grazing for livestock and subsidiary cultivation of cereals in part of the woodlands.

Tannin production from the bark of *Quercus ilex* and *Q. suber* was of regional economic importance with international exchanges until the twentieth century. This production ceased because of the increasing production of manufactured tannins and the importation of tannin. The harvest of tannin from both natural or cultivated trees costs less in other countries.

Other less important forest products are the fruit of *Arbutus unedo*, fungi (truffles) and essential oils from *Myrtus communis*.

In industrialized Mediterranean countries, the importance of the broad-leaved evergreen forest areas as productive lands is obviously decreasing and wood products are no longer important in international trade. The sole exception is cork which supplies a modern processing industry. At the same time the role of evergreen broad-leaved forests as recreative areas is increasing because more people, living in the region or coming from other countries, are attracted to the scenery and the sunny climate, mainly along the coasts. Attempts to make the forest area more attractive to sportsmen have also been successful. All industrialized countries have adopted a forest policy clearly orientated towards conservation, even if such objectives can rarely be expressed in economic terms. The protection of the forest against fire then becomes of primary concern. The development of national parks is part of this conservation policy.

In countries where the population growth rate is high and industrialization slow, a major problem is to make the best use of broad-leaved evergreen forest on hilly and poor land. Uncontrolled multiple use is often inevitable and there remains strong competition between wood production, livestock grazing, and cultivation, all of which can be of vital importance. Laws and regulations have been formulated in order to limit the conversion of forest into new cultivated lands and to prevent or at least control browsing and grazing. However, these regulations often remain ineffective when there is little improvement in the general economic situation. Except on good sites, attempts to replace the native evergreen broad-leaved forest cover by another type of vegetation, more profitable in a modern economic context, have not yet been very successful. This is due to the severe natural restrictions and primarily climate. Much research is still needed to determine for each area a compromise multiple use program (Naveh, 1968).

In both industrialized and less industrialized countries, decisions on the management of broad-leaved evergreen forests must take into account the problems associated with the general utilization of land and water resources. Since water conservation is the main concern for future development of all Mediterranean countries, improved knowledge of the relationships between forests and the water cycle seems a major research objective.

REFERENCES

Anonymous, 1966. *Plaga forestales.* Ministerio de Agricultura, Direccion general de Montes Casca y Pesca fluvial, 96 pp.

Anonymous, 1975. *Soil Taxonomy, a Basic System of Soil Classification for Making and Interpreting Soil Surveys.* Agriculture Handbook, 436. U.S. Dep. Agric., Washington, D.C., 754 pp.

Bagnouls, F. and Gaussen, H., 1952. L'Indice xerothermique. *Bull. Assoc. Geogr. Fr. Paris,* (Janv. Fev.): 222–223.

Bazilevich, N.J., Drozdov, A.V. and Rodin, L.E., 1971. World forest productivity, its basic regularities and relationship with climatic factors. In: *Productivity of Forest Ecosystems. Proceedings of the Brussels Symposium, 1969.* UNESCO, Paris, pp. 345–353.

Binyamini, N. and Avizohar-Hershenzon, Z., 1973. Boletaceae of Israel, 2. *Trans. Br. Mycol. Assoc.,* 60: 99–105.

Biswell, H.H., 1964. Range management in the general economy of Greece. *J. Range Manage.,* 17: 299–304.

Braun-Blanquet, J., 1952. *Les groupements végétaux de la France mediterranéene.* Macabet Frères, Maison la Romaine, 297 pp.

Brisse, H. and Grandjouan, G., 1974. Classification climatique des plantes. *Oecol. Plant.,* 9: 51–80.

Castellani, C., 1970. Volume and yield tables for Italian forests. *Ann. Ist. Sper. Assestamento For. Alpic.,* 436 pp.

Debazac, E.F. and Mavrommatis, G., 1971. Les grandes divisions écologiques de la vegétation forestière en Grèce continentale. *Bull. Soc. Bot. Fr.,* 118: 429–452.

Di Castri, F., Goodall, D.W. and Specht, R.L. (Editors), 1981. *Mediterranean-Type Shrublands.* Ecosystems of the World, 11. Elsevier, Amsterdam, 643 pp.

Duchaufour, P., 1970. *Précis de Pédologie.* Masson, Paris, 481 pp.

Eckardt, F.E., 1967. Mécanisme de la production primaire des écosystèmes terrestres sous climat méditerranéen. *Oecol. Plant.,* 2: 367–397.

Eckardt, F.E., Heim, G., Methy, M. and Sauvenzon, R., 1973. Recherches sur une forêt méditerranéeene à feuillage persistant (*Quercus ilex*). Interception de l'énergie rayonnante, échanges gazeux et croissance.

Emberger, L., 1932. Sur une formule climatique applicable en géographie botanique. *C.R. Acad. Sci. Paris,* 191: 389–391.

Emberger, L., 1952. Sur le quotient pluviothermique. *C.R. Acad. Sci. Paris,* 234: 2508–2510.

Emberger, L., 1955. Une classification biogéographique des climats. *Rec. Trav. Lab. Bot., Géol. Zool., Fac. Sci., Montpellier, Sér. Bot.,* 7: 3–43.

FAO, 1962. *Forest Influences.* FAO, Rome, 394 pp.

FAO–UNESCO, 1970. *Soil Map of the World. Key to Soil Units.* UNESCO, Paris.

Galoux, A., 1971. Flux et transferts d'énergie au niveau des écosystèmes forestiers. In: *Productivity of Forest Ecosystems. Proceedings of the Brussels Symposium. 1969.* UNESCO, Paris, pp. 21–40.

Gindel, I., 1964. Seasonal fluctuations in soil moisture under the canopy of xerophytes and in open areas. *Comm. For Rev.,* 43: 24–34.

Halperrin, L., 1963. Some insects affecting oak. *Horaat Hateva,* 13–14: 43–48.

Herve, P., 1959. Les cóleoptères du sol. *Rev. For. Fr.,* 11: 107–117.

Ilijanic, L. and Graucacin, M., 1972. Water relations of some mediterranean species. *Ber. Dtsch. Bot. Ges.,* 85: 329–339.

Kupferschmidt-Werssbluet, M. and Fahn, A., 1965. Anatomical aspects of leaf anatomy in some plants of the mediterranean maquis. *La Yaaran,* 15: 1–15.

Larcher, W., 1970. Kälteresistenz und Überwinterungsvermögen mediterraner Holzpflanzen. *Oecol. Plant.,* 5: 267–286.

Larcher, W. and Mair, B., 1969. Die Temperaturresistenz als Ökophysiologisches Konstitutionsmerkmal *Quercus ilex* und andere Eichenarten des Mittelmeergebietes. *Oecol. Plant.,* 4: 347–376.

Loissant, P., 1973. French IBP woodlands project near Montpellier. In: *Modeling Forest Ecosystems, Report of International Woodlands Workshop.* International Biological Program/PT Section, 1972. Oak Ridge National Laboratory, Oak Ridge, Tennessee, pp. 123–132.

Loissant, P. and Rapp. M., 1971. Répartition de la matière organique, productivité et cycle des éléments minéraux des écosystèmes de climats méditerranéeens. In: *Productivity of Forest Ecosystems. Proceedings of the Brussels Symposium, 1969.* UNESCO, Paris, pp. 597–617.

Malençon, G. and Marion, J., 1951. L'*Hypoxylon mediterraneum*; ses modalités epidemiologiques en Afrique du Nord. *Congr. AFAS, Tunis,* 4: 538–542.

Molinier, R., 1954. Les climax côtiers de la Mediterranée occidentale. *Vegetatio,* 4: 284–308.

Mooney, H.A., 1969. Dark respiration of related evergreen and deciduous Mediterranean plants during induced drought. *Bull. Torrey Bot. Club*, 96: 550–555.

Naveh, Z., 1968. Multiple use of Mediterranean range lands: a new approach to old problems. *Ann. Arid Zone*, 7: 163–176.

Pavari, A., 1937. L'influenza dei boschi Mediterranei sul clima. *L'Alpe*, 24: 42–57.

Rapp, M., 1969. Production de litière et apport au sol d'éléments minéraux dans deux ecosystèmes méditerranéens: La forêt de *Quercus ilex* et la garrigue de *Quercus coccifera*. *Oecol. Plant.*, 4: 377–410.

Rapp, M., 1973. Le cycle biogéochimique de Soufre dans une forêt de *Quercus ilex* du sud de la france. *Oecol. Plant.*, 8: 325–334.

Rapp, M. and Romane, F., 1968. Contribution à l'étude des bilans en eau dans les ećosystèmes méditerranéens. Egoutement des précipitations sous des peuplements de *Quercus ilex* L. et *Pinus halepensis* Mill. *Oecol. Plant.*, 3: 271–284.

Scachori, A.Y. and Michaeli, A., 1965. Water yield of forest, maquis and grass cover in a semi arid region. A literature review. in: *Methodology of Plant Ecophysiology, Proceedings of the Montpellier Symposium*. UNESCO, Paris.

Silva, F., Azevedo, R., Nogueira, S., Ferreira, J. Cid., 1963. *Estud. divulg. Tec. Serv. Flor. Aqua., Portugal*, 2–8.

Silva, F., Azevedo, R., Nogueira, S. and Ferreira, J. Cid., 1968. Entomofauna do sobieiro. *Estud. Divulg. Tec. Serv. Flor. Aqua.*, Portugal, 20.

Stanhill, G., Israeli, M. and Rosenzweig, D., 1973. The solar radiation balance of scrub forest and pasture on the Carmel mountains, Israel: A comparative study. *Ecology*, 54: 819–828.

Thornthwaite, C.W., 1948. An approach towards a rational classification of climate. *Geogr. Rev.*, 38: 55–94.

Turc, L., 1961. Le bilan d'eau du sol: précipitations, évaporation et écoulement. *Ann. Agron.*, 12: 13–49.

UNESCO–FAO, 1963. *Ecological Study of the Mediterranean Zone. Bioclimatic Maps with Notes*. Arid Zone Research, 21. UNESCO, Paris 60 pp.

UNESCO–FAO, 1969. *Ecological Study of the Mediterranean Zone, Vegetation Map of the Mediterranean Zone*. Arid Zone Research, 30. UNESCO, Paris, 90 pp.

Chapter 7

TEMPERATE BROAD-LEAVED EVERGREEN FORESTS OF THE INDIAN SUBCONTINENT[1]

B. GOPAL and V.M. MEHER-HOMJI

INTRODUCTION

So little is known about the forest ecosystems of the Indian subcontinent that a review of ecosystems can hardly be attempted. The situation is particularly difficult in relation to temperate forests since the existence of a temperate climate comparable to that of northern latitudes is itself questionable. Though the Indian subcontinent lies in the tropical and subtropical belt, areas above 1500 m in the Himalaya and above 1800 m in the south Indian hills resemble the temperate region in their temperature regime (Fig. 7.1). Champion (1936) referred to the vegetation there as temperate. Burtt-Davy (1938) strongly objected to calling these areas temperate and suggested the term tropical montane. Puri (1960) accepted the forest classification of Champion (1936) in principle but preferred to call the climate montane temperate. Meher-Homji (1968, 1969) has also suggested that the climate of the south Indian hills is not truly temperate. However, it is generally agreed to include these areas in the temperate category and the term montane temperate is widely accepted for these forests (Champion and Seth, 1968). Puri (1960), Legris (1963) and Champion and Seth (1968) have reviewed the literature on these forests in some detail.

CLIMATE

The major differences between the temperate and montane climates lies in the duration of daylength through the year and the temperature. The photoperiod varies very little during the year in the south Indian region, and in the Himalaya the seasonal variations in the photoperiod are also not so great as in temperate regions. In the temperate climate the seasonal variations in the mean temperature are large and affect the vegetation. In the montane temperate areas of the Indian subcontinent, diurnal variations are more important and in this respect the Himalayas are closer to the temperate climatic regime. Further differences lie in the amount of precipitation and humidity. The montane regions are comparatively drier than the temperate regions elsewhere and these differences are reflected in the soils. The podzols of temperate regions are unknown on the tropical hill tops of the Indian subcontinent.

The south Indian hills differ greatly from the Himalayan region in the small annual range of mean monthly temperature. The two regions differ also in the rainfall pattern: maximum rainfall occurs in October in the south whilst in the north the maximum occurs during June–July (monsoon). These differences are illustrated by the climatic data for a few representative stations given in Fig. 7.2.

The average length of the dry season is one month at Ootacamund in the Nilgiri Range of south India and nil at Kodaikanal in the Palni Range. These averages are misleading, for in many years it is dry from three to four months at Ootacamund and from one to three months at Kodaikanal. Individual spells without rain can last up to eighty days at Ootacamund and from thirty to sixty days at Kodaikanal (Legris and Blasco, 1969). Detailed maps of climate have been prepared by Legris and Viart (1961), Labroue et al. (1965) and Blasco (1971).

[1]Manuscript completed January 1981.

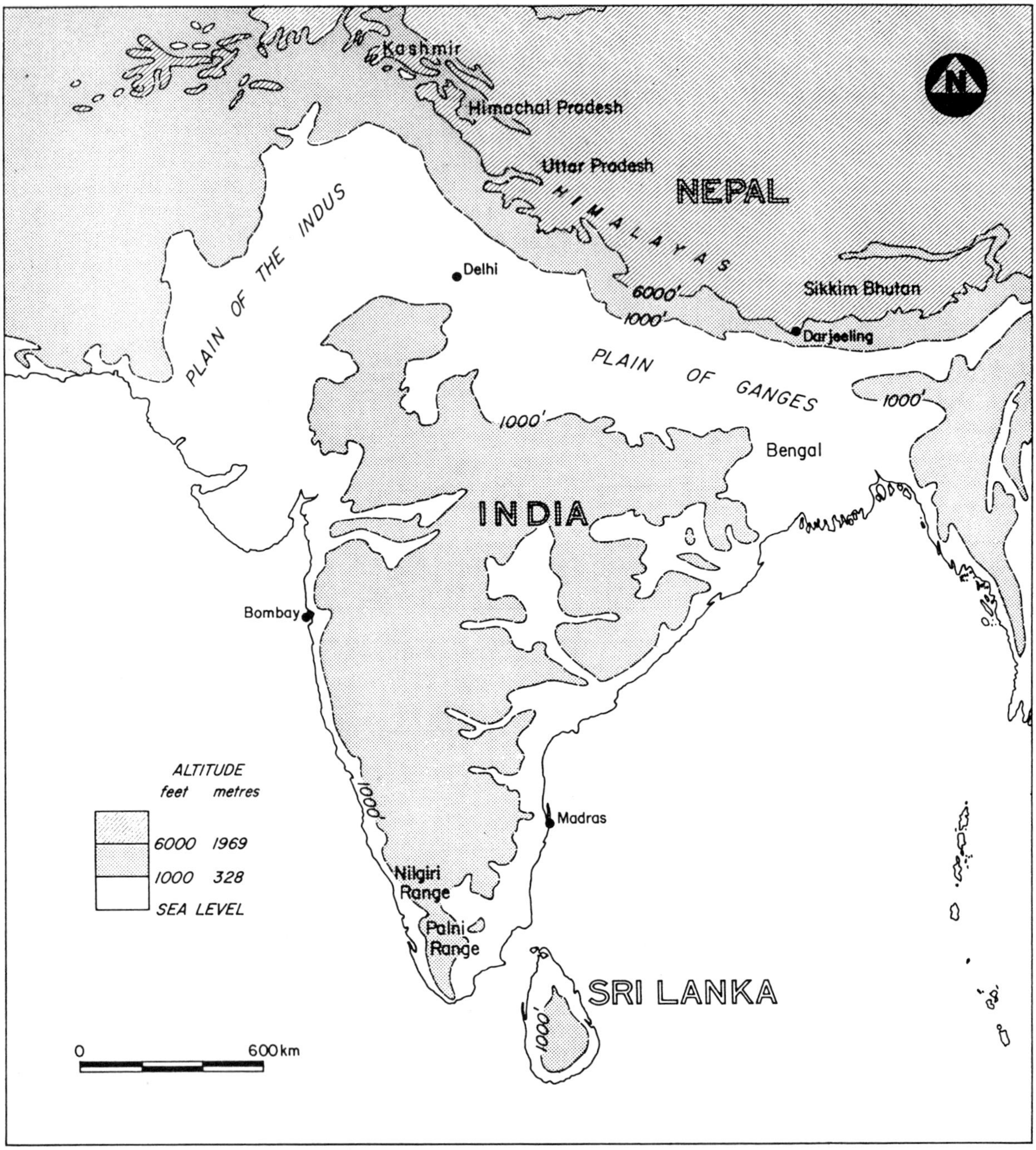

Fig. 7.1. Indian subcontinent.

GEOLOGY AND SOILS

The rocks in the south Indian hills (Nilgiri, Anaimalai, Palni) are mostly gneiss. The humic ferrallitic montane soils are acidic (pH 5–6) and rich in humus with thick, black, upper soil layers. Gneiss also predominates in the Himalaya, but schists, quartzites, granites, shales and conglomerates are also common. The soils are again rich in acidic humus but are mostly loams and sandy loams.

Some studies have been made in the Himalayan region on the soils under coniferous forests (Hoon,

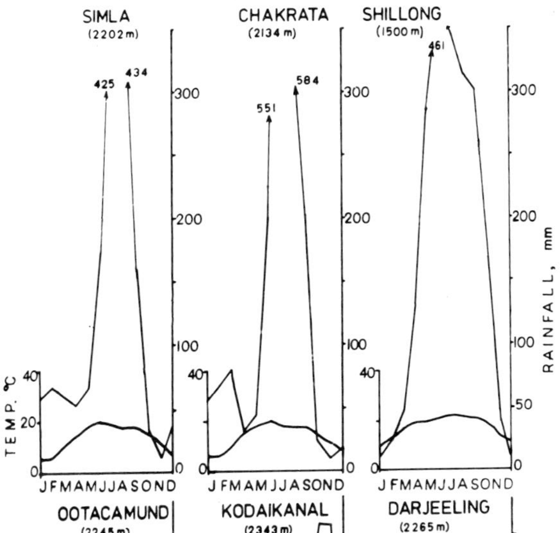

Fig. 7.2. Mean monthly temperature and monthly rainfall at some stations in the broad-leaved evergreen temperate forest regions of the Indian subcontinent.

1939; Puri and Gupta, 1951; Mohan and Puri, 1955; Puri and Maini, 1957). Puri (1960) reported a close relationship between exchangeable calcium content and organic matter in soil profiles in the Bashahar (western Himalaya). Values of pH between 5.4 and 7.1 have been recorded and no important difference has been noted between soils under conifers and *Quercus*. Puri (1954), in a study of soils from *Cryptomeria* and oak plantations in Darjeeling (eastern Himalaya), found high calcium contents and low pH values.

VEGETATION

The temperate forests of India have been divided into three types on the basis of differences in rainfall during the growing season (taken as the period with more than 13°C mean temperature). These are: montane wet temperate, montane moist temperate and Himalayan dry temperate forests. Only the montane wet temperate forests are dominated by broad-leaved evergreen species and they occur both in the Himalaya and in the south Indian hills. The montane moist temperate forests also contain evergreen oaks but are largely mixed conifer–deciduous broad-leaved forests. The dry temperate forests are almost wholly coniferous and rarely contain deciduous species. The Himalayan oaks are evergreen in contrast to those of northern latitudes (Europe and America). The following account of broad-leaved evergreen forests is chiefly concerned with the montane wet temperate forests.

Montane wet temperate forests

The montane wet temperate forests are subdivided into the northern and southern types, the northern type being restricted to the eastern Himalaya and parts of Assam and Bengal. The two types differ from each other in many respects and are better considered separately.

Northern wet temperate forests

Cowan (1929), Chaudhuri (1931), Bor (1938), Deb (1960) and Schweinfurth (1968) have described some of these forests. In the eastern Himalaya and Assam between an altitude of 1800 m and 3000 m, these forests are composed solely of evergreen species in dense closed stands up to 25 m in height. Oaks and lauraceous taxa dominate the tree layer. Due to high rainfall there is a preponderance of epiphytes, ferns, mosses and aroids. The shrub layer and the undergrowth are also composed of evergreen species.

A clear altitudinal zonation occurs in the distribution of tree layer species. Lauraceae abound up to 2100 m beyond which *buk* oak (*Quercus lamellosa*) becomes dominant and gives way to *Q. pachyphylla* above 2500 m. The important taxa in various strata are: tree layer—*Alcimandra* (= *Michelia*) *cathcartii*, *Castanopsis tribuloides*, *Engelhardtia spicata*, *Litsea* sp., *Machilus edulis*, *Magnolia campbellii*, *Mallotus nepalensis*, *Quercus fenestrata*, *Schima wallichii*; shrub layer—*Eurya acuminata*, *Mahonia*, *Symplocos*, *Turpinia* and some Araliaceae. In the buk oak forests at middle altitudes, *Litsaea* and *Machilus* are relegated to secondary strata and the tree layer is dominated by *Acer campbellii*, *Castanopsis*, *Michelia doltsopa* and *Quercus lamellosa*. Species of *Arundinaria*,

Daphne, Rubus, Strobilanthes, Viburnum, etc., occupy the undergrowth. At still higher elevations, the common species in the tree layer are *Acer, Castanopsis, Magnolia campbellii, Quercus lamellosa* and *Q. pachyphylla*. In the Assam Hills (Naga, Aka), *Acer, Bucklandia, Magnolia, Michelia* and *Quercus* form the tree layer. Bamboos (Bambusaceae) are common in the relatively drier regions and evergreen shrubs form dense undergrowth.

At an altitude of 2700 m the oaks are no longer dominant and rhododendrons (*Rhododendron arboreum, R. campanulatum*, etc.) gradually take their place.

Burkill (1913) and more recently Banerji (1958, 1963, 1964), Dobremez (1971), Dobremez and Jest (1971), Stainton (1972) and Ohsawa et al. (1973) have studied the vegetation of Nepal. Here also the temperate forests comprise *Quercus semecarpifolia* mixed with *Rhododendron arboreum* and *Pinus* spp. Deciduous species such as species of *Prunus* and *Pyrus* are also found. Other shrub species are: *Berberis aristata, Buddleja macrostachya, Eurya acuminata, Hypericum patulum, Jasminum humile, Randia tetrasperma* and *Viburnum stellulatum*. The ground flora is excessively rich with species of *Artemisia, Dipsacus, Galium, Pimpinella, Ranunculus, Senecio, Thalictrum*, etc.

The forests have been subjected during recent years to great biotic disturbance from heavy and repeated lopping, clear-felling for shifting cultivation, grazing and burning. In many burnt areas of relatively less rainfall, bamboos have replaced the forests.

Southern wet temperate forests

The forests of the south Indian hills commonly known as shola forests are the best known of the temperate evergreen forests of the region. They are largely confined to steep slopes and sheltered areas. Floristically the sholas are a mixture of tropical and temperate elements. In the Nilgiris at lower altitudes typical tropical and subtropical wet evergreen forests are found as almost a continuation of the sholas but differ in floristic composition and tree height. The shola forests are dominated by members of Lauraceae, Myrtaceae and Styracaceae in the tree layer and by Acanthaceae and Rubiaceae in the undergrowth. The trees are small (20–25 m at most) but form dense closed canopies. Epiphytic ferns, mosses and climbers are abundant. The tree layer consists of *Elaeocarpus* spp., *Euonymus crenulatus, Gordonia obtusa, Ilex denticulata, Litsaea wightiana, Meliosma wightii, Michelia nilagirica, Photinia lindleyana, Symplocos* spp., *Syzygium arnottianum* and *S. montanum. Impatiens* spp., *Psychotria congesta, Strobilanthes* spp., *Vaccinium neilgherrense* and *Viburnum erubescens* form the undergrowth of shrubs and undershrubs. Among other common genera in the ground flora are: *Clematis, Eleagnus, Rosa, Rubus* and *Schefflera*. Detailed accounts of the vegetation of the sholas are given by Ramaswami (1912), Bor (1938), Ranganathan (1938), Shankarnarayan (1958), Meher-Homji (1967, 1969, 1975) and Lakshamanan (1968). Blasco (1971) has published an ecological account of the sholas and savannas.

In exposed areas on slopes and hill tops, the shola forests are almost absent and the vegetation is a shrub-savanna generally referred to as grassland. On the margins of the sholas the vegetation is also different floristically and physiognomically. Here the shrubs are mostly light-requiring and include *Berberis tinctoria, Daphniphyllum glaucescens, Dodonaea viscosa, Eurya japonica, Gaultheria fragrantissima, Hypericum mysorense, Photinia notoniana, Rhododendron nilagiricum, Rhodomyrtus tomentosa, Turpinia cochinchinensis* and *Viburnum coriaceum*.

The shrub-savanna consists of shrubs and ferns such as *Berberis tinctoria, Eupatorium glandulosum, Fragaria nilgherrensis, Lobelia excelsa, Oldenlandia stylosa, Pteridium aquilinum, Rhodomyrtus tomentosa, Rosa leschenaultiana, Strobilanthes kunthiana* and *Ulex europaeus*. The herbs include *Bupleurum distichophyllum* and species of *Anaphalis, Campanula, Cassia, Commelina, Conyza, Heracleum, Justicia, Luzula, Murdannia, Polygala, Potentilla, Spiranthes, Striga, Viola*. The grasses include species of *Andropogon, Arundinella, Bothriochloa, Chrysopogon, Eragrostis, Heteropogon, Isachne, Ischaemum, Themeda* and *Tripogon*. In some situations the grasses dominate and the herbs and shrubs are much reduced.

Montane moist temperate forests

Throughout the entire Himalayan range (Kashmir, Himachal Pradesh, Uttar Pradesh, Nepal, Bhutan, Sikkim and north Bengal), the moist temperate forests occur between 1500 and 3300 m altitude in areas receiving about 1000 to 2500 mm rainfall (Gupta, 1966, 1972; Dobremez et al., 1971; Gupta and Meher-Homji, 1972). These forests are more developed in the western Himalaya and the central Himalaya as compared to the eastern Himalaya where the rainfall is less. Only the oak and oak–conifer forests are evergreen. The oaks form uniform pure stands but are usually less than 20 m high. *Quercus semecarpifolia* occurs on scarp slopes of schists, *Q. dilatata* on flood plains of fine granite, *Q. incana* on quartzite, schists, limestones and sometimes on sandstone (in the Siwaliks) and *Q. ilex* in inner valleys mostly on granites. Oaks occur as an understorey in mixed formations with conifer species of *Abies*, *Cedrus*, *Picea* and *Pinus*, which are up to 50 m tall. The shrub layer consists mostly of evergreens such as *Berberis*, *Castanopsis*, *Daphne*, *Sarcococca* and *Skimmia*. Detailed studies in the western Himalaya have been made by several workers (Kenoyer, 1921; Osmaston, 1922; Dudgeon and Kenoyer, 1925; Coventry, 1929; Gorrie, 1933; Puri, 1950; Puri and Gupta, 1951; Mohan and Puri, 1955). In a study of the Bashahar Himalaya, Mohan and Puri (1955) recognized twenty communities of which only four were composed solely of oaks and four were oak–conifer associations. Between 1500 and 2000 m (sometimes up to 2400 m) *Quercus incana* forms almost pure stands and is considered the climatic climax (Puri, 1960). In biotically disturbed regions the conifer *Pinus excelsa* and *Rhododendron arboreum* invade the *Quercus incana* community to form a mixed association. At relatively higher altitudes, *Q. incana* grows in mixture with *Q. dilatata*, and in moist and disturbed situations (also on dip slopes) the mixed oak-conifer community is found. Here, *Q. dilatata* grows mixed with *Abies*, *Picea* and *Pinus* species. At the same altitude in moist soils *Quercus dilatata* may dominate under disturbed situations with species of *Clematis*, *Desmodium*, *Plectranthus* and *Rosa* present in the shrub undergrowth. Above 2500 m, *Quercus semecarpifolia* forms pure dense communities (climatic climax according to Puri, 1960) in which the shrub layer consists of species of *Berberis*, *Cotoneaster*, *Viburnum*, etc. When heavily disturbed by man through felling, coppicing or grazing, *Abies webbiana*, *Picea smithiana* and *Pinus excelsa* invade the community to form a mixed association.

In very dry regions the evergreen component is *Quercus ilex*, which occurs also as a climax community.

Recently, Saxena et al. (1978) have studied the phytosociology of these forests in the Kumaun Himalaya (near Naini Tal, 29°24′N, 79°28′E). In the mixed oak–conifer forest, *Cupressus torulosa* has a greater stem basal area than *Quercus dilatata* but the two species have the same importance value index (*IVI* = the sum of relative values of density, frequency and abundance) because of the greater density of oaks. In one of the sites located between 2100 and 2227 m altitude, *Quercus dilatata* dominates with *IVI* values ranging from 133 to 271.7 (maximum possible value 300). The best growth of *Q. dilatata* has been observed on north aspects and the poorest on south aspects. On east and south aspects the greatest forest stem basal area is on the hill tops. *Q. incana* also occurs in these forests but only on the upper slopes with south aspects, the lower slope with northeast aspects and on the north aspects at the base of the hills. Among other associates in these forests are *Lyonia ovalifolia* and *Rhododendron arboreum*. The shrub layer comprises *Arundinaria falcata*, *Daphne cannabina*, *Myrsine africana*, *Smilax aspera* and *Wikstroemia canescens*. At other sites located at lower altitudes (1950–2025 m), *Quercus incana* dominates on east and northeast aspects forming at times pure stands. Here the oaks grow best on the lower slopes and at the base of the hills. The forest stem basal area of *Q. incana* has been recorded as high as 0.68 m^2 per 100 m^2.

Basing their observations on the number of saplings in these forests, Saxena et al. (1978) have noted that *Q. dilatata* has a good regeneration potential in contrast to *Q. incana* where regeneration is poor.

ECOLOGICAL NOTES ON SOUTHERN MONTANE TEMPERATE FORESTS

The shola forests and the grasslands of the south Indian hills have been the subject of several studies and some controversial discussion. Ranganathan (1938) considered the grasslands as a distinct climatic climax vegetation, but Champion (1936, 1939) and Bor (1938) regarded them only as biotic climaxes or as a subclimax under the biotic influences of burning and grazing. Most workers since then have accepted them as biotic communities (Shankarnarayan, 1958; Gupta, 1960; Puri, 1960; Chandrasekharan, 1962; Legris, 1963). The basis of their conclusion is largely the success of plantations of exotic species of wattles (*Acacia dealbata*, *A. decurrens* and *A. mearnsii*), *Eucalyptus* (*E. diversicolor* and *E. globulus*), *Cupressus* (*C. funebris* and *C. macrocarpa*), *Alnus nepalensis* and *Pinus khasya* under protection from grazing and burning. However, Meher-Homji (1965, 1967, 1969) has offered another explanation on the basis of the ecological amplitudes of the species in the sholas and grasslands (including shrub-savanna). In a study of successional sequence around Kodaikanal it was noted that the shrub-savanna soils are ameliorated by plantations. When a plantation is clear-felled, a thicket is formed which consists of *Daphniphyllum glaucescens*, *Eurya japonica*, *Photinia notoniana*, *Rhododendron nilagiricum* and *Viburnum coriaceum*. This association never develops into a shola forest though some species of the shola margins, such as *Turpinia* spp. and *Symplocos* spp. may occur in it. With the destruction of the thicket by fire, a shrub-savanna or grassland develops. The failure of the sholas to regenerate and to form a climax community under protection on ameliorated soils can thus be regarded as an indication of the difference between the microclimatic conditions of the shola forests and the open areas.

Blasco (1971) recorded a nocturnal temperature during the winter months as low as $-9°C$ in the open grassy areas while the temperature under the shola canopy was above 0°C. Furthermore, when the relative humidity was below 40% in the open during dry periods, the relative humidity remained quite high under the shelter of trees. The sholas are restricted to slopes and folds of the hills with a high moisture content. The shola species are essentially a tropical element and unable to tolerate low temperature or frost during winter (Meher-Homji, 1967). In contrast, the ligneous species of the shrub-savanna and shola margins consist of subtropical or temperate elements. Many of these are common to the Himalayan flora and therefore can be expected to have a wide ecological amplitude. Their seedlings survive the frost of very cold winters and so readily establish themselves in open areas.

Fire is an important factor in these hilly areas. Some of the shrub-savanna is set on fire by graziers. In dry spells fires pass through the plantations. However, the explanation that fire is the only agent limiting the spread of sholas is inadequate, and the roles of drought and winter cold have to be recognized as additional factors preventing the shola species from invading the open grassy areas. At Anaimudi in Kerala, the highest peak south of the Himalaya, where the incidence of fire and grazing is much less than in the Nilgiri and Palni Hills, the cold and wind seem to play an important role in maintaining the shrub-savanna (Shetty and Vivekanathan, 1971).

Historical proof of the importance of climatic factors in the development of shrub-savanna is provided by the palynological study of Vasanthy et al. (1980). Pollen spectra reveal the presence of extensive areas of shrub-savanna with rhododendron in the period dated more than 6000 years B.C. At that time anthropic activity would be negligible and would not result in large-scale forest destruction.

The presence of an Himalayan element in the south Indian hills has also been a subject of discussion. The consensus favours the view that during the Pleistocene glaciation in the Himalaya many of the species migrated southwards and now persist there (Medlicott and Blanford, 1879; De Terra and Paterson, 1939; Puri, 1947; Meher-Homji, 1972, 1975). The same Pleistocene glaciation has also been considered as responsible for vegetation differences between the eastern and western Himalaya. Glaciation is considered to have been more intense in the western Himalaya. Thus, it is suggested that the oak–laurel community was destroyed in the western Himalaya while it survived in the eastern Himalaya under higher rainfall conditions.

PRODUCTION AND MANAGEMENT

Since these broad-leaved evergreen forests are not of major economic importance, they have been largely ignored scientifically. Except for two recent studies on litter production and decomposition, there have been no studies on the biomass, productivity, nutrient or other budgets of these forests, nor even of the associated microbiota and fauna.

The forests are under Government control, but indiscriminate destruction still continues. The southern montane wet temperate forest is being increasingly planted with species of *Acacia* and *Eucalyptus* for greater economic return. The oaks in the northern forests are used particularly for fuel, and their important roles in the hydrological cycle and conservation are largely ignored. Prompted by their use for fuel supply, foresters have compiled a yield table for *Quercus incana* forests covering various ages and girth classes. Data adapted from the revised yield tables of Seth (1957) are given in Tables 7.1 and 7.2. A 50–60 year-old oak stand of 1 ha has 200 to 300 m^3 of bole wood and branches whose fresh weight is estimated to be 230 to 345 t (specific gravity 1.15). In determining the total biomass it is necessary to add the weight of the small branches, leaves and roots. From these figures it seems that the standing crop of the Himalayan oak forests is comparable to that of other temperate forests. However, no estimate can be made of the production rates in the absence of data on mean annual increments.

Blasco and Tassy (1975) studied the litter production in the Gundar forest near Kodaikanal in the Palni Range of south India (2100 m altitude). The actual yearly production ranged from 3.8 to 7.3 t ha^{-1} between 1969 and 1972, with an average of 5.5 t ha^{-1}. The yearly fluctuations are not corre-

TABLE 7.1

Yield table for *Quercus incana* based on tree girth measurements (Seth, 1957)

Diameter breast height (cm)	Crop height (m)	Number of trees ha^{-1}	Volume of wood (stem and smallwood) (m^3 ha^{-1})		
			main crop	thinning	total
15	12.2	1285	157.2	17.5	174.7
20	13.7[1]	828	208.6	14.3	222.9
	17.1	927	307.6	5.9	314.5
25	16.1	581	220.4	15.1	235.5
	18.9	630	282.4	9.6	292.0
30	18.0	430	295.5	15.3	307.8
	21.0	457	361.0	10.1	371.1

[1]Two sets of data are for two stands of different quality.

TABLE 7.2

Yield table for *Quercus incana* at different ages

Age (yr)	Diameter breast height (cm)	Crop height (m)	Number of trees ha^{-1}	Volume of wood (stem and smallwood) (m^3 ha^{-1})		
				main crop	thinning	total
30	16.8	12.7	455	179.1	33.9	213.0
40	21.1	14.2	322	219.6	27.9	247.5
		17.4	354	307.9	n.a.	
50	24.1	14.9	250	202.9	29.7	232.6
		18.6	287	295.3	n.a.	
60	26.4	15.3	212	199.4	33.6	233.0
		18.9	238	282.7	n.a.	
70	28.2	15.6	189	220.4	35.3	255.7
		20.7	215	319.5	n.a.	
80	29.7	15.9	173	238.7	36.8	275.5
		21.0	195	344.0	n.a.	
90	30.7	25.9	161	250.8	37.8	288.6
		21.0	183	366.9	n.a.	

n.a. = data not available.

lated with the amount of rainfall, length of dry period nor temperature. A significant contribution is made to this forest production by endemic species of which there is a good diversity. Eight principal species belonging to Myrtaceae and Lauraceae contribute about 70% of the mean annual leaf production.

In another study in the Kumaun Himalaya, Saxena et al. (1978) reported a maximum fresh litter biomass on the forest floor in *Quercus dilatata* forests of 475 g m^{-2} at the bottom of a hill and 420 g m^{-2} on the hill top in *Q. incana* forests. The minimum amount of litter biomass was recorded in mixed oak–conifer forests. It was observed that, as the forest stem basal area increased, there was a greater amount of total litter.

Litter fall has been studied only in mixed oak–conifer forests. Of the total fall of 553 g m^{-2} yr^{-1}, *Q. dilatata* contributed 199 g m^{-2} and *Q. incana* 44 g m^{-2}. The maximum litter fall occurred during April and May in both species, but *Q. dilatata* exhibited another peak litter fall in August and September. The data on litter fall in other temperate forests (Jensen, 1974) indicate that the values for Himalayan forests are within the general range for oak forests. Saxena et al. (1978) recorded the rate of litter decomposition for *Q. dilatata* and *Q. incana* by using a litter bag technique. They found that 73% of the leaf litter of *Q. dilatata* and 85% of *Q. incana* are decomposed

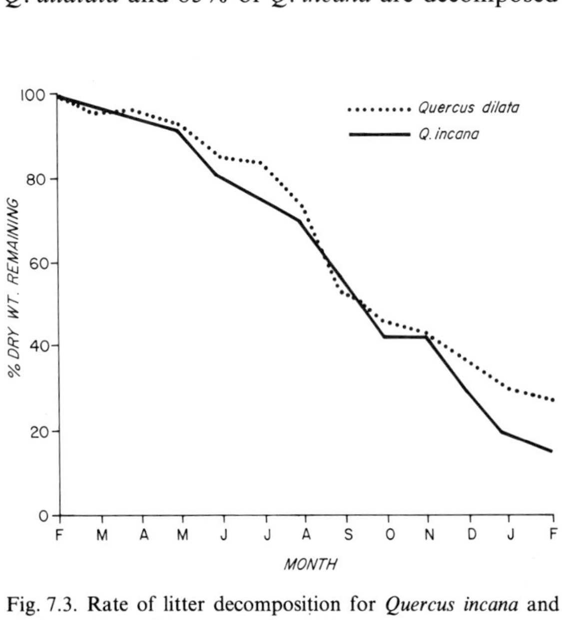

Fig. 7.3. Rate of litter decomposition for *Quercus incana* and *Q. dilata* (data of Saxena et al., 1978).

annually. The data also show (Fig. 7.3) that 39% of the litter decomposition occurs in both species during the rainy season. The dryness of the summer and the low temperature during the winter season result in slow decomposition. An estimate of the decomposition coefficient (Jenny et al., 1949) gave values ranging from 0.67 to 0.80 (estimated total litter fall 617–1443 g m^{-2} yr^{-1}) for *Q. dilatata* forests and 0.68 to 0.79 (estimated litter fall 495–877 g m^{-2} yr^{-1}) for *Q. incana* forests on different aspects and slopes. While the rate of decomposition in these oak–conifer forests lies within the range reported for other oak forests, the decomposition coefficient is higher than reported elsewhere (Lang, 1974; Singh and Gupta, 1977).

CONCLUSIONS

Our knowledge of the structure and functioning of the broad-leaved temperate forests of the Indian subcontinent is very scanty. Most of the areas under these forests, both in the Himalaya and in south India, are under great stress from human activities, particularly cutting, felling, agriculture, plantations, grazing and fire. Regeneration is slow and, in the shola forests of south India, has created a serious problem for the foresters. The raising of plantations of *Eucalyptus* species in south India has been questioned because it is believed that they lower the water table. However, no detailed studies have been undertaken; these are badly needed to form a strong base for conservation and management. The Himalayan forests need further investigation to enable the preparation of proper management plans for conservation and flood control.

REFERENCES

Banerji, M.L., 1958. Botanical explorations in East Nepal. *J. Indian Bot. Soc.*, 55: 243–288.

Banerji, M.L., 1963. Outline of Nepal phytogeography. *Vegetatio*, 11: 288–296.

Banerji, M.L., 1964. Some salient features of East Nepal vegetation. *Candollea*, 19: 215–219.

Blasco, F., 1971. Montagnes du Sud de l'Inde — Forêts, savanes, écologie. *Inst. Fr. Pondichéry, Tr. Sect. Sci. Tech.*, 11: 1–436.

Blasco, F. and Tassy, B., 1975. Etude d'une écosystème forestier montagnard du Sud de l'Inde. *Bull. Ecol.*, 6: 525–539.

Bor, N.L., 1938. The vegetation of the Nilgiris. *Indian For.*, 64: 600–609.

Burkill, I.H., 1913. Notes from a journey to Nepal. *Rec. Bot. Surv.* India, 4: 59–140.

Burtt-Davy, J., 1938. The classification of tropical woody vegetation types. *Imp. For. Inst. Publ.*, 13: 1–86.

Champion, H.G., 1936. A preliminary survey of the forest types of India and Burma. *Indian For. Rec., N.S. Silva*, 1: 1–286.

Champion, H.G., 1939. The relative stability of the Indian vegetational types. *J. Indian Bot. Soc.*, 18: 1–12.

Champion, H.G. and Seth, S.K., 1968. *A Revised Survey of the Forest Types of India.* Manager of Publications, Delhi, 404 pp.

Chandrasekharan, C., 1962. Forest types of Kerala state. *Indian For.*, 88: 837–847.

Chaudhuri, S., 1931. *Working Plan for the Darjeeling Forest Division, North Circle, Bengal (1929–30).* Department of Forests, Calcutta.

Coventry, B.O., 1929. The forest vegetation and evidence of denudation. *Indian For. Rec.*, 14: 49–60.

Cowan, J.M., 1929. The forests of Kalimpong — an ecological account. *Rec. Bot. Surv. India*, 12: 1–74.

Deb, D.B., 1960. Forest type studies in Manipur. *Indian For.*, 86: 94–111.

De Terra, H. and Paterson, T.T., 1939. Studies on the ice age in India and associated human culture. *Publ. Carnegie Inst. Wash.*, 69: 1–354.

Dobremez, J.F., 1971. Biogéographie du centre de Népal. *Bull. Assoc. Géogr. Fr.*, 379–380: 79–90.

Dobremez, J.F. and Jest, C., 1971. Carte écologique du Népal. 1. Région Annapurna–Dhaulagiri (1/250,000). *Doc. Carte Vég. Alpes Grenoble*, 11: 1–49: *Univ. Grenoble Lab. Biol. Vég. CNRS–RCP Paris*, 253: 147–189.

Dobremez, J.F., Vigny, F. and Williams, L.H.J., 1971. Bibliographie du Népal. 3. Sciences naturelles, Botanique. *Cah. Népalais*, 5: 1–100 (CNRS, Paris).

Dudgeon, W. and Kenoyer, L.A., 1925. The ecology of Tehri-Garhwal: A contribution to the ecology of the Western Himalayas. *J. Indian Bot. Soc.*, 4: 233–285.

Gorrie, R.M., 1933. The Sutlej deodar, its ecology and timber production. *Indian For. Rec.*, 17: 1–140.

Gupta, R.K., 1960. Ecological notes on the vegetation of Kodaikanal in South India. *J. Indian Bot. Soc.*, 39: 601–607.

Gupta, R.K., 1966. Bibliography of Himalayan ecology, forestry and phytosociology. *Excerpta Bot.*, B 7: 60–74.

Gupta, R.K., 1972. Bibliography on plant ecology in Bhutan, Sikkim and Tibet. *Excerpta Bot.*, B 12: 226–237.

Gupta, R.K. and Meher-Homji, V.M., 1972. Bibliography on plant ecology in Nepal. *Excerpta Bot.*, B 12: 220–225.

Hoon, R.C., 1939. Study of the soils in hill areas of Kashmir. *Indian For. Rec., N.S.*, 3: 195–261.

Jenny, H., Gessel, S.P. and Bingham, F.T., 1949. Comparative study of decomposition rates of organic matter in temperate and tropical regions. *Soil Sci.*, 68: 419–432.

Jensen, B., 1974. Decomposition of angiosperm tree leaf litter. In: C.H. Dickinson and G.J.F. Pugh (Editors), *Biology of Plant Litter Decomposition, I.* Academic Press, New York, N.Y., pp. 69–104.

Kenoyer, L.A., 1921. Forest formation and succession of the Sat Tal Valley, Kumaon Himalayas. *J. Indian Bot.*, 2: 236–256.

Labroue, L., Legris, P. and Viart, M., 1965. Bioclimats du Sous-continent Indien. *Inst. Fr. Pondichéry, Tr. Sect. Sci. Tech.*, 3: 1–32.

Lakshamanan, C., 1968. The forest types of the Nilgiris and its ecological problems. In: R. Misra and B. Gopal (Editors), *Proc. Symposium on Recent Advances in Tropical Ecology.* Int. Soc. Trop. Ecol., Varanasi, pp. 407–418.

Lang, G.E., 1974. Litter dynamics in a mixed oak forest on the New Jersey Piedmont. *Bull. Torrey Bot. Club.*, 101: 277–286.

Legris, P., 1963. La végétation de l'Inde: Ecologie et flore. *Inst. Fr. Pondichéry, Tr. Sect. Sci. Tech.*, 6: 1–589.

Legris, P. and Blasco, F., 1969. Variabilité des facteurs du climat. Cas des montagnes du Sud de l'Inde et de Ceylon. *Inst. Fr. Pondichéry, Tr. Sect. Sci. Tech.*, 8: 1–95.

Legris, P. and Viart, M., 1961. Bioclimates of South India and Ceylon. *Inst. Fr. Pondichéry, Tr. Sect. Sci. Tech.*, 3: 165–179.

Medlicott, H.B. and Blanford, W.T., 1879. *A Manual of Geology of India.* Calcutta, 2 vols.

Meher-Homji, V.M., 1965. Ecological status of the montane grasslands of the South Indian hills: a phytogeographic reassessment. *Indian For.*, 91: 210–215.

Meher-Homji, V.M., 1967. Phytogeography of the South Indian hill stations. *Bull. Torrey Bot. Club*, 94: 230–242.

Meher-Homji, V.M., 1968. Application of pluviothermic quotient and xerothermic–hygrothermic indices to the Indian subcontinent. *Indian For.*, 94: 351–362.

Meher-Homji, V.M., 1969. Some considerations on the succession of vegetation around Kodaikanal. *J. Indian Bot. Soc.*, 48: 42–51.

Meher-Homji, V.M., 1972. Himalayan plants on South Indian hills: role of Pleistocene glaciation vs. long distance dispersal. *Sci. Cult.*, 38: 8–12.

Meher-Homji, V.M., 1975. On the montane species of Kodaikanal, South India. *Phytocoenologia*, 2: 28–39.

Mohan, N.P. and Puri, G.S., 1955. The Himalayan conifers. III. The succession of forest communities in oak–conifer forests of Bashahar Himalayas. *Indian For.*, 81: 465–487.

Ohsawa, M., Sakhya, P.R. and Numata, M., 1973. Mountaineering of Mt. Makalu II and scientific studies in Eastern Nepal, 1971. *Jap. J. Ecol.*, 23: 218–228. [also *Himalayan Committee of Chiba Univ., Japan*, pp. 99–143, 1975].

Osmaston, A.E., 1922. Notes on the forest communities of the Garhwal Himalayas. *J. Ecol.*, 10: 129–167.

Puri, G.S., 1947. Fossil plants and the Himalayan uplift. *J. Indian Bot. Soc.*, 25: 167–184.

Puri, G.S., 1950. The distribution of conifers in the Kulu Himalaya with special reference to geology. *Indian For.*, 76: 144–153.

Puri, G.S., 1954. Soil climate of some Indian forests. *J. Indian Bot. Soc.*, 33: 394–416.

Puri, G.S., 1960. *Indian Forest Ecology.* Book and Stationery Co., New Delhi, 2 vols.

Puri, G.S. and Gupta, A.C., 1951. The Himalayan conifers. II. The ecology of humus in conifer forests of the Kulu Himalaya. *Indian For.*, 77: 55–63; 124–129.

Puri, G.S. and Maini, J.S., 1957. Succession of forest communities in Chakrata Himalayas. *Proc. Indian Sci. Congr., Abstr.*, pt. III: 287.

Ramaswami, M.S., 1912. A botanical tour in the Tinnevelly hills. *Rec. Bot. Surv. India*, 6: 105–110.

Ranganathan, C.R., 1938. Studies in the ecology of shola grassland vegetation of the Nilgiri plateau. *Indian For.*, 64: 523–540.

Saxena, A.K., Pandey, U. and Singh, J.S., 1978. On the ecology of oak forests in Naini Tal hills, Kumaun Himalaya. In: J.S. Singh and B. Gopal (Editors), *Glimpses of Ecology*. Int. Sci. Publ., Jaipur, pp. 167–180.

Schweinfurth, U., 1968. Vegetation of the Himalaya. In: B.C. Law (Editor), *Mountains and Rivers of India. 21st Int. Geogr. Congr., New Delhi*, pp. 110–136.

Seth, S.K., 1957. Revised volume tables for *Quercus incana*. *Uttar Pradesh For. Dep. Bull.*, 26: 38 pp.

Shankarnarayan, K.A., 1958. The vegetation of the Nilgiris. *J. Biol. Sci.*, 1: 90–98.

Shetty, B.V. and Vivekanathan, K., 1971. Studies on the vascular flora of Anaimudi and the surrounding regions, Kottayam district, Kerala. *Bull. Bot. Surv. India*, 13: 16–42.

Singh, J.S. and Gupta, S.R., 1977. Plant decomposition and soil respiration in terrestrial ecosystems. *Bot. Rev.*, 43: 449–528.

Stainton, J.D.A., 1972. *Forests of Nepal*. John Murray, London, 181 pp.

Vasanthy, G., Caratini, C. and Delibrias, G., 1980. Palynological studies of clayey peats of Palni and Nilgri hills: Palaeoecological significance. In: *Fifth Int. Palynol. Conf., Abstr.*, p. 405.

Chapter 8

TEMPERATE BROAD-LEAVED EVERGREEN FORESTS OF AFRICA SOUTH OF THE SAHARA[1]

D.G.M. DONALD and J.M. THERON

INTRODUCTION

Definition of temperate broad-leaved evergreen forest

Approximately one-eighth of the land surface of Africa south of the Sahara lies south of the Tropic of Capricorn. The remainder is entirely within the tropics. If the division between the warm and cold temperate zones is taken as 6°C (Schimper, 1903), the whole of this land area, with the exception of a small portion of the Orange Free State, lies within the warm temperate to subtropical zones (Weather Bureau, 1954).

Even within this area temperate forest, as normally understood, does not exist. For example, compared with temperate forests in general the forests have a much wider diversity of species, are more complex structurally, lianes occur although they are not numerous, and epiphytes are often abundant. The forests of temperate Africa in some respects have much closer affinities with tropical than with temperate forests.

Of all African forests the forests of the southern Cape coast in the Republic of South Africa are regarded as being most similar to temperate broad-leaved evergreen forests elsewhere, but nevertheless conifers (*Podocarpus* spp.) are common in them.

Forests similar in physiognomy to these Cape forests, and with a surprising number of common species, occur on isolated mountain ranges throughout the continent, where altitude has compensated for latitude to give a temperate climate. This affinity has been recognized by Keay (1959) and White (1978), among others.

Temperate broad-leaved evergreen forest in the African context is regarded as synonymous with White's (1978) concept of afromontane forest. Although Wild (1968), Knapp (1973) and White (1978) have called these forests afromontane, numerous other names have been used, for example: mixed yellowwood and broad-leaved forest (Laughton, 1937), montane evergreen forest (Keay, 1959), montane forest (Phillips, 1959), tropical cloud forest (Walter, 1971), upland forest (Lind and Morrison, 1974), and tropical montane forest (Persson, 1975). Phillips (1959) quoted several other names suggested by various authors: subtropical evergreen forest (Phillips, 1930, 1931a, b), highland evergreen forest (Moreau, 1933, 1935), tropical upper montane forest (Burtt-Davy, 1938), and warm temperate forest (Pitt-Schenkel, 1938). The Yangambi classification uses the name dry montane forest while Greenway calls them upland dry evergreen forest (Chapman and White, 1970).

Boughey (1956) listed 35 synonyms for the afromontane forests in his survey of the nomenclature of the vegetation zones on the mountains of tropical Africa and used the term mountain rain forest.

Many areas formerly covered with natural afromontane forest, which reverted to grassland through mis-management, have been planted to commercial timber plantations. Exotic conifers, mostly *Pinus* and some *Cupressus* species, and broad-leaved species of *Eucalyptus* and *Acacia*, have been extensively planted. In this chapter emphasis is placed on natural forests, and commercial man-made plantations are only briefly covered under management and economics.

The authors of this chapter have had experience

[1]Manuscript completed in December, 1980.

with the afromontane forests in Malawi and South Africa. Of necessity the chapter must be largely a compilation of other people's work, and the authors have drawn freely on the following publications: Wild (1956, 1964, 1968), Phillips (1959), Hedberg and Hedberg (1968), Chapman and White (1970), Knapp (1973), Lind and Morrison (1974), Acocks (1975) and White (1978).

Geographical extent

The afromontane forests of Africa stretch from the southern Cape along the eastern mountain ranges of the continent to Ethiopia, and also occur as an outlier in West Africa. Their distribution is extremely disjointed and has been compared to an archipelago within seven regional mountain systems (White, 1978). Fig. 8.1 indicates the afromontane forest distribution in Africa.

The seven mountain systems from south to north are:

I. Drakensberg—South Africa and Swaziland
II. Chimanimani—Zimbabwe and Moçambique
III. Uluguru–Mulanje—Malawi and Tanzania
IV. Kivu–Ruwenzori—Tanzania, Uganda, Rwanda, Burundi, Sudan and Zaïre
V. Imatong–Usambara—Tanzania, Kenya, Uganda and Sudan
VI. Ethiopian—Ethiopia, Sudan and Somalia
VII. West African—Cameroon, Nigeria, Sierra Leone, Guinea, Liberia and Ivory Coast.

The geographical distribution of the forests within each of these systems will be briefly described.

I. Drakensberg system

Although White (1978) has named these most southerly forests after the major mountain range on which they occur, the greatest area of forest is not associated with the Drakensberg Range. It is found on the southern slopes of the coastal mountains and coastal plateau of the southern Cape from Mosselbaai to Humansdorp. The total area of indigenous forest in the region is approximately 65 000 ha; 66% of this is demarcated forest under state control (Von Breitenbach, 1974). Remnants of these forests occur to the west in kloofs, or gorges, of the coastal mountain ranges where they have been protected against fire. The greatest area of these remnants is at Swellendam, there the Grootvadersbosch Forest covers 425 ha, 68% of which is state-controlled (Taylor, 1955). Further west, forest virtually disappears apart from isolated patches on Table Mountain (Cape Town) and in the protected kloofs of the Jonkershoek Mountains (Stellenbosch), Stetynskloof Mountains (Worcester) and on the coast mountains near Cape Hangklip (Boucher, 1978).

On the east coast, forest remnants are much more numerous than to the west, but nowhere approach the extent of the southern Cape forests. The majority are kloof forests, but larger areas occur, for example, Loerie Forest northwest of Port Elizabeth; the Isidenge, Pirie and Katberg Forests on the Amatolas near King William's Town; the Nquada and Kambi Forests near Umtata in the Transkei; Nkandla, Queni and Weza Forests in Natal; Ndedema and other gorge forests in the Cathedral Peak area; and Wood Bush Forest near Tzaneen and Soutpansberg Forest northeast of Louis Trichardt in Transvaal. Afromontane forests in the northeastern Transvaal have recently been described by Scheepers (1978). These forests lie on the slopes of the Great Eastern Escarpment on the watershed between the Letaba and Koedoes Rivers at altitudes between 1200 and 1500 m.

The forests north of the Transkei are situated on the southern slopes of the Drakensberg Range. The Drakensberg mountain system is the only one of the seven where afromontane forest approaches sea level. In the southern Cape, at latitude 34°S, the forests occur from about 200 m above sea level; with decreasing latitude up the east coast the altitude of the forest increases until it reaches about 1830 m in the Cathedral Peak area, at 29°S. In the remaining mountain systems afromontane forests are seldom present below 1800 m.

Forests similar to afromontane, but more tropical in nature, occur at lower altitudes in the Cape (Alexandria Forest), Transkei (Ntsubane Forest), and Natal (Ngomi Forest). These are intermediate between the true coastal or lowland forests and the high-altitude or afromontane forests of the east coast mountains—submontane.

II. Chimanimani system

The afromontane forests of this system are in the eastern highlands of Zimbabwe, and in adjoining Moçambique. The forests are small and scattered,

Fig. 8.1. Distribution of the "islands of the afromontane archipelago in the seven regional mountain systems. (Adapted from White, 1978, with the permission of the publishers, W. Junk, The Hague).

occurring as patches in the Chimanimani Mountains, on the Vumba and on Mount Inyanga. Chirinda Forest, which is one of the larger evergreen forests, is submontane, transitional to afromontane forest. It occurs up to 1250 m—that is, at a lower altitude than the afromontane forests.

III. Uluguru–Mulanje

This mountain system commences at Mount Mulanje in southern Malawi and continues up the western side of Lake Nyasa (Lake Malawi). It includes Mount Zomba, Dedza Mountain, the Viphya and Nyika Plateaux in Malawi and the

Livingstone, Poroto, Rungwe, Rubeko, Uluguru and Nguru Mountains of Tanzania. Afromontane forests occur throughout, but in the south of Mount Mulanje they are dominated by conifers, such as *Widdringtonia nodiflora*, which form an emergent layer over the evergreen broad-leaved forest.

On the Mulanje and Zomba Mountains the forests are present between 1525 and 2135 m. They are no longer continuous but form scattered fragments, often large, in the more protected portions of the plateaux. Montane forest occurs on the plateau of Mount Dedza above 1980 m. Patches of forest virtually identical to that on Dedza occur on Chongoni Mountain at approximately the same altitude. True afromontane forest is confined to higher parts of the Viphya and Nyika Plateaux. They consist of small patches naturally protected from fire. On the Nyika Plateau conifers again form an emergent strata over a broad-leaved evergreen understorey. The conifer here is *Juniperus procera*. These forests are at altitudes between 2135 and 2230 m.

Much more extensive submontane forests occur on both plateaux at altitudes below 1980 m and down to about 1830 m. They are also present on the Zambian side of the Nyika Plateau. These are floristically richer and structurally more complex than the montane forests, and are considered more closely associated with tropical lowland forests than with temperate forests. The extensive forests of the Misuku Hills in the north of Malawi are also submontane.

The Tanzanian afromontane forests occur in the southern highlands particularly around Mount Rungwe. *Juniperus procera*-dominated forest occurs at Ndumbi Forest on the Kipengere Range.

Two remote regions of forests are present on the Kilombero Scarp and in the Image Forest, 70 km northeast of the town of Iringa. These are regarded as protection forests because of their remoteness, although they contain exploitable timber. As in most of these afromontane forests, conifers occur frequently, in this case *Podocarpus* species.

On the Rubeho, Nguru and Uluguru Mountains, forests occur at high altitudes. They are valued for their protective function and form an important link with the adjoining mountain system.

IV. Kivu–Ruwenzori mountain system

Two small areas of relict afromontane forest remain in Tanzania; on Kungwe Mountain on the eastern side of Lake Nyasa (Malawi) and in Mbizi Forest some 20 km west of the northern end of Lake Rukwa. Most of this system is in Uganda and Zaïre, along the mountains associated with the rift valley. Forests are found on the Virunga and Ruwenzori Mountains at altitudes of 2000 to 2400 m. Throughout the high-altitude regions of East Africa, rainfall and mist in the dry season significantly influence the distribution of the afromontane forest. Conifers tend to be better represented on drier than on wetter slopes.

V. Imatong–Usambara mountain system

The forests of this group occur in northern Tanzania on the Usambara Mountains, on Mounts Meru and Kilimanjaro, the Masailand Mountains, and on the rift wall above Lake Manyara (Soda). In Kenya they are on Mount Kenya, Mount Elgon, the Aberdare Mountains, and the Mau–Elgeyo–Cherangany Mountain ranges; in Uganda on Mount Elgon and Mount Debasien; and on the Imatong Mountains of the Uganda/Sudanese border.

VI. The Ethiopian mountain system

Virtually all of the afromontane broad-leaved forests lie within Ethiopia itself. These forests are in the southwest of the country at altitudes from 1200 to 2000 m where rainfall is plentiful. At 1500 to 2200 m and under slightly drier conditions *Podocarpus* spp. become common. Above 2000 m to approximately 3200 m and on even drier sites *Juniperus procera* replaces *Podocarpus*. Examples of the broad-leaved forest are to be found at Bellata Forest 40 km WSW of Jimā and at Sai Forest near Gore. Forest dominated by *Podocarpus* occurs on the Mendebo Mountains, 150 km ESE of Sashemenē (Shashamanna). *Juniperus*-dominated forest is found on the highest edge of the rift escarpment, 20 km ESE of Lake Langano and in the Menegasha Park west of Addis Ababa.

Conifer-dominated outliers of dry afromontane forest occur on the Red Sea Hills in Sudan and on the Ahl Mescat Mountains in the Somali Republic. Since little is known of their present condition and this summary deals primarily with broad-leaved forest, they are not considered further here.

VII. West Africa

The major area of afromontane forest in West Africa is in Cameroon, with an additional small area in Nigeria. The Cameroon forests on Mount Cameroun are at an altitude of approximately 2000 m and on the higher altitudes on this range to Mount Oku. Afromontane vegetation is restricted to the northern plateau in the Jarawa Hills near Jos and in the Shebshi Mountains in Nigeria.

Several scattered forest patches occur in the far west in the Loma Mountains and Tingi Hills of Sierra Leone; in the Nimba Mountains of Liberia and Guinea, on the Ziama Massif of Guinea and on the Don Massif of the Ivory Coast.

Climatic, topographic, geological and soil conditions

Climate

There is a great diversity of climate within the afromontane forest regions of southern Africa (Jackson, 1961). Rainfall, for example, has a winter maximum in the southwestern Cape, is fairly constant throughout the year in the southern Cape and has a summer maximum in the eastern Cape, Natal, Transvaal, Zimbabwe, Moçambique and Malawi. In East Africa two rainy seasons, March/April and November/December, associated with the monsoons, occur; and in West Africa, which also has a monsoon-type rainfall, dry seasons are very short. Rainfall figures throughout the afromontane region are invariably higher than the surrounding regions since the amount of rain received is frequently governed by the relief.

Since rainfall is greater on the side of the high-altitude mountains facing the rain-bearing winds, the afromontane forest is present at lower altitudes there than on the drier sides. On Mount Kenya, for example, the broad-leaved afromontane forest comes down to 1800 m on the wetter south side but only descends to 2700 m on the drier northern side. On Mount Meru the altitudinal difference is some 600 m while on Kilimanjaro it is only some 150 m. The large East African mountains on the rift valley, which are appreciably further from the coast, also demonstrate this phenomenon. On Mount Virunga the forest extends to just below 2000 m on the wet western side while on the drier eastern side it is not found below 2400 m. This altitudinal difference is only 250 m on Mount Ruwenzori.

The afromontane broad-leaved forest regions frequently experience long periods of dense mist, enabling them to exist even in areas with pronounced dry periods. Rainfall is usually greater than 2000 mm per annum in all the high mountain broad-leaved evergreen forests. Only in the South Cape forests is a rainfall less than 1200 mm yr^{-1} experienced.

In the low-altitude forests of South Africa moisture availability, rather than temperature is the climatic factor limiting the presence of broad-leaved forest. Further north, although moisture still plays a major role, temperature becomes increasingly important. On the high mountains and plateaux of East Africa, the vegetation zones are determined primarily by a temperature decrease of *c.* 1.6°C with every 300 m increase in altitude. Vegetation is not present above 5000 m, where often there is permanent snow. The afromontane heath present below this thermal limit of vegetation gives way at lower altitudes to a shrubby vegetation in many ways reminiscent of the Cape fynbos vegetation (see Specht, 1979). Below this are the afromontane broad-leaved forests, then submontane and eventually lowland forest.

The altitude at which these temperature changes are induced varies from mountain to mountain depending upon topography, the size of the massif, distance from the sea and rainfall. In Kenya, Uganda and Tanzania the zonation is reasonably constant, the afromontane broad-leaved forest lying between 1800 m (1500 m on the south face of Mount Meru) and 2400 m. Frost occurs in all these high-altitude forests but seldom does serious damage. The seasonal temperature range at high altitudes on the large equatorial mountains of Africa is much less than the daily range (Richards, 1963).

Wind seldom affects the forests directly, but may have a marked indirect effect by modifying precipitation and evapotranspiration. Warm dry winds such as the berg wind of South Africa facilitate the penetration of fire along the forest margins. Table 8.1 gives climatic conditions of typical centres in each of the seven mountain systems.

Topography

Aspect, by affecting the radiation received, is only important in the most southerly end of the continent. The warmer northern aspects are less

TABLE 8.1

Climatic data for selected afromontane forests adapted from Jackson (1961)

Station	Latitude	Longitude	Altitude (m)	Mean annual rainfall (mm)	Distribution of rainfall	Daily temperature (°C)							
						January		March		July		September	
						mean	max.	mean	max.	mean	max.	mean	max.
Deepwalls	33°57′S	22°10′E	500	1000	constant	15	20	15	17.5	5	10	10	12.5
Qudeni	28°40′S	30°50′E	1700	900	summer max.	20	25	17.5	22.5	10	17.5	15	20
Mariepskop	24°30′S	31°E	1800	1400	summer max.	20	25	17.5	25	10	17.5	15	22.5
Eastern Highlands	19°S	32°30′E	2000	1400	summer max.	20	25	20	22.5	12.5	17.5	17.5	22.5
Mount Mulanje	16°10′S	35°40′E	1950	1800	summer max.	20	22.5	20	22.5	15	17.5	17.5	22.5
Nyika Plateaux	10°30′S	33°30′E	2100	1200	summer max.	20	25	20	25	15	22.5	20	27.5
Usambara Mountains	4°20′S	38°30′E	2000	1400	summer max.	20	25	20	25	17.5	20	17.5	22.5
Ruwenzori	0°40′N	30°E	2000	1400	constant	20	27.5	20	27.5	20	25	20	25
Mount Elgon	1°40′N	34°40′E	2000	1400	monsoon	22.5	27.5	22.5	27.5	20	25	20	27.5
Jimā Mountains	7°N	37°E	1800	1400	monsoon	20	30	22.5	30	20	25	20	27.5
Loma Mountains	8°40′N	11°W	2000	2600	summer max.	20	25	20	25	20	20	20	20
Mount Cameroun	4°10′N	9°10′E	2000	3200	monsoon	20	20	20	20	17.5	17.5	17.5	17.5

favourable to afromontane broad-leaved forest because of the greater evapotranspiration. The importance of aspect in relation to rain-bearing winds has already been discussed. Forest only occurs on the southern or sea side of the Cape coastal mountain ranges. The vegetation is generally shorter and more xerophytic or may even be reduced to fynbos when associated with shallow soils. Steep slopes and uneven rough terrain are common features in the afromontane broad-leaved forests. Many forest remnants occur in steep gullies, ravines and similar areas where they are naturally protected from fire and difficult to exploit. Flat accessible terrain on which the forest is easily exploited and which can be used for agriculture is the first to be deforested, and by and large already has been.

Geology and soils

The mountain ranges and coastal plateau of the southern Cape consist virtually entirely of sandstone of the Table Mountain Group of the Cape Supergroup. The sandstone has been metamorphosed to quartzite in many areas.

On plateaux sites where Table Mountain Sandstone drift has been deposited over shale or granite, the derived soils are usually less than a metre deep with heavy clay to sandy clay subsoils. These impede drainage and inhibit root penetration. Chemically the soils are acid, low in bases and poor in phosphate. Transported materials, largely colluvial, may give rise to deeper sandy soils on lower slopes, but chemically they are still extremely impoverished.

The Drakensberg Range consists of geologically younger rocks and is composed of various Groups of the Karoo Supergroup overlying the Cape Supergroup deposits. Dwyka, Ecca, Beaufort and Stormberg Groups are found. The soils are mostly ferrallitic, moderate to highly leached, chemically poor but with a higher clay content than the soils of the southern Cape. Nevertheless, soil depth and physical properties are usually favourable for plant growth.

The Chimanimani Mountains in eastern Zimbabwe consist of very old rocks, mainly Precambrian quartzites and schists and Late Cambrian Umkondo shales. The soils derived from the quartzites are white sands with less than 1% clay, less than 5 ppm available phosphorus and less than 15 meq kg^{-1} exchangeable bases and hence are very infertile. The shale soils are reddish in colour with about 20% clay, 55 ppm phosphorus and 10 meq kg^{-1} exchangeable bases. Wild (1964) considered the very infertile quartzite soils can inhibit the development of forest and force an edaphic fynbos-like climax.

Southern Malawi consists of Middle Precambrian rocks, mostly schists and gneisses. Granite and syenite intrusions form Mounts Mulanje, Zomba and Dedza. Rocks in northern Malawi are of Upper Precambrian origin, consisting of schists, quartzites and phyllites. Granite intrusions form the main massif of the Nyika Plateau. Chapman and White (1970) could find no evidence to suggest a correlation between the vegetation and the kind of underlying rocks.

The soils on which the broad-leaved evergreen forests of Malawi occur have been described by Chapman and White (1970). They are essentially leached soils but differ from similar soils outside the forest in their much higher organic content both as a surface layer and incorporated into the top 10 to 30 cm.

The following profile taken from Chapman and White (1970) is from a well-drained site at an altitude of 1980 m on the Nyika Plateau:

litter layer	partly decomposed *Ocotea* leaves up to 5 cm thick
0–30 cm	very dark grey brown (10 YR 3/2)[1] sand to sandy loam; very highly organic; roots extremely abundant in the form of a dense mat
30–150 cm	brown (7.5 YR 5/4) sandy clay loam, fairly stony with many roots, some up to 8 cm in diameter, down to 150 cm

[1]Munsell = soil colour notations.

In Table 8.2 the physical and chemical properties of this profile are shown. Soils from the other forest areas are basically similar although those of Dedza and Mulanje (Chambe Plateau) have clay subsoils. Dedza soils are also much less acid (pH 6.2–6.5).

As in the rest of Africa rainfall is the major soil-forming factor in East Africa. The soils in the afromontane forests appear to be largely humic ferrallitic to humic ferrasols although podzolization has been reported from afromontane forest in the Usambara Mountains in Tanzania (Lind and Morrison, 1974).

TABLE 8.2

Chemical and physical properties of a profile from the Nyika Plateau, altitude 1980 m (from Chapman and White, 1970)

Properties	Soil depth (cm)			
	0–10	10–20	20–30	30–150
Coarse and fine sand %	89	81	76	60
Silt %	11	7	10	18
Clay %	—	12	14	22
Texture	S	LS	SL	SCL
pH	3.6	3.8	4.5	4.9
Organic carbon %	18	11	8.9	0.87
Nitrogen %	1.5	1.2	0.96	0.06
C/N ratio	12	9.1	9.8	15
Available P (ppm)	26	8	1	11
Reserve P (ppm)	—	11	5	—
Available K (meq kg^{-1})	8.0	2.5	1.6	1.5

In West Africa the afromontane forests of Sierra Leone, Liberia, Ivory Coast, Guinea and Nigeria are found on Precambrian formations, mostly consisting of schists, gneiss, amphibolite and granite. This so-called basement complex also gives rise to humic ferrallitic soils in the high-altitude, high-rainfall conditions where the forests occur. In Cameroon the soils of the afromontane broad-leaved forest are derived from volcanic basalts. They can be regarded as immature ferrallitic soils. Fournier (UNESCO, 1963) called them "brown tropical soils of the humid regions". They are more fertile than the ferrallitic soils, having a high exchange capacity and high base saturation. The clay fraction differs from that of the ferrallitic soils in being mostly 2:1 lattice instead of 1:1 lattice type (illites instead of kaolinite). The line of mountains along the fault line in southwestern Cameroon are all of volcanic origin and it is on this line that the afromontane forest occurs. Recent volcanic eruptions (Mount Cameroun, 1909, 1922 and 1959), have done severe damage to afromontane broad-leaved forest in the Cameroons (Furon, 1963).

Soils of volcanic origin occur in East Africa along the rift valley in Uganda, Kenya and Tanzania. Several of these soils have been described from experimental projects associated with afromontane forest in East Africa (Scott, 1962a, b; Scott and Friend, 1962). In such projects on the Sitoten

Mountains of Kenya, afromontane forest areas were converted to tea plantations; on the Aberdare Mountains of Kenya and also on the eastern edge of the rift valley, bamboo thicket was converted to coniferous plantations; and on Mount Mbeya in Tanzania a cultivated catchment was taken out of cultivation, and compared both with a secondary forest catchment brought back into cultivation and a primary afromontane forest catchment.

All these forest soils exhibit pumice layers over buried organic horizons. On Mount Mbeya the underlying rocks are schists and gneiss while in the Kenya projects the soils have developed from deep lava flows and are often over 3 m deep. In all three soils, the volcanic deposits are much older than on Mount Cameroun. The soils are basically humic ferrallitic, have red B horizons and, being highly leached, have a low base content.

Floristic features

One of the most remarkable features of the afromontane forests is their floristic composition. Although separated by great distances, each "island" of this archipelago of high mountains has a flora completely different from that of the surrounding lowland forest but with marked similarity to the other "islands" in the chain. Furthermore, the majority of the flora of these afromontane forests are endemic to these moist, high-altitude habitats, at least at the species level.

Table 8.3 from White (1978) gives the geographical distribution of 90 species of trees of the afromontane forests of southern Africa. White recognized three types of distribution: (1) afromontane endemic wides; (2) relatively restricted endemic species; and (3) local endemics. Wides include tree species present in at least four out of the seven regional mountain systems. Fifty-eight species in Table 8.3 are considered wides. The following eight species occur in all seven systems: *Apodytes dimidiata*, *Ilex mitis*, *Nuxia congesta*, *Prunus africana*, *Rapanea melanophloeos*, *Hypericum revolutum*, *Maesa lanceolata*, *Myrica salicifolia*. Restricted endemics (38 species) occur in two or three of the mountain systems. Fifty-six species are local endemics occurring in only one mountain system.

The total number of tree species increases with decreasing latitude from the low-lying southern Cape forests, along the east coast of South Africa into Zimbabwe and Malawi until a maximum is reached in the afromontane forests of East Africa. Numbers decrease further north into Ethiopia and west into West Africa.

Several species which are confined to the Drakensberg mountain system are considered to be derived from the southern or Cape flora rather than from tropical origins. *Platylophus trifoliatus* and *Virgilia oroboides*, for example, are confined to the southwestern and southern Cape. Other species which occur in the northern areas of the Drakensberg system and further north in other systems have failed to reach the southern Cape forests, for example, *Ptaeroxylon obliquum* and *Xymalos monospora*.

White (1978) analyzed the distribution of 51 afromontane species in the Drakensberg mountain system. Twenty-five occur in the southwestern Cape, 32 in the southern Cape, 44 in the eastern Cape, 46 in the Natal Drakensberg and 43 in the Transvaal Drakensberg. When compared with the other mountain systems the Drakensberg has 53 of the 90 species listed in Table 8.3, Chimanimani has 55, Uluguru–Mulanje has 68, Imatong–Usambara has 63, Kivu–Ruwenzori has 53, while Ethiopia and West Africa have significantly fewer, 34 and 18 respectively.

The endemics of the Drakensberg system have much wider ranges within the system than endemics in other systems, which may be confined to a single mountain range. This is explained by the greater ease with which tree species have migrated down the coastal mountain ranges to the low-altitude Cape forests. In the north the larger mountains are often completely isolated from the adjoining systems by areas of low altitude, high temperature and, frequently, low rainfall. In some systems individual mountains are isolated in a similar manner.

White (1978) believed that afromontane forest can be characterized by the presence of the following species: *Apodytes dimidiata*, *Ilex mitis*, *Nuxia congesta*, *Prunus africana*, *Rapanea melanophloeos*, *Halleria lucida*, *Kiggelaria africana*, *Nuxia floribunda*, *Ocotea bullata* (*O. kenyensis*), *Xymalos monospora* and the two conifers *Podocarpus latifolius* and *P. falcatus*. The first five of these occur in all seven mountain systems. All but four (*Kiggelaria*, *Nuxia floribunda*, *Podocarpus latifolius* and *Xymalos*) are found up into Ethiopia, while all but five

TABLE 8.3

Distribution of the afromontane tree species within the afromontane archipelago (adapted from White, 1978, with permission of the publishers, W. Junk, The Hague)

Species	System[1]: I	II	III	IV	V	VI	VII
Acacia abyssinica		×	×	×	×	×	
Afrocrania volkensii		×	×	×	×		
Agauria salicifolia			×	×	×	×	×
Albizia schimperana		×	×	×		×	
Aningeria adolfi-friederici		×	×	×	×	×	
Anthocleista grandiflora	×	×	×	×	×		
Aphloia theiformis	×	×	×	×			
Apodytes dimidiata	×	×	×	×	×	×	×
Arundinaria alpina			×	×	×	×	×
Aulacocalyx diervilloides		×	×	×	×		
Bersama swynnertonii		×					
Bridelia brideliifolia			×	×	×		
Calodendrum capense	×	×	×	×			
Canthium mundianum	×						
C. ventosum (*C. inerme*)	×						
Casearia battiscombei		×	×	×	×		×
Cassine crocea	×						
C. peragua	×						
Cassipourea gummiflua	×	×	×	×	×		×
Catha edulis	×	×	×	×	×	×	
Chrysophyllum gorungosanum		×	×	×	×		
Cola greenwayi		×	×	×	×	×	
Combretum krausii	×						
Cryptocarya latifolia	×						
C. liebertiana	×	×	×	×			
C. myrtifolia	×						
Cunonia capensis	×						
Curtisia dentata	×	×					
Cussonia spicata	×	×	×	×	×		
Cylicomorpha parviflora			×	×			
Diospyros whyteana	×	×	×				
Entandrophragma excelsum			×	×	×		
Fagaropsis angolensis		×	×	×	×	×	
Ficalhoa laurifolia			×	×	×		
Galiniera coffeoides			×	×	×	×	
Garcinia gerrardii s.l.	×	×	×	×	×		
G. kingaensis		×	×	×	×		
Gnidia glauca			×	×		×	×
Gonioma kamassi	×						
Hagenia abyssinica			×	×	×	×	
Halleria lucida	×	×	×	×	×	×	
Hypericum revolutum	×	×	×	×	×	×	×
Ilex mitis	×	×	×	×	×	×	×
Juniperus procera		×	×	×	×	×	
Kiggelaria africana	×	×	×	×			
Lepidotrichilia volkensii			×	×	×	×	
Leucosidea sericea	×	×					
Linociera peglerae	×						
Macaranga kilimandscharica			×	×	×	×	
M. mellifera		×	×				
Maesa lanceolata	×	×	×	×	×	×	×
Maytenus peduncularis	×						

(*continued*)

TABLE 8.3 *(continued)*

Distribution of the afromontane tree species within the afromontane archipelago (adapted from White, 1978, with permission of the publishers, W. Junk, The Hague)

Species System[1]:	I	II	III	IV	V	VI	VII
Mitragyna rubrostipulata			×	×	×	×	
Myrianthus holstii		×	×	×	×		
Myrica salicifolia s.l.	×	×	×	×	×	×	×
Neoboutonia macrocalyx			×	×	×		
Nuxia congesta	×	×	×	×	×	×	×
N. floribunda	×	×	×	×	×		
Ochna holstii	×	×	×	×	×		
Ocotea bullata s.l.	×	×	×	×	×	×	
O. usambarensis			×	×	×		
Olinia sp.	×	×	×	×	×	×	
Oreobambos buchwaldii		×	×	×			
Platylophus trifoliatus	×						
Podocarpus falcatus s.l.	×		×	×	×	×	
P. henkelii s.l.	×		×	×			
P. latifolius s.l.	×	×	×	×	×		×
Prunus africana	×	×	×	×	×	×	×
Pterocelastrus sp.	×	×	×				
Rapanea melanophloeos s.l.	×	×	×	×	×	×	×
Rhus chirindensis	×	×					
Ritchiea albersii		×	×	×	×	×	×
Schefflera abyssinica			×	×	×	×	×
S. goetzenii		×	×	×	×		
S. myriantha			×	×	×	×	
S. umbellifera	×	×	×				
Scolopia mundii	×						
Seemannaralia gerrardii	×						
Strombosia scheffleri		×	×	×	×		×
Suregada procera	?	×	×	×	×	×	
Syzygium masukuense		×	×				
Ternstroemia polypetala			×				
Trichocladus ellipticus	×	×	×	×	×	×	
Turraea holstii			×	×		×	
T. robusta				×	×	×	
Warburgia salutaris	×	×		×	×		
Widdringtonia nodiflora	×	×	×				
Xymalos monospora	×	×	×	×	×		×
Zanthoxylum capense (syn. *Fagara capensis*)	×	×					
Zanthoxylum davyi (syn. *F. davyi*)	×	×					

[1]Systems: I = Drakensberg system; II = Chimanimani system; III = Uluguru–Mulanje system; IV = Imatong–Usambara; V = Kivu–Ruwenzori system; VI = Ethiopian system; VII = West African system.

(*Halleria*, *Kiggelaria*, *Nuxia floribunda*, *Ocotea* and *Podocarpus falcatus*) reach West Africa.

Gillet (1956) considered that the afromontane forests of Ethiopia were similar to those of Kenya–Uganda. He analyzed the distribution of 246 species from three forests in South Ethiopia and concluded that the southern Ethiopian forests are much closer floristically to Kenya and Uganda than to northern Ethiopia.

Historical elements

The tree species of the afromontane forests have a very disjunct distribution. This can be clearly seen from the large gaps between, and also within, the mountain systems in Fig. 8.1. How these distributions arose is a matter of scientific conjecture. There are basically two main explanations: one suggesting that long-range dispersal is responsible

(Hedberg, 1961, 1969) and the other invoking a climatic change (Moreau, 1966; Morton, 1972).

The tree seed and fruiting bodies of the afromontane flora do not appear particularly adapted for long-range dispersal; only 42% of the tree species of the southern Cape forests have such seeds or fruiting bodies. The seeds and fruiting bodies of the remaining 58% of tree species seem suited only for local distribution or only for distribution immediately under the parent tree. Table 8.4 gives the seed and/or fruit type for the major tree species found in the southern Cape forests. Morton (1972) presented similar data from the afromontane forests on the Loma Mountains in Sierra Leone. He concluded that 40% of the flora could conceivably enjoy long-range dispersal , but that the remainder could not.

Evidence of past climatic change has been fully summarized by Coetzee and Van Zinderen Bakker (1970), but will not be reported here. They concluded that periods with temperatures cooler than those of the present day existed at three distinct times in the past, between 50 000 and 43 000 years ago, between 30 000 and 12 000 years ago and about 4700 years ago. The temperature reduction for the first two periods is estimated as between 4 and 9°C. Morton (1972) considered this sufficient to lower the base level of the afromontane forests by approximately 1000 m. The hypothesis that this temperature reduction would create a high land

TABLE 8.4

Distribution range of South Cape forest trees

Species	Size of seed (cm)		Size of fruit (cm)		Dispersal range	
	<1	1–5	<1	1–5	short	long
Apodytes dimidiata	×		×		×	
Calodendrum capense		×		×	×	
Canthium mundianum	×		×		×	×
C. inerme (syn. *C. ventosum*)	×			×	×	×
Cassine crocea		×		×	×	
C. peragua	×		×		×	
Cunnonia capensis	×		×			×
Curtisia dentata	×		×		×	×
Cussonia spicata	×		×		×	
Diospyros whyteana	×		×		×	
Gonioma kamassi	×			×	×	×
Halleria lucida	×			×	×	
Ilex mitis	×		×		×	
Kiggelaria africana	×			×	×	×
Linociera foveolata	×			×	×	×
Maytenus peduncularis	×			×	×	
Nuxia floribunda	×			×	×	×
Ocotea bullata		×		×	×	
Olinia ventosa (syn. *O. cymosa*)	×		×		×	
Platylophus trifoliatus		×	×		×	
Podocarpus latifolius		×		×	×	
P. falcatus		×		×	×	
Prunus africana	×		×		×	×
Pterocelastrus tricuspidatus	×		×		×	×
P. rostratus	×		×		×	×
Rapanea melanophloeos	×		×		×	×
Rhus chirindensis	×		×		×	×
Scolopia mundii	×		×		×	
Trichocladus ellipticus	×		×		×	
Zanthoxylum capense	×		×		×	
Z. davyi	×		×		×	

bridge to allow migration of the afromontane species down the east coast and over to the west appears feasible. Perhaps high land suitable for afromontane species was not completely continuous, although it most likely was in South Africa. Nevertheless, the distances to be covered by long-range distribution would have been reduced to much more acceptable levels.

If this climatic change hypothesis is correct, it would be logical to expect relict populations of afromontane species to be present on high-altitude ground between the main mountain systems. Morton (1972) and White (1978) have shown the existence of such relict populations in Zimbabwe, Zambia, the Shaba Province of Zaïre, and Angola. Not all of these are necessarily stages in the migratory routes of the afromontane species. Wild (1956) considered some to be rather the dead ends of the earlier migratory process. An example of such a dead end would be the typically afromontane forests in the kloofs of the Magaliesberg Mountains in South Africa.

Destructive agencies including fire

Large areas of African vegetation are fire climax, and fire has been a major factor of the environment since man learnt to control fire. Prehistoric man is known to have used fire in the southern Cape as far back as the early Late Pleistocene; fire traces and prepared hearths have been positively dated from this period (Deacon and Brooker, 1976). Even so, human populations were small and it is unlikely that the prehistoric peoples made serious inroads on the forest.

As these peoples were displaced by the migration of the Bantu peoples so the destruction of the forest increased. The Bantu were culturally more developed: living a settled existence, keeping cattle and practising a type of shifting agriculture. In many areas they engaged actively in the smelting of iron (Chapman and White, 1970). All these activities led to an increased use of fire and took their toll of the forest.

This Bantu migration commenced between 2200 and 2000 years B.P. in East Africa (Hamilton, 1974) and spread down the east coast of the continent into Transvaal and Natal around 2000 B.P. (H. Deacon, pers. comm., 1978). Populations increased slowly but by 1575 A.D. Natal was known to be well populated (Edwards, 1967). It remained so until the rise of the Zulu nation under Shaka in 1818.

Much of Natal was severely depopulated under the Zulu kings Shaka and Dingaan, and it was not until the death of Dingaan in 1841 that people began to return. The Europeans had also arrived by this time and there was a veritable population explosion. The black population density was estimated as 2 km^{-2} in 1816 (prior to the depopulation), and rose to 23 km^{-2} in 1916 and up to 36 km^{-2} in 1946 (Edwards, 1967). As a result there was a significant increase in the need for wood (firewood, hut poles and hut wattles) and land, for shifting cultivation and grazing. An average-sized native hut requires hundreds of saplings and straight branches for the construction of its hemispherical framework (Killick, 1963). Most of these were cut from the afromontane forests prior to the establishment of exotic *Acacia* and *Eucalyptus* plantations. The bark of some species, such as *Ocotea bullata*, was in such demand for medicinal purposes that the species virtually disappeared from some forests. Many areas of afromontane forest disappeared towards the end of the nineteenth century.

The arrival of Europeans further increased the destruction of the forest but also began its conservation. In Tanzania for example "the natural closed montane forests of the country had already been completely cleared or extensively replaced by secondary forest types" prior to the arrival of the Germans towards the end of the nineteenth century (Parry, 1962). The first major task of the German administration was to establish forest reserves to conserve what forest was left.

Today, virtually the only afromontane forest that is still relatively intact is that which is protected either in forest reserves or national parks. Outside the reserves the forests have already been or are rapidly being destroyed, largely by uncontrolled fires (Chapman and White, 1970). The afromontane forests, because they exist under wet to moist conditions, do not burn easily. Nevertheless, most authors consider fire as the major factor in the continued reduction of the forest area (Taylor, 1961; Killick, 1963; Edwards, 1967; Morton, 1968; Osmaston, 1968; Chapman and White, 1970; Hamilton, 1974; Acocks, 1975).

Reverse succession goes from climax forest to

scrub forest to montane grassland which is maintained as a fire sub-climax by frequent burning. If fire can be excluded for long periods, a secondary forest can again occupy the site, but this seldom occurs except on small experimental areas (Granger, 1976). The very sharp boundaries between the grassland and the remaining forest virtually throughout the continent are clear evidence that fire is preventing secondary succession.

Current changes

In post-war times considerable areas of afromontane forest have been cleared for agriculture under economic pressure, particularly to grow coffee and tea. Large areas in Tanzania, Uganda and Kenya are so involved (Pereira, 1962a). Clearing for banana plantations has been widespread on Mount Cameroun (Richards, 1963). The remaining forests are strictly protected, since in general they are largely in catchment areas where their conservation value is fully appreciated (Pereira, 1962b; Parry, 1962; Chapman and White, 1970; Lind and Morrison, 1974). It is unlikely therefore that the present area of afromontane forests currently under government protection will change significantly. Neither will they make anything other than a purely local impact on the global forest resource.

GROSS CHARACTERISTICS OF TEMPERATE BROAD-LEAVED FORESTS

Forest structure and composition

The tree and understorey compositions of the afromontane forests are best dealt with by describing some of the forests throughout their range and, where possible, specific forests.

Drakensberg mountain system

South Cape forests. The species composition of these forests varies somewhat with altitude, aspect, drainage and utilizable soil depth. The example is taken from enumerations of a management experimental area in compartment A11 of the Deepwalls Forest (33°57′S, 23°10′E). The forest lies on the seaward side of the Outeniqua Mountains, some 48 km from the town of Knysna. The forest is classified as medium-moist forest — that is, it occurs on sites with a temperate humid climate with moderately deep loam soils, frequently over dense clay subsoils. The medium-moist forest is the most widespread and also one of the most productive of the South Cape forest types.

The continuous upper canopy of the forest is approximately 20 m high, with emergents—chiefly *Podocarpus falcatus* — up to 30 m. There is an understorey of tree species at 6 to 12 m and a dense shrub layer 3 to 6 m tall. Because of the density of the shrub layer, the ground flora is sparse. In moist forest the shrub layer is poorly developed and the ground flora is appreciably richer. Table 8.5 summarizes the percentage composition of the 45-ha compartment. Every tree over 9.7 cm diameter at 1.37 m was measured and recorded by species and girth class. The average number of stems per hectare greater than 9.7 cm was 285, but because of the shrub layer this relatively open forest was an exceptionally dense community. Although Table 8.5 reflects the strata in which the species occur, immature specimens of the largest trees may be in the lower strata; the percentages are by species, not by strata.

The preponderance of *Olea capensis* subsp. *macrocarpa* in this forest type is considered to be due largely to selective exploitation of the other co-dominant species. *O. capensis* subsp. *macrocarpa* is an exceptionally hard and heavy timber and very difficult to work (Von Breitenbach, 1974). Many species which are infrequent in this medium-moist forest are much more frequent in moist and wet forest, for example, *Ilex mitis* and *Platylophus trifoliatus*, with *P. trifoliatus* being replaced by *Cunonia capensis* at higher altitudes. Other species normally associated with medium-moist forest but not occurring in the compartment are *Canthium inerme* (syn. *C. ventosum*), *Rhus lucida*, *Rothmannia capensis* and *Trimeria rotundifolia*. All occur in the lower strata.

The form of the trees is attractive for commercial purposes, the dominants having long, straight, clear boles. Weak buttressing occurs in some species but is not a feature of these forests.

The shrub layer in the medium-moist forest is characterized by a dense thicket of *Trichocladus crinitus*. Other shrubs present are *Carissa bispi-*

TABLE 8.5

Species distribution by strata, A11 Deepwalls Forest

Species	Percentage of total stem number	Strata
Olea capensis subsp. *macrocarpa*	24.40	upper
Podocarpus latifolius	15.23	upper
Gonioma kamassi	9.22	lower
Pterocelastrus tricuspidatus	9.06	upper
Apodytes dimidiata	8.33	upper
Curtisia dentata	8.97	upper
Canthium obovatum	3.20	upper/lower
Cassine crocea	2.98	upper/lower
Ocotea bullata	2.07	upper
Halleria lucida	2.04	lower
Olea capensis subsp. *capensis*	1.91	lower
Nuxia floribunda	1.89	upper
Diospyros whyteana	1.83	lower
Maytenus peduncularis	1.74	lower
Lachnostylis hirta	1.46	lower
Maytenus acuminata	1.40	lower
Canthium mundianum	1.11	lower
Platylophus trifoliatus	0.94	upper
Podocarpus falcatus	0.56	upper
Ochna arborea	0.42	lower
Ilex mitis	0.40	upper
Linociera foveolata	0.38	lower
Olinia ventosa (syn. *O. cymosa*)	0.29	upper
Burchellia bubalina	0.14	lower
Rapanea melanophloeos		upper
Gardenia thunbergia	0.03	lower
Ekebergia capensis		upper
27 species	100%	—

nosa, *Cassinopsis ilicifolia* and *Rhamnus prinoides*. In wet forest, tree ferns (*Alsophila capensis*) form an important part of this stratum while on the drier sites more thorny, often scrambling, shrubs are prevalent, such as *Dovyalis rhamnoides*, *Maytenus heterophylla* and *Scutia myrtina*.

The ground flora consists mainly of ferns and herbs, with a few grasses; a typical list would be: *Blechnum punctulatum*, *Polystichum lucidum* and *Thelypteris bergiana*; *Aristea ensifolia*, *Carex aethiopica*, *Dietes vegeta* and *Kobresia lancea*; *Brachypodium flexum* and *Oplismenus hirtellus* (Von Breitenbach, 1974).

This forest type regenerates freely but growth of the tree seedlings virtually stagnates until an opening in the canopy stimulates more rapid growth. A tree of *Podocarpus falcatus* with a trunk 2 m in diameter was felled in this forest and had taken just over 100 years to achieve a diameter growth of 15 cm.

Lianes occur frequently in the lower tree and shrub layers but only rarely in the upper; *Cynanchum ellipticum*, *C. obtusifolium*, *Melothria punctata* and *Pyrenacantha scandens* are the commoner important species.

Epiphytes such as the lichen *Usnea*, the club moss *Lycopodium gnidioides*, and the tree orchids *Tridactyle bicadata* and *Angraecum* spp., are found in medium-moist forest (Von Breitenbach, 1974). Bryophytes and epiphytes are common in wet forest.

Natal forests. The following description is based on Qudeni forest in Natal (28°40′S, 30°38′E). The forest is situated on steep southern slopes within an altitude range of 1000 to 1500 m (Edwards, 1967). Qudeni, of approximately 2000 ha, is a demarcated State Forest Reserve. Edwards calls this forest "Mist Belt Mixed *Podocarpus* Forest", but it is very similar to the South Cape forests. At altitudes above 1400 m and up to 2000 m Edwards differentiates a further climax "Mountain *Podocarpus* Forest". Although these forests have fewer species and are not so well developed as the mist-belt forest they do not differ any more than the various forest types within the South Cape forests, and in this account the two are taken as environmental expressions of the same climax forest. The upper levels of Qudeni forest approach "Mountain *Podocarpus* forest".

Structurally the forests are similar to those of the South Cape, though significantly richer in species. The general canopy level varies in height from 15 to 20 m with numerous species as emergents, individuals of which reach 20 to 30 m in height. The emergent species frequently form part of the general canopy. *Calodendrum capense*, *Celtis africana*, *Combretum kraussii**, *Cussonia sphaerocephala**, *Homalium dentatum**, *Kiggelaria africana*, *Olea capensis* subsp. *macrocarpa*, *Podocarpus latifolius*, *Prunus africana**, *Ptaeroxylon obliquum**, *Rapanea melanophloeos* and *Zanthoxylum davyi** (syn. *Fagara davyi*) are frequent to abundant. All but the six species asterisked occur in the South Cape forests. The following species also occur in the general canopy but not as emergents, Group 1 occurs in the South Cape and Group 2 does not.

Group 1	Group 2
Apodytes dimidiata	*Clerodendrum glabrum*
Curtisia dentata	*Cryptocarya woodii*
Ekebergia capensis	*Dais cotinifolia*
Halleria lucida	*Linociera peglerae*
Maytenus peduncularis	*Pleurostylia capensis*
Pterocelastrus echinatus	*Rawsonia lucida*
Rhus chirindensis	*Seemannaralia gerrardii*
Sclopia mundii	*Xymalos monospora*
S. zeyheri	

The lower canopy is often broken and varies appreciably in height from 5 to 13 m. The species present include:

Upper layer	Lower layer
Burchellia bubalina	*Bersama tysoniana*
Cassine papillosa	*Cassine eucleiformis*
Diospyros whyteana	*C. peragua*
Eugenia zeyheri	*Peddiea africana*
Maytenus acuminata	*Psychotria capensis* (syn. *Grumilia capensis*)
Ochna arborea	*Trimeria grandiflora*

The shrub layer is richer than for the South Cape forests, is discontinuous and can be approximately 3 m tall. The following species occur:

Canthium ciliatum	*Indigofera natalensis*
Carissa bispinosa	*Machaya bella*
Cassinopsis ilicifolia	*Maytenus mossambicensis*
Desmodium repandum	*M. undata*
Doryalis lucida	*Micrococca capensis*
D. rhamnoides	*Pavetta natalensis*
D. zeyheri	*P. tristis*
Duvernoia adhatodoides	*Piper capensis*
Excoecaria simii	*Tricalysia capensis*

The ground flora, as in the South Cape forests, varies appreciably with the availability of light and moisture. Species commonly found include: *Begonia sutherlandii*, *Carex zuluensis*, *Clivia miniata*, *Dietes butcheriana*, *Haemanthus katherinae*, *Hypoëstes verticillaris*, *Impatiens duthiei* and *Plectranthus* spp. Grasses are not common, but *Brachypodium flexum*, *Oplismenus hirtellus*, *Panicum aquinerve* and *Setaria chevalieri* occur.

Lianes and scrambling shrubs are not common but the following species may be present: *Buddleja auriculata*, *B. pulchella*, *Cassine tetragona*, *Clematis brachiata*, *Dalbergia obovata*, *Popowia caffra*, *Quisqualis parviflora*, *Rhoicissus rhomboidea*, *Scutia myrtina* and *Secamone alpini*. Epiphytes, being favoured by the frequent mists are numerous and include the ferns *Asplenium* spp., *Pleopeltis schraderi*, *Polypodium polypodioides*; lichens (*Usnea*); *Lycopodium* spp. and bryophytes.

Transvaal forests. The Mariepskop forests occupy some 2954 ha of the Mariepskop Forest Reserve in the Pilgrims Rest District of the Transvaal Drakensberg Mountains (24°30′S, 31°E). The Drakensberg Escarpment changes direction at this point from N–S. to NW–SE. The forests attain their maximum development on the south and southeast slopes, particularly in deep kloofs (Van der Schijff and Schoonraad, 1971). The following data are taken from reports by these authors. The usual strata — emergents, top canopy, lower tree stratum, shrub stratum and ground flora — occur.

The commonest species in the top canopy are:

Apodytes dimidiata	*Kiggelaria africana*
Bersama tysoniana	*Nuxia congesta*
Brachylaena discolor subsp. *discolor*	*N. floribunda*
B. discolor subsp. *transvaalensis*	*Ochna arborea*
Celtis africana	*Olea capensis* subsp. *macrocarpa*
Combretum kraussii	*O. woodiana*
Cryptocarya liebertiana	*Podocarpus falcatus* (emergent)
C. woodii	*P. latifolius*
Curtisia dentata	*Protorhus longifolia*
Cussonia sphaerocephala	*Prunus africana*
C. spicata	*Pterocelastrus echinatus*
Diospyros whyteana	*Schrebera alata*
Faurea macnaughtonii	*Syzygium guineense* subsp. *gerrardii*
Ficus petersii	*Trichocladus grandiflorus*
Halleria lucida	*Xymalos monospora*
Ilex mitis	

In the lower tree stratum the following trees and large shrubs occur:

Calpurnia aurea	*Rhus chirindensis*
Croton sylvaticus	*Rothmannia capensis*
Oxyanthus gerrardii	*Tricalysia capensis*
Psychotria capensis	*Trimeria rotundifolia*
P. kirkii	*Zanthoxylum davyi*
Rawsonia lucida	

The shrub layer is variable, occasionally forming dense thickets but often virtually absent depending upon the light and moisture available. Species found are: *Bowkeria cymosa*, *Canthium obovatum*, *Duvernoia adhatodoides*, *Mackaya bella*, *Pavetta lanceolata*, *Piper capense*, *Plectranthus* spp., *Sclerochiton haveyanus* and *Vangueria cyanescens*. *Canthium obovatum* is a large tree, frequently in the upper canopy, in the South Cape.

The ground flora is equally variable. Where shade is dense, species such as *Carex spicato-paniculata*, *Clivia caulescens* and *Oplismenus hirtellus* occur. In more open areas the following species can be found:

Alchemilla rehmannii	*Impatiens duthiei*
Asplenium aethiopicum	*I. sylvicola*
A. erectum	*Laportea pedunculata*
A. inaequaelaterale	*Peperomia retusa*
A. rutifolium	*Pseudobromus africanus*
Begonia spp.	*Streptocarpus micranthus*
Cardamine africana	*Thalictrum rhynchocarpum*
Dryopteris inaequalis	*Thelypteris pozoi*
Hypoëstes verticillaris	

Lianes are more conspicuous with more species present than in the South Cape forests. These include the following species:

Behnia reticulata	*Rhoicissus revoilii*
Canthium gueinzii	*R. rhomboideus*
Cnestis natalensis	*R. tomentosa*
Helinus integrifolius	*Riocreuxia torulosa*
Jasminum abyssinicum	*Secamone alpini*
J. fluminense	*S. gerrardi*
Landolphia capensis	*Smilax kraussiana*
Quisqualis parviflora	*Stephania abyssinica*

Epiphytes consist mainly of pteridophytes and bryophytes, but orchids also occur. Common pteridophytes are:

Asplenium anisophyllum	*L. verticillatum*
A. sandersonii	*Pleopeltis macrocarpa*
A. splendens	*Polypodium polypodioides* subsp. *ecklonii*
Elaphoglossum acrostichoides	*Vittaria isoetifolia*
Loxogramme lanceolata	
Lycopodium dacrydioides	

The orchids present include: *Angraecum sacciferum*, *Bulbophyllum sandersonii*, *Mystacidium capense* and *Tridactyle tricuspus*.

Several *Streptocarpus* species can occur as epiphytes but are equally at home on rock ledges.

The Uluguru–Mulanje mountain system

Two forests have been selected from this system: a forest on Mount Mulanje in southern Malawi, and a forest on the Nyika Plateau in northern Malawi. The data for both are taken from Chapman and White (1970).

Chambe Plateau Forest, Mount Mulanje. This forest occurs on fairly level ground on the plateau at an altitude of 1950 m (16°S, 35°40′E). The emergent tree, *Widdringtonia nodiflora*, reaches a height of 27 m, well above the somewhat scattered evergreen broad-leaved canopy, which occurs at about 14 m. Species in this stratum are:

Afrocrania volkensii	*Nuxia congesta*
Aphloia theiformis	*Ochna holstii*
Cassipourea congensis	*Olinia rochetiana*
Cussonia sphaerocephala	*Pittosporum viridiflorum*
Ekebergia capensis	*Podocarpus milanjianus*
Garcinia mlanjiensis	*Polyscias fulva*
Ilex mitis	*Rapanea melanophloeos*
Kiggelaria africana	*Xymalos monospora*

The lower tree stratum is dense with an average height of about 9 m. The following species are present:

Cassine aethiopica	*Nuxia floribunda*
Clausena anisata	*Oxyanthus speciosus*
Diospyros whyteana	*Pterocelastrus tricuspidatus*
Faurea forficuliflora	*Rothmannia fischeri*
Lasianthus kilimandscharicus	*Trichocladus ellipticus*
Maytenus acuminata	

The shrub layer includes:

Canthium captum	*Pavetta* sp.
Carissa sp.	*Piper capense*
Choristylis rhamnoides	*Rhamnus prinoides*
Dracaena laxissima	*Tricalysia acocantheroides*

The ground flora is well developed, ferns being particularly numerous on wetter sites. The following herbaceous species occur:

Brachystephanus africanus	*Panicum monticola*
Carex brassii	*Pilea ceratomera*
Desmodium repandum	*Plectranthus elegans*
Dietes prolongata	*Pseudobromus sylvaticus*
Isachne zethiopica	

Regeneration of tree species is common in the lower strata. Lianes are not common but include *Asparagus plumosus*, *Landolphia* sp. and *Urera hypselodendron*. Epiphytes are abundant particularly on wet sites. The ferns include *Asplenium*, *Elaphoglossum*, *Hymenophyllum*, *Polypodium* and *Trichomanes*. Three club mosses and fourteen orchids have been recorded.

The emergent conifer in this forest, *Widdring-*

tonia nodiflora, is of southern origin. It reaches its most northerly limit on Mount Mulanje and also its greatest development. In South Africa it is known as a small tree, seldom exceeding 8 m tall and associated with fynbos and dry scrub forest.

Nyika Plateau forest. The afromontane forests of the Nyika Plateau are of two types: the *Juniperus procera*-dominated forest and the broad-leaved forest. On the Nyika there appears to be little difference in the moisture status of the two, although this is the major differentiating factor in East Africa. The *Juniperus* forest has virtually disappeared, there being only a few patches remaining, such as Uyaghaya Valley. There is little difference between the two, apart from an upper canopy of *Juniperus* which can be closed and 30 m in height on protected sites. The following description is given for a broad-leaved evergreen forest in the Uyaghaya Valley at an altitude of 2135 m on the open slopes of the high plateau (10°30′S, 33°30′E).

The forest consists of two strata with the upper canopy seldom higher than 15 m. There are no emergents. Species in the upper canopy are:

Afrocrania volkensii	*Nuxia congesta*
Aphloia theiformis	*Olinia rochetiana*
Apodytes dimidiata	*Podocarpus milanjianus*
Cassipourea congensis	*Polyscias fulva*
Cussonia spicata	*Prunus africana*
Ekebergia capensis	*Rapanea melanophloeos*
Ilex mitis	*Xymalos monospora*
Kiggelaria africana	

Olea capensis, which occurs under *Juniperus*, was not found in this forest.

Species in the lower tree stratum are:

Canthium cf. *huillense*	*Galiniera caffeoides*
Cassine aethiopica	*Hagenia abyssinica*
Clausena anisata	*Maytenus acuminata*
Diospyros whyteana	*Peddiea africana*
Dodonea viscosa	*Philippia benguelensis*

This is the most southerly occurrence of *Hagenia abyssinica*, an important high-altitude species in East Africa. Both these strata are fairly dense and continuous; their development can be significantly affected by exposure to wind.

The shrub layer is somewhat open because large antelope graze these forests. The grasses *Panicum monticola* and *Pseudobromus* sp. occur in the openings, while the shrubs are browsed to knee height. *Dicliptera* spp. and other Acanthaceae make up the remainder of the stratum. The lianes *Embelia schimperi* and *Toddalia asiatica* are frequent. Epiphytes consisting of ferns, mosses and orchids are plentiful, lichens are common in the tree crowns.

Imatong–Usambara, Kivu–Ruwenzori mountain systems

The afromontane broad-leaved evergreen forests on the east coast (4°40′S, 38°30′E) have been selected as the main example for these systems. The following information is from Lind and Morrison (1974). These forests occur at an altitude of 1700 to 2400 m in a very wet climate, the annual rainfall exceeding 2200 mm. Pitt-Schenkel has described eight forest types in this forest (Lind and Morrison, 1974), the most important being the *Ocotea* forest. Structurally there are four strata: an upper and a lower tree stratum, a shrub stratum and the ground or herb stratum.

Species in the upper canopy which may be up to 45 m tall are:

Aningeria adolfi-friederici	*Ilex mitis*
Balthasaria schliebernii	*Mammea africana*
Casearia battiscombei	*Ochna holstii*
C. sp. cf. *engleri*	*Ocotea kenyensis*
Chrysophyllum albidum	*O. usambarensis*
C. gorungosanum	*Podocarpus usambarensis*
Dombeya leucoderma	*Polyscias kikuyuensis*
Ekebergia capensis	*Prunus africana*
Entandophragma excelsum	*Rapanea rhododendroides*
Ficalhoa laurifolia	*R. usambarensis*

The lower tree stratum varies from 9 to 21 m in height and contains the following species:

Afrocrania volkensii	*Macaranga kilimandscharica*
Albizia gummifera	*Myrica salicifolia*
Allophylus abyssinicus	*Neoboutonia macrocalyx*
Aphloia theiformis	*Ochna stuhlmannii*
Apodytes dimidiata	*Olea hochstetteri*
Cassipourea malosana	*Pauridiantha holstii*
Craibia brownii	*Podocarpus gracilior*
Croton machrostachys	*Strombosia scheffleri*
Cussonia spicata	*Syzygium guineense*
Dasylepis leptophylla	*Trichocladus ellipticus*
Dracaena steudneri	*Tabernaemontana holstii*
Faurea saligna	*Urophyllum holstii*

Species in the shrub layer, some of which are better known as small trees, are:

Alsophila sp. (syn. *Cyathea*)	
Catha edulis	*Myrsine africana*
Clausena anisata	*M. deminutum*
Croton sp. cf. *scheffleri*	*Pavetta hymenophylla*
Dodonea viscosa	*Piper capense*
Dracaena deremensis	*Pittosporum lanatum*
Ensete ventricosum	*Psychotria riparia*
Lasianthus kilimandscharicus	*Syzygium cordatum*
Memecylon cogniauxii	*Teclea nobilis*
	Xymalos monospora

The ground in this forest is devoid of vegetation, most likely because of the density of the shrub layer, but in other areas species of Acanthaceae and *Impatiens* are plentiful.

Lianes are not common, but epiphytes are numerous on boles, branches and in the crowns of the trees. Mosses, ferns, lichens and orchids form the bulk of this population, but fleshy herbs such as *Peperomia* spp. and *Canarina* spp. also occur.

Forests, similar in both floristics and structure occur on all the higher mountains of East Africa (Jackson, 1956; Ross, 1956; Lind and Morrison, 1974): on Mounts Meru and Kilimanjaro in Tanzania; on the Aberdares, Mounts Kenya and Elgon in Kenya; on Mounts Elgon, Ruwenzori and Virunga and the Imatong Mountains in Uganda. On slightly drier sites (rainfall > 1400 mm as opposed to > 2000 mm), the conifers *Juniperus* and *Podocarpus* become more numerous and frequently dominant. The understorey is still broad-leaved and evergreen. The species present include *Buddleja polystacha*, *Diospyros abyssinica*, *Dombeya* sp., *Euclea microcarpa*, *Halleria lucida*, *Olea africana*, *O. welwitschii*, *Peddiea fischeri* and *Teclea simplicifolia*.

Above an altitude of about 2400 m, a thicket of mountain bamboo (*Arundinaria alpina*) replaces the forest on all the highest mountains, except Kilimanjaro. The altitude of the lower and upper limits of this belt varies from mountain to mountain. The bamboo appears to be a climax consociation although there is evidence that it can be seral to broad-leaved evergreen forest, where the latter is destroyed either by fire or cultivation (Kerfoot, 1964).

Above the bamboo zone the forest has only three strata. The upper tree stratum is usually dominated by *Hagenia abyssinica* but with *Afrocrania volkensii*, *Dombeya goetzenii*, *Hypericum lanceolatum*, *Prunus africana* and *Rapanea rhododendroides* locally abundant. The canopy is open, and occasionally 18 m high but more often less than 10 m. The shrub stratum is rather poor in species, but *Lobelia gibberoa*, *Sambucus africana* and *Senecio amblyphyllus* are common. The ground flora is exceptionally rich in species; Lind and Morrison (1974) listed 30 common species from Hagenia Forest on Mount Elgon. Epiphytes occupy every available space, often forming dense cushions of moss on the trunks and branches. *Usnea*-like lichens are plentiful in the tree crowns (Jackson, 1956).

The Ethiopian mountain system

Although floristically poorer than the East African afromontane forests, those of Ethiopia also have two tree strata. In the upper canopy species such as *Aningeria adolfi-friederici*, *Ekebergia ruppeliana*, *Prunus africana* and *Syzygium guineense* occur, while *Galiniera caffeoides* and *Teclea nobilis* are common in the lower canopy. On the drier sites, as in East Africa, the conifers *Podocarpus* sp. and *Juniperus procera* become the dominant trees. Lianes are rare but epiphytes are abundant particularly on wetter sites and on the higher areas where *Erica arborea*, *Hagenia abyssinica* and *Hypericum lanceolatum* are the dominant trees. Tree ferns (*Alsophila* sp.) occur in the wettest areas of these forests (Beals, 1968).

West Africa

The afromontane forests of West Africa are floristically impoverished compared to those of East Africa. Structurally they are similar, having four strata, an upper and a lower tree stratum, a shrub stratum and a ground layer. Richards (1963) studied the afromontane forests at Mann's Spring on the southwestern face of Mount Cameroun between 1800 m and 2256 m. The following information is taken from his report.

The forests form a nearly continuous belt in this area. They are open and irregular, and the trees, although tall, are short trunked with wide crowns. They are not commercially attractive. The upper canopy, which can attain a height of 37 to 40 m, consists mainly of *Syzygium staudtii* with *Schefflera abyssinica* and *S. mannii*. The lower tree layer can reach 18 m and consists of *Allophylus bullatus*, *Crassocephalum mannii*, *Discopodium penninervium*, *Pittosporum mannii*, *Prunus africana* and *Canthium glabriflorum*. This last also occurs in the rain forest at lower altitudes, being one of the few species so to overlap.

In the shrub layer *Maesa lanceolata*, *Mimulopsis solmsii*, *Oreacanthus mannii*, *Pavetta hookeriana*, *Piper capense* and *Solanum mauritianum* occur. The tree fern *Alsophila manniana* is abundant below 1800 m. Woody lianes such as *Oreosyce africana* and *Stephania abyssinica* occur, but are rare. Epiphytes are abundant, particularly the bryophytes *Lepidopilum* and *Plagiochila* on the trunks and branches of the crowns. *Usnea* sp. is common in the upper crowns. The abundance of epiphytes becomes very striking in the highest forest around 2550 m. Johannson (1974) has done a detailed study of the epiphytes of the West African rain forests including the lower afromontane forests.

The ground layer consists of ferns, grasses, herbs and mosses; Richards (1963) listed 28 species from a first sample plot and an additional three from a second. Stinging Urticaceae are frequently a feature of this layer.

Faunal component

Mammalia

Although Africa as a whole has a large mammalian fauna, the populations in afromontane forest are not large, either by species or numbers (Bigalke, 1978). Table 8.6 lists the major orders, families and genera which occur in the afromontane forests. This Table and the following notes are largely taken from Bigalke.

The golden moles *Amblysomus* and *Chrysochloris* both have forest species, the former occurring in high-rainfall forests of South Africa and the latter further north; *C. stuhlmanii* occurs in the afromontane forests of Mount Elgon and Ruwenzori. The golden moles burrow in the soil using strongly developed fore-claws. They feed on insects and worms, but are not common. Of the forest shrews, *Myosorex varius* occurs in the South Cape forests. *M. geata* occurs up to 2300 m on the Uluguru Mountains of Tanzania.

Elephant shrews (*Rhynchocyon*) occur in moist forest where an adequate organic layer has developed. *R. cirnei* is found north of the Zambezi and can possibly extend its range to afromontane forest. Elephant shrews occur in forest in western Uganda but have not negotiated the Nile into Kenya (Kingdon, 1971).

TABLE 8.6

Orders, families and genera of mammals occurring in afromontane forest of Africa south of the Sahara

Order / Family / Genus	Common name	Number of species in forest
Insectivora		
Chrysochloridae		
Amblysomus	golden mole	3
Chrysochloris	golden mole	1
Soricidae		
Myosorex	forest shrew	3
Macroscelididae		
Rhynchocyon	elephant shrew	1
Chiroptera		
Pteropidae		
Eidolon	bat	1
Epomophorus	bat	1
Epomops	bat	2
Rousettus	bat	1
Primates		
Pongidae		
Gorilla	gorilla	1
Galagidae		
Galago	galago	3
Cercopithecidae		
Cercopithecus	monkey	2
Papio	baboon	2
Rodentia		
Sciuridae		
Funisciurus	striped squirrel	2
Heliosciurus	sun squirrel	2
Muridae		
Beamys	pouched rat	1
Hybomys	forest mouse	1
Lophuromys	harsh-furred mouse	2
Malacomys	long-footed rat	1
Praomys	wood mouse	3
Thamnomys	forest mouse	1
Cricetidae		
Dendromus	climbing mouse	1
Otomys	vlei rat	1
Muscardinidae		
Graphiurus	dormouse	1
Hystricidae		
Hystrix	porcupine	1
Carnivora		
Mustelidae		
Mellivora	badger	1
Viverridae		
Herpestes	mongoose	2
Nandinia	civet	1
Felidae		
Felis	wild cat	2
Panthera	leopard	1

(*continued*)

TABLE 8.6 *(continued)*

Orders, families and genera of mammals occurring in afromontane forest of Africa south of the Sahara

Order Family Genus	Common name	Number of species in forest
Proboscidea		
Elephantidae		
Loxodonta	elephant	1
Hyracoidea		
Procaviidae		
Dendrohyrax	tree dassie	3
Procavia	dassie	1
Artiodactyla		
Suidae		
Potamochoerus	bush pig	1
Bovidae		
Cephalophus	duiker	5
Tragelaphus	bush buck	2
Taurotragus	eland	1

Bats of four genera occur in forest areas. *Epomophorus wahlbergi* is widespread. *Eidolon helvum* is common in the north but rare in South Africa. *Epomops* ranges from Angola to Zambia but is largely a lowland forest species. *Rousettus aegyptiacus* has a very wide distribution occurring virtually from the Mediterranean to the Cape. It prefers forest areas but is not restricted to this habitat.

The primates are commoner in savanna and tropical lowland forest than in the afromontane forests; only three genera are consistently found in them. The mountain variety of *Gorilla gorilla*, which occurs on the Zaïre/Uganda boundary, particularly on Mounts Ruwenzori and Virunga, is the largest of the three. This is the only remaining habitat for the gorilla in East Africa.

Galagos (*Galago* spp.) occur in afromontane forests of East and West Africa. They are small, agile, furry animals with long hind-legs and bushy tails. Being nocturnal they are difficult to study. *Galago crassicaudatus* is widely distributed, particularly in East Africa on the Uluguru Mountains and Kilimanjaro. *G. demidovii* occurs in Uganda and in West Africa in Cameroon, Sierra Leone and Liberia. *G. zanzibaricus* is found in the Usambara Mountains of Tanzania. It also occurs in lowland forest and along the coast (Kingdon, 1971).

Cercopithecus pygerythrus, the vervet monkey, is widely distributed in afromontane broad-leaved evergreen forests in South Africa, up the east coast to Kenya and also in West Africa. It is not confined to this habitat, being a particularly adaptable animal. Other *Cercopithecus* species appear mainly to be confined to lowland forest, but *C. albogularis* occurs in the broad-leaved evergreen forests of Malawi, Uganda and parts of South Africa.

Baboons (*Papio* spp.) prefer more open woodland than afromontane forest but occasionally seek small forest remnants in which to sleep. Some troops of baboons live successfully in forest (Kingdon, 1971).

Squirrels, rats and mice have several genera which occur in afromontane forests. *Heliosciurus rufobrachium* occurs in the eastern highlands of Zimbabwe. Two species, *Lophuromys flavopunctatus* and *Praomys jacksonii*, occur in montane forest in Zambia. The afromontane forests of eastern Zaïre and Uganda, particularly Mounts Kigezi and Ruwenzori, have several squirrels and rats found only there: *Funisciurus carruthersi*, *Heliosciurus ruwenzorii* and *Lophuromys woosnami*, for example (Kingdon, 1971). *Praomys verreauxi* and *Thamnomys dolichurus* occur in the South Cape forests (Von Breitenbach, 1974). A species of *Beamys* has been reported in the forests of Malawi. The most widespread of the forest mice are *Hybomys univittatus*, *Malacomys longipes* and *Praomys stella*. The population density of forest rodents is particularly low. Fleming (1975, in Bigalke, 1978) estimated it to be between 3 and 25 ha^{-1}, much less than in grassland or open savanna.

Mellivora capensis, the badger, occurs in broad-leaved evergreen forest in addition to many other habitats. It is a very adaptable and virtually omnivorous animal, occurring throughout the continent. The mongooses, *Herpestes sanguineus* and *H. pulverulentus*, are equally adapted to different habitats including forest.

The palm civet (*Nandinia binotata*) is mainly a lowland forest species, but a relict population occurs in afromontane forest in Zambia.

Several species of cats occur in afromontane forest. *Felis libyca*, the wild cat, is widespread while *F. serval*, the serval cat, is only occasional in broad-leaved evergreen forest, being more at home in moist savanna. Leopard (*Panthera pardus*) is the biggest of the surviving cats to occur in afromontane forest. Nowhere common, it is still widespread wherever duiker and dassie are plentiful (Kingdon, 1971).

Elephants exert a considerable destructive pressure on afromontane forests, even when in small numbers. In the South Cape where the population has variously been estimated as between seven and twelve animals on an area of approximately 2500 hectares (Von Breitenbach, 1974), severe damage has been done to the research plots at Deepwalls Forest, particularly to regeneration and pole stage trees. Elephants are widely distributed in Africa, living mostly in savanna and lowland forest and visiting the afromontane forests during seasonal migrations, even though altitudinal differences are often great (Chapman and White, 1970; Bigalke, 1978). There is a subspecies which lives in forest, *Loxodonta africana cyclotis*.

The Cape dassie (*Procavia capensis*) is not a forest dweller, but because the afromontane forests frequently include rocky terrain these dassies are often associated with them. Several species of tree dassie (*Dendrohyrax*) occur in afromontane forests. *D. validus* varieties are found on the Usambaras, on Mount Meru and on Kilimanjaro in Tanzania, while *D. arboreus* occurs on Ruwenzori and other mountains in Kenya and Uganda and as far south as the eastern Cape Province. *D. dorsalis* occurs in Uganda, Cameroon and Sierra Leone (Kingdon, 1971; Bigalke, 1978). Bush pigs (*Potamochoerus porcus*) are widely distributed in afromontane forest throughout the continent. They are nocturnal and omnivorous (Bigalke, 1978).

The antelopes are represented in the afromontane forests by *Cephalophus* and *Tragelaphus*. *Cephalophus natalensis*, the red duiker, occurs in afromontane forest from Somalia to Natal. *C. monticola* occurs in West Africa and also in the South Cape forests. *C. sylvicultor* is found in Kenya and Zambia. All the duikers are browsers and grass is not normally part of their diet. They are considered to be part of a relict forest population (Bigalke, 1978).

Tragelaphus scriptus, bush buck, is widely distributed in afromontane forest from the South Cape, the eastern Cape, Natal and Transvaal Drakensberg, Zimbabwe, Malawi, to East Africa and West Africa. It is not confined to afromontane forest; indeed, any forest which is thick and with adequate water makes a suitable habitat. The animal is thought to have an altitudinal limit of 4000 m, well within the afromontane forest range (Bigalke, 1978).

Large antelope such as kudu (*Tragelaphus strepsiceros*) and eland (*Taurotragus oryx*) although not normally inhabitants of afromontane forest, have been known to browse in small relict patches (Chapman and White, 1970).

Avifauna

Moreau (1966) has given an excellent detailed description of the African avifauna. The birds of the afromontane forests in Africa south of the Sahara are a distinct, largely relict population (Winterbottom, 1978); few species are of lowland origin.

The following details of the bird population of the South Cape high forests are from Phillips (1931a, b), McLachlan and Liversidge (1970), and Von Breitenbach (1974). Species occurring throughout the afromontane archipelago are marked with an asterisk.

Water birds. The black duck (*Anas sparsa**) occurs in forest streams, otherwise water birds are unknown in the forests.

Raptors. *Aviceda cuculoides**, the cuckoo falcon, *Hieraaetus spilocaster*, the hawk eagle, *Stephanoaetus coronatus**, the crowned eagle, *Buteo buteo trizonatus**, mountain buzzard, *Accipiter rufiventris**, the red-breasted sparrow-hawk, *A. melanoleucus**, the black goshawk, and *A. tachiro**, the African goshawk, are the forest predators.

Game birds and other ground birds. *Pternistis afer notatus*, the red-necked francolin, is a bird of the forest margin rather than the interior. *Sarothrura elegans**, the buff-spotted fluff-tail, occurs in the thick forest.

Non-passerines. Two members of the dove family live in the forest: *Columba arquatrix**, the rameron pigeon, and *Aplopelia larvata*, the cinnamon dove.

*Tauraco corythaix**, the Knysna loerie, is probably the best known forest bird, because of its distinctive noisy call.

Two cuckoos are found in the forests during the summer months: *Cuculus solitarius**, the red-chested cuckoo, and *Chrysococcyx cupreus**, the emerald cuckoo. The former is apparently equally at home in exotic plantations.

Two owls, *Ciccaba woodfordii**, the wood owl, and *Bubo africanus**, the spotted eagle owl, occur in the forests.

Tockus alboterminatus, the crowned hornbill, is a noisy bird which has been reported in the forests east of Knysna, but is commoner at lower latitudes.

Two woodpeckers, *Campethera notata*, the Knysna woodpecker, and *Mesopicos griseocephalus*, the olive woodpecker, are found in the forest.

Only one cuckoo-shrike, the grey cuckoo-shrike (*Coracina caesia**), inhabits the forests. The black-headed oriole (*Oriolus larvatus*) lives in the canopy of the high forest. Its melodic whistle is heard more than the bird is seen.

Passerines. Two bulbuls occur in the forest but are commoner along the margins and more open scrub: *Phyllastrephus terrestris**, the terrestrial bulbul, and *Andropadus importunus*, the sombre bulbul.

The thrush family has four representatives: *Turdus olivaceus*, the olive thrush, *Cossypha dichroa*, the chorister robin, *C. caffra**, the Cape robin, and *Pogonocichla stellata*, the starred robin. The Cape robin is not common in high forest but has been recorded in the dense shrub layer of the medium-moist forest. In South Africa the starred robin moves from the high-altitude forests into the lower lying valleys in the winter.

The warblers (Sylviidae) are found in scrub and open forest rather than high forest.

Several fly-catchers are forest inhabitants, for example, *Muscicapa adusta**, *Batis capensis*, *Trochocecus cyanomelas* and *Terpsiphone viridis**. The last is migratory in the South Cape forests, occurring only in the summer months. It is resident in Natal and Transvaal.

Chlorophoneus olivaceus, the olive bush-shrike, occurs in high forest in the South Cape but is frequently found in drier shrub on the east coast. Many other birds are not true forest dwellers but visit the forest at certain seasons to feed on specific fruits, such as *Onychognathus morio*, the red-winged starling. Others which live in scrub and forest margins are occasionally found in the high forest, such as *Zosterops pallidus virens*, the Cape white-eye.

Two forest species which do not occur in the South Cape forests are found in the afromontane forests of eastern Cape and further north. They are *Poicephalus robustus**, the brown-necked parrot, and *Lioptilus nigricapilus*, the bush black-cap. One species, *Prinia robertsi*, the forest prinia, is confined to afromontane forest edges and high-altitude scrub in Zimbabwe.

Of the 35 species reported as resident in South African forests only 3 are confined within South Africa's borders. The Knysna woodpecker is found from the South Cape to Natal; the bush black-cap and the chorister robin reach the Soutpansberg in the Transvaal. The remainder occur in afromontane forest from the South Cape, up the east coast mountains to Kenya and even Somalia. Twenty-two of these (marked by an asterisk in previous paragraphs) occur in afromontane forest of West Africa in Cameroon and as far west as Sierra Leone.

Herpetofauna

The herpetofauna of Africa has been reasonably well sampled, but a great deal of work must still be done. Poynton and Broadley (1978) have summarized current knowledge.

There are few amphibians and even fewer reptiles which live permanently in the afromontane forests. The most striking feature of the forest anurans is that they invariably lay their eggs out of the water, for example, *Arthroleptella* in wet vegetation, *Leptopelis* in moist humus and *Breviceps* even in the soil. The tadpoles then find their way to water or develop within the egg, never entering water at all. One, *Bufo anotis*, lays its eggs in small water-filled cavities in the roots of large trees.

The following genera are considered to be forest amphibians: *Anhydrophryne*, *Arthroleptella*, *Arthroleptis*, *Breviceps*, *Heleophryne*, *Leptopelis*, *Natalobatrachus* and *Probreviceps*. Most of these, apart from species of *Leptopelis* which are arboreal, live on or in the ground.

The few forest reptiles can be split into those which live in the trees and those which live in or on the ground. Geckos (*Lygodactylus* spp.) and lizards (*Mabuya* spp.) live on tree trunks while chameleons (*Chamaeleo* spp.) and snakes inhabit the upper crown. Tree snakes include such species as *Atheris nitschei rungweensis* and *Dendroaspis jamesoni jamesoni*.

Ground reptiles apart from a limbless skink, *Melanoseps* sp., are all snakes. These include *Bitis gabonica gabonica*, *Naja melanoleuca*, *Natriciteres variegata sylvatica* and *Polemon christyi*. Burrowing snakes of the genera *Leptotyphlops* and *Typhlops* are common in the Knysna forests. Most of the snakes which occur in the South Cape high forests are equally at home in the fynbos or the forest margin, for example *Dispholidus typus*, the tree snake, and *Philothamnus hoplogaster*, the green water snake.

The forest microfauna is not considered here mainly because information is limited and neither author is competent to make a contribution. The available data have been ably summarized in a recent publication (Werger, 1978).

Pests and diseases

Very little information on the pests and diseases of the afromontane forests exist, and what there is, is largely South African. The following information refers to the South Cape forests and is taken from Phillips (1931a, b), Laughton (1937), Doidge (1950), and the personal experience of the senior author in these forests from 1954 to 1959.

There are few serious insect pests in the South Cape forests. Stem borers such as *Promises longicepes* (Cerambycidae) attack seedling and coppice shoots of many species, particularly *Ocotea bullata*. Aphididae and Psyllidae attack the flowers of many species, such as *Apodytes dimidiata*, *Olea capensis* subsp. *macrocarpa* and *Olinia ventosa*. Larvae of the moth *Leto venus* are stem borers on *Virgilia oroboides*, a pioneer species of forest margins and in scrub. When trees are felled, bark beetles (Scolytidae) attack the stems in large numbers, but no evidence of these insects affecting healthy trees has been seen.

Fungal attacks are more widespread but being part of the forest dynamics do not appear to create serious problems. The only serious parasitic fungus is *Ganoderma applanatum* which causes heart-rot and will kill the tree. It attacks most species but is particularly troublesome on *Olea capensis* subsp. *macrocarpa*. *Ganoderma oroflavum* has been reported on *Curtisia dentata* and *Podocarpus* spp., while *Fomes geotropus*, *F. hornodermis*, *F. robinsoniae* are recorded from *Apodytes dimidiata*, *Ekebergia capensis*, *Gonioma kamassi*, *Ocotea bullata*, *Olea* spp. and *Platylophus trifoliatus*.

Leaf fungi are numerous but apart from killing the attacked leaves, and often only parts thereof, they make little impression. Complete defoliation is unknown. A list of some of the commoner fungi, mostly Ascomycetes, follows:

Asterina africana on *Kiggelaria africana*
A. hendersoni on *Ilex mitis*
A. reticulata on *Olinia ventosa*
Asterodothis solaris on *Olea* spp., particularly
O. capensis subsp. *macrocarpa*
Coryneliospora fructicola on *Rapanea* and *Cassine* spp.
Irene nuxiae on *Nuxia floribunda*
Lembosia phillipsii on *Ocotea bullata*
Meliola comata on *Olea capensis* subsp. *macrocarpa*
M. ganglifera on *Curtisia dentata*
M. oleicola on *Olea* spp.
Parenglerula macowaniana on *Cassine* and *Maytenus* spp.
Schiffnerula radians on *Cunonia capensis*.

Seedlings are much more susceptible to these leaf fungi than are mature trees. Occasionally *Pestalotia ocoteae* destroys flowers and immature fruits of *Ocotea bullata*. Damping-off fungi kill large numbers of seedlings of all species. *Fusarium* spp., *Pythium* spp. and *Rhizoctonia solani* are probably the commonest damping-off fungi in the South Cape forests.

FOREST FUNCTIONAL PROCESSES

Organic production, decomposition and nutrient circulation

Very little information is available about the flow of organic matter and nutrients through forest ecosystems. In Kenya Lind and Morrison (1974) found that within the afromontane areas increasing rainfall is more favourable to vegetative growth and organic matter production than increasing temperature. The favourable effect of temperature on plant growth and organic production is counter-balanced by increasing microbial activity resulting in organic matter destruction. It was estimated that, for a 300 m increase in altitude, the organic matter in the top soil of the afromontane forest increased by 0.08%. In January 1977 the authors of this chapter initiated a three-year study of litter fall in the South Cape forests. Litter is collected monthly from four traps in each of four forest types — that is, dry, medium-moist, moist and wet. The results show a relatively constant litter fall amongst the forest types — that is, 4054 ± 46.0 kg ha^{-1} yr^{-1}. However, variation within each of the forest types was relatively large. Considerable monthly variations occur with a peak in June (639 kg ha^{-1}), followed by two smaller peaks in August (486 kg ha^{-1}) and November–December (423 kg ha^{-1} $month^{-1}$).

Litter traps were also erected underneath a mature *Podocarpus falcatus* and in a dry forest type where the undergrowth had been removed. At these sites the annual litter fall was 5899 ± 548.0 kg ha^{-1} and 3747 ± 375.1 kg ha^{-1} respectively.

The annual average percentage of nitrogen in the litter of the dry forest type was 1.10%, with only small monthly variations. In the medium-moist, moist and wet forest types, the mean annual nitrogen percentages were 1.27, 1.32 and 1.32% respectively. For these three sites the monthly mean ranged from 1.07% (December) to 1.75% (October), averaging 1.50% for the period July to October. The nitrogen content of the litter collected below the *Podocarpus falcatus* canopy had a mean of 1.36% and ranged from 1.06% (April) to 1.83% (July), averaging 1.61% for the period July to October. In the dry forest type where undergrowth had been removed, the mean was 1.29%, with little monthly variation apart from a peak in September (1.83%) and October (1.87%).

Water circulation

The continuous cover of afromontane forest regulates streamflow very effectively. Dagg and Pratt (1962) found that at Sambret (Kenya) stormflow amounted to only 1.3% of the rainfall. The most severe storm recorded — 75 mm in 50 min — produced a run-off of only 2.5%. The Penman estimate of potential open water evaporation is approximately 1500 mm yr^{-1}, and from December to February it exceeds the average rainfall (McCulloch, 1962). However, Kerfoot (1962) found that trees exploited the porous volcanic soil (at Sambret) to depths of 9 m, and it seems that even during the dry season water is fully available. Since the average annual rainfall on the studied catchment varies from 1750 to 2000 mm, the annual surplus water is 250 to 500 mm.

In a similar study carried out by Pereira et al. (1962) at Mbeya Peak (Tanzania), exact figures were not obtained. The forest showed no sign of water-stress in the late dry season and seemed to draw on deep water storage. The streamflow increased with an increase in total storm rainfall.

Quantitative hydrological research in South Africa has mainly been confined to fynbos and grassland catchment areas. Phillips (1931a, b) refers to the relatively low rate of evapotranspiration in the South Cape forests as a result of the relatively low temperatures and high air humidity. However, during a berg wind there is an appreciable increase due to increased temperature, decreased humidity and movement of air.

PAST, PRESENT AND POTENTIAL USE

Wood production

Until a few decades ago most afromontane forests were a source of timber for general construction purposes, household implements and furniture, as well as minor forest products such as firewood and poles. The South Cape forests supplied material for railway sleepers, wagon-, house- and ship-building. Because these forests were over-exploited and the relicts are now protected to a large extent, wood production has decreased considerably.

In the South Cape the medium-moist and moist forest types (approx. 50% of the indigenous forest area) today produce a large variety of high-grade timber for special purposes, such as prestige furniture and cabinet work, face veneers, panelling, flooring and curios. The bulk of the most valuable timber is of the following fourteen species which are found in climax high forest associations:

Acacia melanoxylon (exotic)
Apodytes dimidiata
Canthium obovatum
Cunonia capensis
Curtisia dentata
Ekebergia capensis
Ocotea bullata
Olea capensis subsp. *macrocarpa*
Olinia ventosa
Platylophus trifoliatus
Podocarpus falcatus
P. latifolius
Pterocelastrus tricuspidatus
Rapanea melanophloeos

In the medium-moist and moist forest types the current annual volume increment culminates with values of 3 and 6 m^3 ha^{-1} respectively, when the basal area is approximately 23 m^2 ha^{-1}. In both forest types the mean annual increment is greatest at a basal area of about 32 m^2 ha^{-1}. On reaching the climax stage the basal area is about 46 m^2 ha^{-1} in both types, the average stems ha^{-1} are 800 and 640, and the wood volume *c.* 250 and 280 m^3 ha^{-1} in the medium-moist and moist types respectively.

In order to allow the forest to recover after the severe exploitation which has taken place, a transition management system is followed. The wood removed is less than the increment and the less valuable species are removed rather than the more valuable species which previously were over-exploited. The initial annual removals average approximately 1.4 m^3 ha^{-1}, of which an estimated 10% is utilizable. The yields are expected to increase gradually until the year 2000 when they should reach a sustained maximum of 2.8 m^3 ha^{-1} yr^{-1} or 56 000 m^3 yr^{-1} for the total 20 000 ha of productive forest — that is, being equal to the increment. At least 75% of these removals should be utilizable. In addition timber yields are also obtained from the occasional removal of over-mature trees from protection forests (Von Breitenbach, 1968).

By the year 2050 the species composition of the forest should reach its optimum. Von Breitenbach expects the annual utilizable yields of the major species to be approximately as follows:

Olea capensis subsp. *macrocarpa*	10 600 m^3
Ocotea bullata	6400
Podocarpus latifolius	4250
P. falcatus	4250
Apodytes dimidiata	4250
Rapanea melanophloeos	4250

Assessments were carried out over a period of twenty years in five permanent sample plots in compartment A11, Deepwalls, South Cape. Although no records exist, some timber extraction may have taken place in this area many years before the assessments were initiated. The species composition is given in Table 8.5. The current annual increment of these plots increased with an increase in soil moisture from 1.18 m^3 ha^{-1} to 2.80 m^3 ha^{-1}, averaging 1.67 m^3 ha^{-1}.

Moll and Woods (1971) studied the rate of growth of five canopy species at Xumeni, Natal. During a period of 37 years the current annual increment of diameter at breast height averaged 0.16 $\pm$ 0.01 cm for all the species. *Podocarpus henkelii* had the lowest increment rate, that of *Xymalos monospora* was slightly less than the average, while *Podocarpus falcatus* and *Kiggelaria africana* slightly exceeded the average value. *Zanthoxylum davyi* (syn. *Fagara davyi*) had the biggest increment of the species studied.

In Tanzania *Ocotea usambarensis* can be established by means of the *taungya*[1] scheme. On the best sites a rotation length of 60 years and a final stocking of 120 stems ha^{-1} is recommended. It has been estimated that a gross volume of 105 m^3 ha^{-1} from later thinnings plus 175 m^3 ha^{-1} from clear-fellings, that is, a mean annual increment of 4.7 m^3 ha^{-1}, should be feasible (Willan, 1965).

Reliable information on wood production in broad-leaved afromontane forests in other countries is not available.

Other products

Grazing by domestic stock occurs in afromontane forests in some countries, for example Ethiopia and Somalia. Since it is generally not properly controlled, it often leads to the deterioration of the forests (Beals, 1968; Hemming, 1968).

Game is attracted to forests because of the shelter provided. In this respect afromontane forests serve a useful purpose. Had it not been for the South Cape forests, the Knysna elephants would no longer exist (Von Breitenbach, 1968).

Many afromontane forests are protected to ensure the supply of water to lower-lying areas, such as the Aberdare Mountain range which is one of Kenya's largest areas of heavy and reliable rainfall (Pereira, 1962b).

Firewood, charcoal and poles are essential forest products in all African countries. Although not as heavily exploited as the lowland rain and savanna forests, the afromontane forests also contribute to these minor forest products in most countries. However, no details are available of the volumes or revenue obtained from such products originating from afromontane forests (Persson, 1975).

Firewood is the most important minor forest product in the South Cape. During the transition management period the volume of non-utilizable

[1] *Taungya* involves the raising of a forest crop in conjunction with a temporary agricultural crop.

removals could build up to an annual total of approximately 25 500 m^3. It will decrease slowly to an expected annual minimum of about 14 000 m^3. Firewood production, however, depends entirely on the demand, which at present is considerably less than the previous figures. Poles and droppers are occasionally prepared. Other minor products, such as charcoal and *Gonioma kamassi* bark for alkaloid extraction, have received minimal attention (Von Breitenbach, 1968).

Aesthetics

Some of Africa's most beautiful scenery is found within and in the vicinity of afromontane forests. Following past large-scale destruction their aesthetics are now partly conserved and improved in some countries by adequate protection. The removal of exotics, which have invaded some of the forests, is practised, for example *Acacia melanoxylon* in the South Cape.

Recreation

The beauty, undisturbed peace, abundant plant and animal life of forests attract people. The tourist trade has been promoted in countries such as South Africa, Zimbabwe, Malawi and Kenya by the creation and maintenance of picnic and scenic spots, camping sites, and forest drives and walks in afromontane forest areas.

In the South Cape forests the area set aside for recreation management totals approximately 2000 ha or 5% of the forest (Von Breitenbach, 1968). The increase in number of visitors is regarded as very satisfactory (Department of Forestry, 1976).

MANAGEMENT

Forest legislation and administration

Apart from South Africa no information is available on the present legislation and administration of forests in African countries.

The South African Department of Forestry controls approximately 188 000 ha of indigenous forest. The biggest area of high forest occurs in the South Cape. Registered woodcutters extracted limited quotas of wood from these forests until 1938 when they were de-registered and compensated. From then until 1968 the forests were protected to allow them to recover from the over-exploitation which had taken place — that is, only dead and dying trees were removed and sold. Since 1968 management is based on sound silvicultural principles with the object of ensuring optimum perpetual yields and the highest possible profits, but taking into consideration their protective, scientific and aesthetical values (Garnett, 1973).

Legislation exists for the expropriation of land for forestry purposes and for protection of forests and trees on private land. Provision is made for compensation. The Minister may set aside any State Forest or portion thereof as nature reserves, protection forests or wilderness areas in order to preserve forests and secure their direct and indirect uses (Republic of South Africa, 1968).

A section for the management of indigenous forests was established within the Department of Forestry and appreciable progress has been made in the South Cape forests. It is intended to expand the activities of the section to other indigenous forests (Garnett, 1973).

The indigenous forests of the South Cape area are administered as four management groups, each consisting of the indigenous forests within several forest reserves. Large units are preferred because a better balance of work is achieved and it warrants a specialized operational centre. The indigenous forest area within a particular forest reserve is divided into blocks which are subdivided into compartments, varying from 1 to 80 ha with the optimum size being 4 to 40 ha. Compartments are allotted to specific management classes which are described on p. 161 (Von Breitenbach, 1968).

Planted and non-planted forests

Apart from South Africa the most recent data available on plantation and indigenous forest areas for other countries are given by Persson (1975). The data for most of the countries are for 1970 and 1971. Since then the plantation areas of many countries have increased considerably and the indigenous forest areas of some countries have decreased. Furthermore, these areas are given for each country as a whole, so that planted and non-planted forest areas within the afromontane regions are not available.

Information about South Africa's present commercial timber resources is given in the report of the Department of Forestry (1978) in which plantation areas are shown by species, magisterial districts and drainage areas. However, the districts and drainage areas are relatively large, with the result that wide climatic ranges exist within them. Further subdivision, which is not available, would be required to determine the plantation area within South Africa's montane region.

The major species planted in montane regions of Africa are: *Eucalyptus grandis*, *Acacia mearnsii*, *Pinus patula*, *P. elliottii*, *P. radiata* and *Cupressus lusitanica* (mainly in East Africa) (Persson, 1975). Extensive provenance trials with cold-resistant eucalypts are being conducted in the highveld areas of Transvaal and Natal. The species are:

Eucalyptus bicostata (10)
E. bridgesiana (4)
E. camaldulensis (12)
E. dalrympleana (8)
E. delegatensis (13)
E. elata (3)
E. fastigata (11)
E. globulus (8)
E. macarthurii (4)
E. maidenii (7)
E. melliodora (5)
E. nitens (14)
E. regnans (10)
E. tereticornis (7)
E. viminalis (24)

The figures in parentheses indicate the number of provenances under test.

Management systems and objectives

Most afromontane forests are managed for a multiplicity of purposes, for example, protection of fauna, flora, soil and water sources; production of wood and minor forest products; and recreation. Therefore, multiple-use management systems have to be followed. The South Cape forests are divided into five management classes, viz. production, protection, reconstruction, recreation, and research.

The production class is managed on a silvicultural selection system, combining forest rehabilitation and timber production. The object is to produce high-grade timber under near-natural all-aged forest conditions. Periodical operations, such as removals, plantings and tending on a ten-year working cycle, aim at a basal area-controlled normality of all-aged stocking.

The protection class includes all dry and wet type forests not classed under research, reconstruction or recreation. These forest types are unstable — that is, the dry becomes drier and the wet becomes marshy when unduly disturbed.

Objectives of the reconstruction class are to rehabilitate destroyed forest areas by introducing pioneer and climax species, and to eliminate alien vegetation. These areas will eventually be added to the production or protection areas.

The recreation class involves the opening of portions of forest for the use of the public, such as picnic spots, camping sites, forest walks and forest drives.

The research class includes permanent sample plots, biological reserves and areas where rare species are preserved (Von Breitenbach, 1968).

In other countries multiple-use management systems are mainly determined by the local needs. For example, in Sudan and Ethiopia afromontane forests are important sources of timber with the result that conservation is neglected (Bari, 1968; Beals, 1968), whereas in Zimbabwe the afromontane forests are protected to a large extent.

McCulloch and Dagg (1965) made the following proposals for protection forestry in East Africa: steep land should be retained under natural or plantation forestry and grazing should be prohibited. The occasional flatter areas existing within a region of sharply divided topography should not be cultivated.

Sloping land in high- or medium-rainfall areas may be suitable for high-value plantation and other crops, which may justify the expense of elaborate conservation works, and thereby only slightly affecting ground-water supplies. Sloping land receiving a mean annual rainfall of less than 1000 mm can support high productive agriculture, provided adequate soil conservation measures are taken, and generally water supplies will increase as compared to where there is a forest cover.

Flat and gently sloping land should mainly be used for agricultural purposes.

Management techniques

In the South Cape production forests, enrichment plantings are carried out in forest openings where natural regeneration is poor. This occurs mainly in moist forest types as a result of fungal attacks and fierce weed competition. The main species planted are *Podocarpus falcatus* and *Ocotea bullata*, which have been selectively over-exploited in the past.

Seeds, collected from selected trees, are sown in boxes, and plants are lined out into large polythene tubes. If there is not enough seed available, small natural regenerated seedlings, which have very little chance of surviving *in situ*, may also be collected from the forest and lined out.

Seedlings are kept in a shelter-house for the first three to six months, then left in the open until they are approximately 50 cm high. They are returned to the shelter-house for a further three to six months, to prepare them for shaded conditions, before being planted in the forest. Only the best plants are planted. Fertilizer is applied to improve the growth rate and root pruning is practised to develop a good root system (Geldenhuys, 1975). Planting holes are 25 cm deep, 40 cm wide and 3.5 to 4.5 m apart. They are made from the middle of the gap outwards, but no pits are made at the fringes of gaps. Care must be taken that the transplants are not planted deeper than they stood in the nursery. The planting seasons are March–April and October–November.

Nurse-crops, such as *Virgilia oroboides*, are sometimes used. *Ocotea bullata* transplants and selected coppice shoots are protected against browsing by meshwire tubes supported by fence-droppers. The tubes are removed when the plants have reached a tall sapling stage and are out of browse range.

Tending of artificial and natural regeneration includes the following activities:

Trichocladus crinitus (*onderbos*) and overgrowing weeds are slashed. Climbers and twiners should also be slashed as they can kill or deform potential top-storey trees. Where more than 20% of the planted trees have died they should be replaced. Coppice shoots must be selectively reduced to one or two per stump. Special thinnings to half of the original number may be required at a tall sapling stage. Exotic species, such as *Acacia melanoxylon*, which have invaded forest margins, are being eliminated systematically.

Marking of trees to be felled is done under close supervision. In order to minimize damage to the remaining stand, removal of trees must be done with great care, using one of the following techniques:

(1) Conventional felling of small trees during summer.

(2) Big trees without merchantable timber are killed standing, using 2,4,5-T during summer. The branches of treated trees will rot and break off before the bole does.

(3) Dominant or co-dominant trees with merchantable timber are topped prior to felling. This should be done in winter in order to minimize danger of fungal and insect attacks.

Marketable timber is removed from the compartment to the roadside by mules and sledges to ensure the least possible damage to the site (Von Breitenbach, 1968).

Willan (1965) described the natural regeneration projects of *Ocotea usambarensis* on south Kilimanjaro and the west Usambara Mountains in Tanzania, where this species is dominant.

A solution of sodium arsenite is applied in frills to defective *O. usambarensis* trees. Although the section above the frill is killed, coppice and sucker shoots develop. Bramble is killed by spraying a mixture of 2,4-D and 2,4,5-T. Both these operations are carried out three years before exploitation takes place.

Within a year after extraction any surviving brambles are dug out, logging tracks are blocked with brushwood to reduce erosion, remaining weed trees in the over-storey are poisoned, gap plantings are carried out where natural regeneration is insufficient, and *O. usambarensis* coppice shoots are thinned to leave the best two per stump. Ten years later *Macaranga* trees, a useful nurse crop, which are likely to suppress *Ocotea usambarensis* are removed.

Thinnings of *O. usambarensis* are carried out at ages 15, 20, 40 and 60 years to approximately 1000, 500, 250 and 125 stems ha^{-1}. The final felling is at 75 years of age when the average diameter at breast height is 60 cm.

Olea welwitschii seedlings germinate profusely on east Meru, but most die during the dry season. Seed is easily collected and raised in a nursery, therefore, *taungya* planting is a practical proposition. Problems encountered include buck browsing and a strong tendency to multiple leaders (Willan, 1965).

In Tanzania controlled early burning of glades within forest reserves is preferred to complete protection. If left over a number of years, a large quantity of very inflammable material is accumulated and the resultant burn will then be exceedingly fierce. Afromontane forests are usually too moist at all times of the year to be in danger of internal burning. Fire-traces are made around forest patches and along the boundaries where possible. Burning back from these fire-traces at the beginning of the dry season has proved a very effective system to prevent fire damaging the edges of afromontane forests.

No grazing is allowed in afforested areas. This is achieved by constant patrolling or fencing. Elsewhere grazing is acceptable, provided that the area is not over-stocked and cattle are properly herded. Graziers are expected to assist in early burning operations (Hamlin, 1962).

Nature of ownership and management control

The major portion of remaining indigenous afromontane forests occurs in state-owned forest reserves and national parks (Hedberg and Hedberg, 1968).

Of the countries known to have relatively large plantation areas in afromontane regions, the majority of those in South Africa, Swaziland and Zimbabwe are owned by the private sector, whereas those in Malawi, Tanzania and Kenya are mainly owned by government. All these plantations are in relatively big units.

Afforestation programs

Hardly any information is available on current afforestation programs in afromontane regions. In South Africa and Zimbabwe *Acacia mearnsii*, *Eucalyptus* spp. and *Pinus* spp., are planted fairly extensively, while in Swaziland and Malawi, *Pinus* spp. are planted for the production of pulp. In Tanzania, in the vicinity of Sao Hill *Acacia mearnsii* and *Eucalyptus grandis* are planted in order to supplement the softwood supplies to a proposed pulp mill (H. Lohbeck, pers. comm., 1978).

Research emphasis

In the developing countries serious restraints on research are introduced by deficient funds, facilities and qualified staff. A considerable amount of botanical research has been carried out in afromontane forests of some countries, such as South Africa and Malawi, but a great need remains for ecological and silvicultural research in most countries.

In the South Cape, research has been done and is continuing on all aspects of ecological research, for the broadening and strengthening of scientific foundations of forest management. Silvicultural studies include the following: germination requirements of all species; methods to reconstruct the forest in the shortest possible time with the minimum effort; the regeneration and increment of different species as influenced by site and management; and methods of killing large non-utilizable trees. The herbarium, and the management of preservation areas of rare species, also fall under the research division (Von Breitenbach, 1968).

According to Parry (1962), there was a great need at that time in Tanzania for research on protection and management of forest catchments.

A considerable amount of research on land-use hydrology has been carried out in all vegetation types in East Africa. The main objective of these studies is to ascertain whether the continued conservation of forests with little economical value is justified by the benefits derived from their protection. It is suggested that alternative forms of land use might be of greater economic value (McCulloch and Dagg, 1965).

ECONOMIC STATUS

Information on the economic status of plantations in afromontane regions is not available because figures for different countries are not given by climatic regions occurring within them.

The total value of the removals from the South Cape forests during the period 1776 to 1939 may be estimated at a minimum of R100 million[1]. It is expected that by 1985 revenue will equal expenditure and by the year 2000 the revenue will become stationary, with an estimated surplus of approximately 18% over expenditure. This must be regarded as a very rough approximation and no allowance has been made for changes in money values (Von Breitenbach, 1968).

During the 1930s sawn *Ocotea bullata* was sold at R5 to R9 m^{-3} while the prices of other species

[1]In 1981 R1.00 = $US 1.22.

varied from R0.70 to R1.75 m^{-3} (Laughton, 1937).

At the 1977 auctions in the South Cape 1555 m^3 timber was sold for a total value of R376 725. The average and highest prices per cubic metre paid for prima logs were:

Species	Average price ($R\ m^{-3}$)	Highest price ($R\ m^{-3}$)
Ocotea bullata	1278	3700*
Podocarpus latifolius	361	440
P. falcatus	283	350
Acacia melanoxylon	249	410

*For a log 5.1 m in length and a mid-diameter of 72 cm, which was sold for R7696. The total value of the tree from which this log was cut was R9259 (Department of Forestry, 1977).

CONSERVATION STATUS

Exploitation

All accessible forests in Africa were exploited injudiciously until a few decades ago. In some countries exploitation is still continuing at an alarming rate with little or no thought given to sustained yields or to the indirect benefits obtained from forests (Hedberg and Hedberg, 1968). For approximately 150 years the South Cape forests supplied the major portion of timber in South Africa. It has been estimated that between 1776 and 1939 (when extraction was temporarily ended) 1.5 to 3 million m^3 of timber was removed, reaching a peak at the beginning of the nineteenth century when 20 000 m^3 was removed annually (Von Breitenbach, 1968).

Adequacy of protected areas

Exact figures of protected areas of afromontane forests are not available. Apart from southern Africa no recent information on protection is available.

The protected areas have increased considerably during the past few decades, but many authors who presented papers at a symposium held in 1966 on the conservation of vegetation in Africa south of the Sahara, expressed the need for increased protection of afromontane forests (Hedberg and Hedberg, 1968). Edwards (1974) referred to the dangers of assessing conservation status upon a too superficial viewpoint of the vegetation, for example only in terms of physiognomic types and of the dominant species. He also emphasized the need for reserves of adequate size for ecological and genetic reasons. Furthermore, many countries are not yet conscious of the urgent need for the thorough conservation of protected areas, with the result that different degrees of protection exist between different countries. Grazing, uncontrolled fires, encroachment and theft are jeopardizing the future existence of many so-called protected afromontane forests. Serious restraints are also introduced by a shortage of funds and of qualified personnel, and the pressure for more agricultural land for food production for increasing populations.

Generally, forests are protected in Africa for the production of indigenous timber, the establishment of plantations with exotics, the securing of water sources, the prevention of soil erosion and the conservation of larger mammals. As a result of the selective exploitation of the more valuable species, the composition of species in many forest reserves has been drastically changed, even in reserves regarded as being protected. The conservation of areas of all different vegetation types should receive much more attention in virtually all African countries.

The relicts of afromontane forests in South Africa are adequately protected, for example in Tsitsikamma Coastal Park (Cape), Giants Castle Reserve (Natal) and Blyderivier Nature Reserve (Transvaal) and in numerous forest reserves in these provinces. In Zimbabwe the montane forests are also adequately protected, for example Vumba, Chimanimani and Inyanga National Parks, Banti Forest Reserve and Chirinda Forest. In Moçambique, however, the patches of montane forest should be protected in small reserves and a national park could be proclaimed to adjoin Chimanimani National Park in Zimbabwe. Afromontane forest patches on the Nyika Plateau are adequately protected in Zambia and Malawi. Other afromontane forests in Malawi are enclosed in nine forest reserves of which Mount Mulanje is the most important. Some vulnerable forest patches outside reserves need protection from fire. In all these countries small botanical reserves could be proclaimed to afford the maximum degree of protection (Huntley, 1978).

According to Huntley (1974), none of the small patches of relic afromontane forest in Angola was protected at that time, but attempts were being made to proclaim them as conservation areas. He regarded the afromontane ecosystems as the most seriously threatened of all in Angola, due to excessive burning and felling.

In Kenya, Uganda and Tanzania conservation is a familiar idea because of the financial gain from its practice. In these countries the policy is that there should be as much forest as possible in high-rainfall areas and on hilltops and mountains in arid areas to conserve water supplies. There remains a need for botanical research. In Kenya, the Aberdare and Mount Kenya National Parks contain large areas of well-protected afromontane forest, but it has been suggested that the forests of Marsabit and Elgon Mountains should also be protected more effectively (Lucas, 1968). In Uganda, the Forest Department has set aside nature reserves within its many forest reserves, which are protected against all interference. It has been proposed that more afromontane forests should be included in these nature reserves, for example, on Ruwenzori, Virunga and Elgon Mountains (Osmaston, 1968).

Parry (1962) referred to the destruction of the natural closed montane forests of Tanzania. By the end of the nineteenth century these forests had already been cleared or extensively replaced by secondary forest types. Fortunately many forest reserves were created during the first half of this century. Only small patches of high forest now remain outside these reserves. Polhill (1968) suggested that nature reserves should be sited within some of these reserves, for example, Meru, Kilimanjaro, Usambara and Livingstone Mountains, in order to protect afromontane forests more securely.

According to Letouzey (1968), afromontane forests in Cameroon were not adequately protected, but it has been suggested that reserves, which include montane forests, should be demarcated on Mount Cameroun. Due to dense populations afromontane forests in Rwanda and Burundi have been degraded and invaded by secondary species. Ten years ago Burundi had no reserves in the strict sense of the word, but felling was not allowed in moist afromontane forests. In Rwanda the Albert National Park includes afromontane forests. However, in both countries the afromontane forests of the Congo/Nile watershed urgently required adequate protection (Deuse, 1968; Lewalle, 1968). In Zaïre the afromontane forests are adequately protected in the Albert National Park (Germain, 1968).

In Sudan no montane forest was protected ten years ago (Bari, 1968). Although it has been proposed that such forests should be enclosed on the Didinga and Imatong Mountains, the feasibility thereof has been questioned. In Somalia the so-called protected afromontane forest types are overgrazed with the result that there is hardly any natural regeneration. Here too adequate protection appears to be unobtainable (Hemming, 1968). In Ethiopia no afromontane broad-leaved forest was protected ten years ago, but the opinion was expressed that Belleta and Sai Forests should be protected (Beals, 1968).

Guinea has a fine reserve, which includes montane grassland and upland gallery forest on the Nimba Mountains (Schnell, 1968). There are also reserved areas on these mountains within the borders of Liberia and Ivory Coast (Voorhoeve, 1968; Adjanohoun et al., 1968).

In Sierra Leone the Loma Mountains and Tingi Hills Forest Reserves are not effectively protected against fires (Morton, 1968). Some afromontane forests in Nigeria are apparently included in forest reserves, but it has been proposed that strict natural reserves, which include afromontane forests, should be proclaimed on Mambila Plateau and Vogel Peak (Charter, 1968).

Conservation management

The policy of most countries in respect of afromontane forests is one of conservation, but apart from Von Breitenbach (1968) the forestry literature hardly refers to conservation management.

Approximately 45% of the South Cape forests qualify for protection because timber utilization in these areas would result in the deterioration of the forests. The conservation objectives are to preserve the flora and fauna and to secure the indirect uses of the forests.

The only active management entails patrolling to exclude encroachment, grazing, theft and fire. Silvicultural management may be required to improve the indirect uses, for example, elimination of weed trees to preserve the aesthetic appearance; removal of utilizable overmature trees in extremely dense areas to improve streamflow; careful culling of overmature trees liable to be windblown, to prevent large gaps in the canopy. Fynbos species which are inflammable can invade these gaps, thereby significantly increasing the fire risk.

Preservation areas have been demarcated to preserve species such as *Faurea macnaughtonii* and *Prunus africana* in their remaining habitats. In such areas the only interference allowed is to remove weed trees, extract fallen trees and interplant suitable species.

Biological reserves are declared in representative areas of different forest types, including the various seral stages, undisturbed climax stands and other areas of special biological and ecological interest. These reserves are closed to all human interference, except for qualified scientists who may enter with special permission.

Conservation is concentrated on the entire forest, however, rather than on that of selected micro-communities or individual plants or animals. Only in exceptional circumstances, for example, where a species is nearing extinction, is selective preservation management carried out. The management of such areas remains subordinate to, and an integral part of, indigenous forest management (Von Breitenbach, 1968).

In some countries national parks are fenced, montane forests are regularly patrolled and firebreaks encircling them are cleared annually (Parry, 1962; Hedberg and Hedberg, 1968). In Burundi exploitation in natural forests has been forbidden since 1968 (Persson, 1975). In Tanzania regeneration of some of the valuable indigenous species, such as *Ocotea usambarensis*, has been carried out with considerable success (Willan, 1965). The continent of Africa is at present in a state of flux. No country can afford to be complacent if these unique ecosystems are to be preserved for posterity.

REFERENCES

Acocks, J.P.H., 1975. Veld types of South Africa. *Bot. Surv. S. Afr., Mem.*, 40: 128 pp.

Adjanohoun, E., Aké Assi, L. and Guillaumet, J.L., 1968. Conservation of vegetation in Ivory Coast. In: I. Hedberg and O. Hedberg (Editors), *Conservation of Vegetation in Africa South of the Sahara. Acta Phytogeogr. Suec.*, 54: 76–81.

Bari, E.A., 1968. Conservation of vegetation in Sudan. In: I. Hedberg and O. Hedberg (Editors), *Conservation of Vegetation in Africa South of the Sahara. Acta Phytogeogr. Suec.*, 54: 59–64.

Beals, E.W., 1968. Conservation of vegetation in Ethiopia. In: I. Hedberg and O. Hedberg (Editors), *Conservation of Vegetation in Africa South of the Sahara. Acta Phytogeogr. Suec.*, 54: 137–140.

Bigalke, R.C., 1978. Mammals. In: M.J.A. Werger (Editor), *Biogeography and Ecology of Southern Africa*. W. Junk, The Hague, pp. 981–1048.

Boucher, C., 1978. Cape Hangklip area. II. The Vegetation. *Bothalia*, 12: 455–497.

Boughey, A.S., 1956. The nomenclature of the vegetation zones on the mountains of tropical Africa. *Webbia*, 11: 413–424.

Burtt-Davy, J., 1938. The classification of tropical woody vegetation types. *Imp. For. Inst. Publ.*, 13: 1–86.

Chapman, J.D. and White, F., 1970. *The Evergreen Forests of Malaŵi*. Commonwealth Forestry Institute, Oxford, 190 pp. [plus photographs].

Charter, J.R., 1968. Conservation of vegetation in Nigeria. In: I. Hedberg and O. Hedberg (Editors), *Conservation of Vegetation in Africa South of the Sahara. Acta Phytogeogr. Suec.*, 54: 91–94.

Coetzee, J.A. and Van Zinderen Bakker, E.M., 1970. Paleological problems of the Quaternary of Africa. *S. Afr. J. Sci.*, 4: 78–84.

Dagg, M. and Pratt, M.A.C., 1962. Relation of storm-flow to incident rainfall. *East Afr. Agric. For. J.*, 27: 31–35.

Deacon, H.J. and Brooker, M., 1976. The Holocene and upper Pleistocene sequence in the southern Cape. *Ann. S. Afr. Mus.*, 71: 203–214.

Department of Forestry, 1976. *Indigenous Forest Management and Conservation*. Pretoria, cyclostyled report, 13 pp.

Department of Forestry, 1977. Aanvraag na inheemse hout steeds lewendig. *For. News*, 4: 15.

Department of Forestry, 1978. *South Africa's Commercial Timber Resources 1976–1977*. Government Printer, Pretoria, 45 pp. [plus map].

Deuse, P., 1968. Conservation of vegetation in Rwanda. In: I. Hedberg and O. Hedberg (Editors). *Conservation of Vegetation in Africa South of the Sahara. Acta Phytogeogr. Suec.*, 54: 125–127.

Doidge, E.M., 1950. The South African fungi and lichens. *Bothalia*, 5: 1–1094.

Edwards, D., 1967. A plant ecological survey of the Tugela River Basin. *Bot. Surv. S. Afr., Mem.*, 36: 285 pp.

Edwards, D., 1974. Survey to determine the adequacy of existing conserved areas in relation to vegetation types. A preliminary report. *Koedoe*, 17: 3–38.

Furon, R., 1963. *The Geology of Africa*. Hafner, New York, N.Y., 377 pp.

Garnett, T.F.F., 1973. *Die Department van Bosbou en sy werksaamhede*. Pamphlet 107. Die Staatsdrukker, Pretoria, 39 pp.

Geldenhuys, C.J., 1975. *Raising Indigenous Trees*. Pamphlet 150. Government Printer, Pretoria, 33 pp.

Germain, R., 1968. Conservation of vegetation in Congo–Kinshasa (Zaire). In: I. Hedberg and O. Hedberg (Editors), *Conservation of Vegetation in Africa South of the Sahara. Acta Phytogeogr. Suec.*, 54: 121–125.

Gillet, J.B., 1956. The relation between the highland floras of Ethiopia and British East Africa. *Webbia*, 11: 459–466.

Granger, J.E., 1976. *Vegetation Changes, Some Related Factors and Changes in the Water Balance Following 20 Years of Fire Exclusion in Catchment 9; Cathedral Peak Forestry Research Station.* Thesis, Department of Botany, University of Natal, Durban, 612 pp.

Hamilton, A.C., 1974. The history of the vegetation. In: E.M. Lind and M.E.S. Morrison (Editors), *East African Vegetation.* Longman, London, pp. 188–209.

Hamlin, H.W., 1962. Protection of forest reserves from fire and grazing. *East Afr. Agric. For. J.*, 27: 111.

Hedberg, I. and Hedberg, O. (Editors), 1968. *Conservation of Vegetation in Africa South of the Sahara. Acta Phytogeogr. Suec.*, 54: 320 pp.

Hedberg, O., 1961. The phytogeographical position of the afroalpine flora. *Rec. Adv. Bot.*, 1: 914–919 [University Press, Toronto, Ont.].

Hedberg, O., 1969. Evolution and speciation in a tropical high montane flora. *Biol. J. Linn. Soc.*, 1: 135–148.

Hemming, C.F., 1968. Conservation of vegetation in Somali Republic North. In: I. Hedberg and O. Hedberg (Editors), *Conservation of Vegetation in Africa South of the Sahara. Acta Phytogeogr. Suec.*, 54: 141–145.

Huntley, B.J., 1974. *Vegetation and Flora Conservation in Angola.* Cyclostyled report, 11 pp. [plus tables and maps].

Huntley, B.J., 1978. Ecosystem conservation in Southern Africa. In: M.J.A. Werger (Editor), *Biogeography and Ecology of Southern Africa.* W. Junk, The Hague, pp. 1333–1384.

Jackson, J.K., 1956. Vegetation of the Imatong Mountains. *J. Ecol.*, 44: 341–374.

Jackson, S.P., 1961. *Climatological Atlas of Africa.* Government Printer, Pretoria, 55 maps.

Johannson, D., 1974. Ecology of vascular epiphytes in West African rain forests. *Acta Phytogeogr. Suec.*, 59: 130 pp.

Keay, R.W.J., 1959. *Vegetation Map of Africa (1:10,000,000).* University Press, Oxford, 11 pp. [plus map].

Kerfoot, O., 1962. Root systems of forest trees, shade trees and tea bushes. *East Afr. Agric. For. J.*, 27: 24.

Kerfoot, O., 1964. The vegetation of the southwest Mau Forest. *East Afr. Agric. For. J.*, 30(4): 295–318.

Killick, D.J.B., 1963. An account of the plant ecology of the Cathedral Peak area of the Natal Drakensberg. *Bot. Surv. S. Afr., Mem.*, 34: 145 pp. [plus photographs].

Kingdon, J., 1971. *East African Mammals.* Academic Press, London, 446 pp.

Knapp, R., 1973. *Die Vegetation von Afrika.* Gustav Fischer-Verlag, Stuttgart, 626 pp.

Laughton, F.S., 1937. The silviculture of the indigenous forests of the Union of South Africa. *Sci. Bull.*, No. 157 (For. Ser. 7): 168 pp.

Letouzey, R., 1968. Conservation of vegetation in Cameroun. In: I. Hedberg and O. Hedberg (Editors), *Conservation of Vegetation in Africa South of the Sahara. Acta Phytogeogr. Suec.*, 54: 115–121.

Lewalle, J., 1968. Conservation of vegetation in Burundi. In: I. Hedberg and O. Hedberg (Editors), *Conservation of Vegetation in Africa South of the Sahara. Acta Phytogeogr. Suec.*, 54: 127–130.

Lind, E.M. and Morrison, M.E.S., 1974. *East African Vegetation.* Longman, London, 257 pp.

Lucas, G.L., 1968. Conservation of vegetation in Kenya. In: I. Hedberg and O. Hedberg (Editors), *Conservation of Vegetation in Africa South of the Sahara. Acta Phytogeogr. Suec.*, 54: 152–166.

McCulloch, J.S.G., 1962. Measurements of rainfall and evaporation. *East Afr. Agric. For. J.*, 27: 27–30.

McCulloch, J.S.G. and Dagg, M., 1965. Hydrological aspects of protection forestry in East Africa. *East Afr. Agric. For. J.*, 30(4): 390–394.

McLachlan, G.R. and Liversidge, R., 1970. *Robert's Birds of South Africa.* John Foelcker Bird Book Fund, Cape Town, 3rd ed., 643 pp.

Moll, E.J. and Woods, D.B., 1971. The rate of forest tree growth and a forest ordination at Xumeni, Natal. *Bothalia*, 10(3): 451–460.

Moreau, R.E., 1933. Pleistocene climatic changes and the distribution of life in East Africa. *J. Ecol.*, 21: 414–435.

Moreau, R.E., 1935. A synecological study of Usambara, Tanganyika Territory with particular reference to birds. *J. Ecol.*, 23: 1–43.

Moreau, R.E., 1966. *The Bird Faunas of Africa and Its Islands.* Academic Press, London, 424 pp.

Morton, J.K., 1968. Conservation of vegetation in Sierra Leone. In: I. Hedberg and O. Hedberg (Editors), *Conservation of Vegetation in Africa South of the Sahara. Acta Phytogeogr. Suec.*, 54: 72–74.

Morton, J.K., 1972. Phytogeography of the West African mountains. In: D.H. Valentine (Editor), *Taxonomy, Phytogeography and Evolution.* Academic Press, London, pp. 221–236.

Osmaston, H.A., 1968. Conservation of vegetation in Uganda. In: I. Hedberg and O. Hedberg (Editors), *Conservation of Vegetation in Africa South of the Sahara. Acta Phytogeogr. Suec.*, 54: 148–151.

Parry, M.S., 1962. Progress in the protection of stream-source areas in Tanganyika. *East Afr. Agric. For. J.*, 27: 104–106.

Pereira, H.C., 1962a. The development of tea estates in tall rain forest. *East Afr. Agric. For. J.*, 27: 16–21.

Pereira, H.C., 1962b. The land-use problem and the experimental environment. *East. Afr. Agric. For. J.*, 27: 76–79.

Pereira, H.C., Dagg, M. and Hosegood, P.H., 1962. The waterbalance of the cultivated and control catchments. *East Afr. Agric. For. J.*, 27: 118–122.

Persson, R., 1975. *Forest Resources of Africa.* Research Notes No. 18, Royal College of Forestry, Stockholm, 294 pp.

Phillips, J., 1959. *Agriculture and Ecology in Africa.* Faber and Faber, London, 412 pp.

Phillips, J.F.V., 1930. Some important vegetation communities in the Central Province of Tanganyika Territory. *J. Ecol.*, 18: 193–234.

Phillips, J.F.V., 1931a. Forest-succession and ecology in the Knysna Region. *Bot. Surv. S. Afr., Mem.*, 14: 327 pp. [plus maps and photographs].

Phillips, J.F.V., 1931b. Ecological investigation in South, Central and East Africa: Outline of a progressive scheme. *J. Ecol.*, 19: 474–482.

Pitt-Schenkel, C.J.W., 1938. Some important communities of warm temperate rain forest at Magamba, West Usambara, Tanganyika Territory. *J. Ecol.*, 26: 50–81.

Polhill, R., 1968. Conservation of vegetation in Tanzania. In: I. Hedberg and O. Hedberg (Editors), *Conservation of Vegetation in Africa South of the Sahara. Acta Phytogeogr. Suec.*, 54: 166–178.

Poynton, J.C. and Broadley, D.G., 1978. Herpetofauna. In: M.J.A. Werger (Editor), *Biogeography and Ecology of Southern Africa.* W. Junk, The Hague, pp. 925–948.

Republic of South Africa, 1968. *Forest Act No. 72.* Government Printer, Pretoria, 42 pp.

Richards, P.W., 1963. Ecological notes on West African vegetation. III. The upland forests of Cameroons Mountain. *J. Ecol.*, 51: 529–554.

Ross, R., 1956. Some aspects of the vegetation of Ruwenzori. *Webbia*, 11: 451–457.

Scheepers, J.C., 1978. Vegetation of the Westfalia Estate on the north-eastern Transvaal Escarpment. *Bot. Surv. S. Afr., Mem.*, 42: 230 pp.

Schimper, A.F.W., 1903. *Plant Geography.* Clarendon Press, Oxford, 839 pp.

Schnell, R., 1968. Conservation of vegetation in Guinea. In: I. Hedberg and O. Hedberg (Editors), *Conservation of Vegetation in Africa South of the Sahara. Acta Phytogeogr. Suec.*, 54: 69–72.

Scott, R.M., 1962a. Summary of soil survey observations on the Sambret Valley. *East Afr. Agric. For. J.*, 27: 22.

Scott, R.M., 1962b. Soil survey of the experimental valleys (of the Mbeya Peak). *East Afr. Agric. For. J.*, 27: 112–114.

Scott, R.M. and Friend, M.T., 1962. Soil Survey of the intensely measured catchments A, B & C (in the Aberdares). *East Afr. Agric. For. J.*, 27: 84.

Specht, R.L. (Editor), 1979. *Heathlands and Related Shrublands, A. Descriptive Studies.* Ecosystems of the World, 9A. Elsevier, Amsterdam, 497 pp.

Taylor, H.C., 1955. Forest types and flora composition of Grootvadersbosch. *S. Afr. For. J.*, 26: 33–46.

Taylor, H.C., 1961. The Karkloof forest: a plea for its protection. *For. S. Afr.*, 1: 123–134.

UNESCO, 1963. A review of the natural resources of the African continent. *Nat. Resour. Res.*, I: 437 pp.

Van der Schijff, H.P. and Schoonraad, E., 1971. The flora of the Mariepskop complex. *Bothalia*, 10(3): 461–500.

Von Breitenbach, F., 1968. *Southern Cape Indigenous Forest Management Manual, II.* Department of Forestry, George, 149 pp.

Von Breitenbach, F., 1974. *Southern Cape Forests and Trees.* Government Printer, Pretoria, 328 pp.

Voorhoeve, A.G., 1968. Conservation of vegetation in Liberia. In: I. Hedberg and O. Hedberg (Editors), *Conservation of Vegetation in Africa South of the Sahara. Acta Phytogeogr. Suec.*, 54: 74–76.

Walter, H., 1971. *Ecology of Tropical and Subtropical Vegetation.* Oliver and Boyd, Edinburgh, 539 pp.

Weather Bureau, 1954. *Climate of South Africa, 1. Climate Statistics.* Government Printer, Pretoria, 160 pp.

Werger, M.J.A. (Editor), 1978. *Biogeography and Ecology of Southern Africa.* W. Junk, The Hague, 2 vols., 1439 pp.

White, F., 1978. The Afromontane region. In: M.J.A. Werger (Editor), *Biogeography and Ecology of Southern Africa.* W. Junk, The Hague, pp. 463–513.

Wild, H., 1956. The principal phytogeographical elements of the Southern Rhodesian flora. *Proc. Trans. Rhodes. Sci. Assoc.*, 44: 53–62.

Wild, H., 1964. The endemic species of the Chimanimani Mountains and their significance. *Kirkia*, 4: 125–157.

Wild, H., 1968. Phytogeography of South Central Africa. *Kirkia*, 6: 197–223.

Willan, R.L., 1965. Natural regeneration of high forest in Tanganyika. *East Afr. Agric. For. J.*, 31(1): 43–53.

Winterbottom, J.M., 1978. Birds. In: M.J.A. Werger (Editor), *Biogeography and Ecology of Southern Africa.* W. Junk, The Hague, pp. 949–979.

Chapter 9

TEMPERATE BROAD-LEAVED EVERGREEN FORESTS OF JAPAN[1]

T. SATOO

INTRODUCTION

Definition of temperate broad-leaved evergreen forest

The distribution pattern of forest vegetation in the Japanese Archipelago is broadly independent of water supply because of the abundant precipitation. Temperature is an important factor because of the extent of the Archipelago from north to south. The forests can be divided into zones related to the temperature gradient. The first to describe such a zonation was Tanaka (1885). Later, Mayr (1891), Honda (1899–1900, 1900a, b, 1912) and Kira (1949) published modified versions. The classification by Honda (1912) is most generally used in Japan. He divided Japanese forests into four zones: tropical (or *Ficus superba*), warm-temperate (or *Cyclobalanopsis*), cool-temperate (or *Fagus crenata*), and subarctic (or *Abies sachalinensis*) zones. The temperate broad-leaved evergreen forests of Japan correspond to the warm-temperate zone of Honda (1912) as shown in Fig. 9.1. They also correspond to the "*Pinus thunbergii*" zone of Tanaka (1885), to the "subtropische Waldzone der immergrünen Eichen und Lorbeerbäume [subtropical forest zone of evergreen oaks and laurel trees]" of Mayr (1891) and to the laurel forest zone of Kira (1949).

Similar types of forest are also found in the lowlands of the southernmost part of the Korean Peninsula (Honda, 1912) and in the montane forests of Taiwan (Honda, 1900a, b; Suzuki, 1952). Plant-geographically this zone corresponds to the Association of *Machileto-Shiietum sieboldii*, belonging to the Alliance of Machilion thunbergii by Nakano (1942), and to the Alliance of Shiion sieboldii, and is part of the Class-group of Lauréceo-Fagacea sempervirens (Suzuki, 1975) or a part of the Class Camellietea japonicae (Miyawaki and Ohba, 1963). Wang (1961) described this type of forest as "Evergreen broad-leaved forest of evergreen oaks, *Schima* and laurels", of which there are various subdivisions. He gave the distribution of this forest type as lowland and montane areas of the southeastern parts of continental China, montane forests of the islands of Taiwan and Hainan, southern Japan and the southernmost parts of the Korean Peninsula. Yoda (1967) described the broad-leaved evergreen forest zones, *Schima–Castanopsis* zone and evergreen oak zone, in eastern Nepal. According to him, Kawakita (1956) has described a *Schima–Castanopsis* zone also in central Nepal. These broad-leaved evergreen forests (see also Gopal and Meher-Hoinji, Ch. 7) have similarities to those in Japan in composition of families and genera. Though part of the warm-temperate zone, the eastern parts of the Eurasian continent have abundant precipitation in summer due to the monsoon while the western parts of the Mediterranean area have a very dry summer. Consequently, the temperate broad-leaved evergreen forests of East Asia are seen as belonging to the "Laurisilvae" as distinct from the "Durisilvae" of the dry Mediterranean area.

Geographical extent

The temperate broad-leaved evergreen forests of Japan occur approximately between the 13°C and 21°C isotherms (Fig. 9.1). On average the southern

[1]Manuscript completed December, 1980.

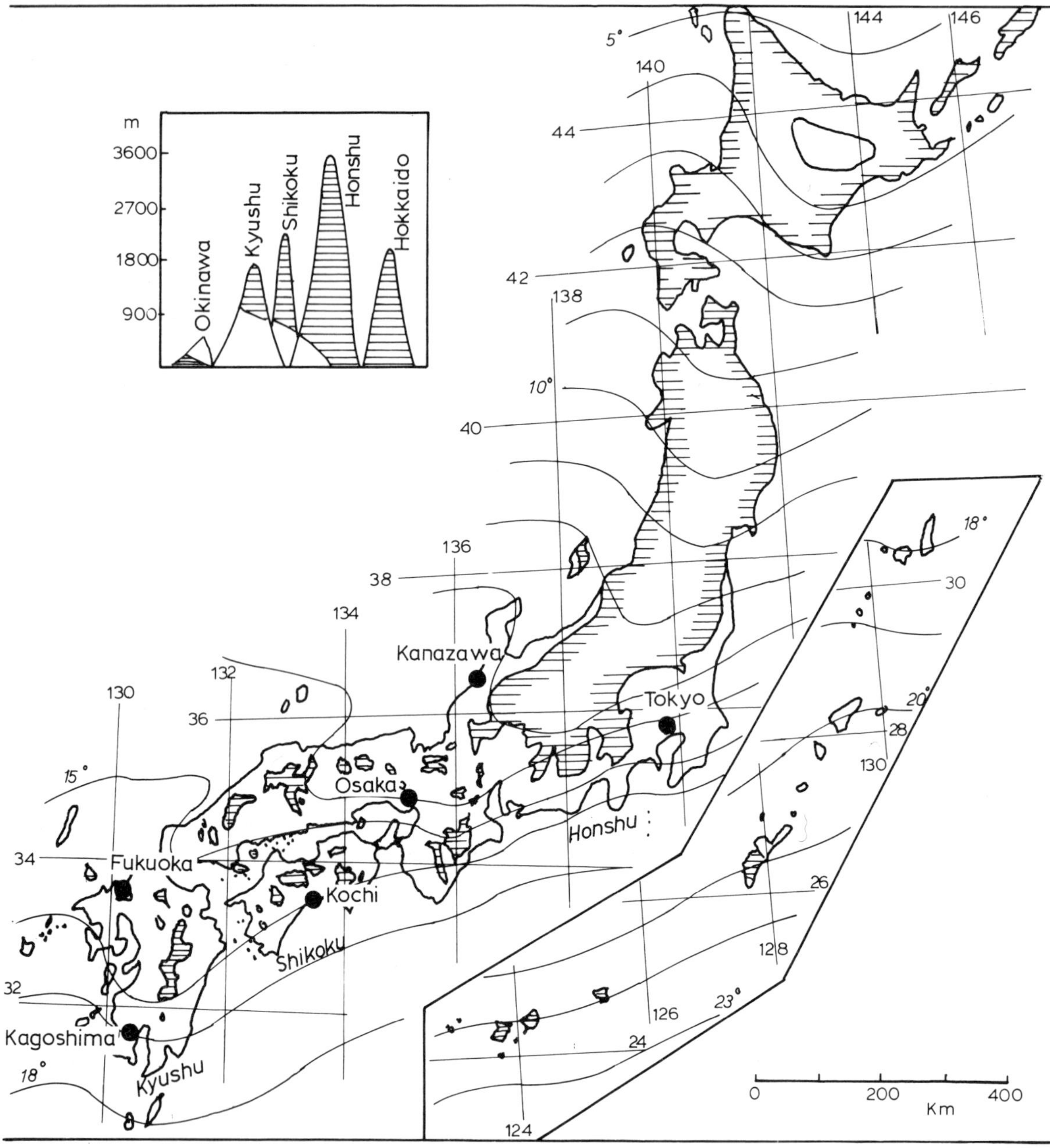

Fig. 9.1. Warm-temperate forest zone of Japan, except hatched areas (from Honda, 1912).

limit extends to 26°30′N and the northern limit is 35°30′N. Under the influence of the warm ocean currents on both sides of the Archipelago, the northern limit is 37°30′N in coastal regions and about 34°30′N inland at the higher elevations.

The temperate broad-leaved evergreen forest zone includes the lowlands of the southern half of Honshu, Shikoku and Kyushu and many islands in the northern half of the Okinawa Islands. Altitudinally, there is no lower limit in Japan, except for some islands of Okinawa. The mean upper limit is 850 m above sea level for Kyushu, 750 m for Shikoku and 600 m for southern Honshu and becomes gradually lower northwards to zero at the northern limit. Indices other than mean annual temperature have been considered to indicate the

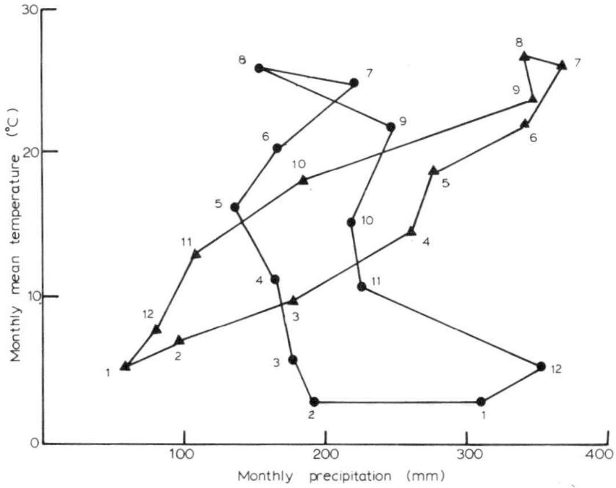

Fig. 9.2. Examples of hythergraphs from the Pacific side and the Sea of Japan side of Japan within the distribution range of warm-temperate broad-leaved evergreen forests. Triangles: Kochi on the Pacific side; dots: Kanazawa on the Sea of Japan side (see also climatic data in Table 9.1).

boundaries of plant zones (e.g. the "warmth and coldness indices" of Kira, 1949). However, Honda (1912) stated that "It is most convenient to use mean annual temperature as the limits of zones, because it is easy to compare with climate of other parts of the world. Though there are discrepancies between isotherms and limits of zones, it is to be discussed when detailed description of the zones are given ... I believe that it is more adequate and more convenient to use isotherms for mean annual temperature in a general description of forest zones."

Climatic, geological, topographic and soil conditions

The Japanese Archipelago is situated between a large land mass and a large water mass: the Eurasian continent and the Pacific Ocean respectively. The climate is influenced greatly by the monsoon. In winter, because of the high atmospheric pressure over the continent and the low atmospheric pressure over the ocean, there is a prevailing cold and dry northwest wind. In summer, warm and moist southeast winds prevail. In winter, the air mass from the continent takes up moisture from the Sea of Japan and this causes very heavy snowfall on the slopes facing the Sea of Japan. The air becomes very dry as it descends the slopes facing the Pacific Ocean. In summer, precipitation is greater on the slopes facing the Pacific than on slopes facing the Sea of Japan but the difference is not great. This difference of climate is clearly seen when comparing eastern and western hythergraphs. The hythergraphs of Kochi on the Pacific coast and of Kanazawa on the coast of the Sea of Japan are shown in Fig. 9.2 as examples. There is a large difference between the two cities, though both have an annual precipitation of about 2600 mm. During the summer typhoons from tropical parts of the Pacific Ocean bring tremendous rainfalls with storms. In June and July, heavy rainfall is also brought by the "Baiu" front. Rainfalls have been as much as 1104 mm day^{-1} at Saigo, Nagasaki and 150 mm h^{-1} and 49 mm in 10 minutes at Asizuri, Kochi. These heavy rains cause serious floods, landslides and soil erosion.

Climatic data at five cities within this forest zone are given in Table 9.1, the locations of these cities being shown in Fig. 9.1. Generally speaking, the summers are very hot and the winters rather cold with snow and frosts. Even at Kagoshima in southern Kyushu there may be frost and snow. Compared to other parts of the world, except the east coast of North America, having similar mean annual temperatures, the summer is hotter and the winter colder. However, in most of this zone, oranges, tea and the palms *Trachycarpus fortunei* and *T. wagnerianus* are planted. In the southernmost parts, sugar cane and banana can be cultivated and the cycad *Cycas revoluta* and the tree fern *Cyathea boninsimensis* grow.

Though the islands are generally small, volcanoes and tectonic lines are common. Many kinds of rocks belonging to various groups, systems and series can be found. The topography is very complicated and mountainous; there are many valleys and small plains among the mountain ranges. Slopes are generally steep. Most of the forest is in mountainous regions since almost all flat land is used for agriculture and industries.

Most forest soils are brown forest soils of various types, mostly differentiated by moisture supply, but black soils of volcanic origin, Andosols, are found especially in Kyushu. On the coastal hills truncated, immature soils are found. The red-colored paleosoils have a patchy distribution. The alluvial soils are mostly devoted to rice cultivation.

TABLE 9.1

Climatic data of six cities within the temperate broad-leaved evergreen forest zone

	Tokyo	Osaka	Kochi	Kagoshima	Fukuoka	Kanazawa
Mean temperature (°C)						
year	14.7	15.5	15.8	16.8	15.1	13.5
Nov.–April	7.7	8.3	9.4	10.5	8.6	6.2
May–Oct.	21.7	22.7	22.3	23.1	21.6	20.7
January	3.7	4.5	5.2	6.6	5.1	2.5
August	26.4	27.8	26.3	27.1	26.0	25.9
mean maximum in hottest month	30.7	32.8	31.1	31.8	31.4	30.9
mean minimum in coldest month	−0.9	0.5	0.2	1.8	1.2	−0.5
Frost						
first	Nov. 18	Nov. 14	Nov. 22	Nov. 27	Nov. 14	Nov. 19
last	April 2	April 7	March 31	March 23	April 12	April 13
Snow						
first	Dec. 26	Dec. 23	Dec. 30	Jan. 8	Dec. 22	Nov. 29
last	March 20	March 20	March 4	Feb. 22	March 8	April 3
Mean precipitation (mm)						
year	1526	1360	2668	2336	1704	2559
Nov.–April	519	457	798	737	537	1415
May–Oct.	1043	903	1870	1599	1167	1149
number of days with rain over 100 mm in 10 years	4	12	38	29	25	4

Floristic features

Temperate broad-leaved evergreen forests of Japan give an appearance of rain forest with many lianas, epiphytes and ferns present. Whilst there are many species of evergreen trees and shrubs, evergreen members of the Fagaceae and Lauraceae constitute an important component of the tree layer.

Historical elements

Japanese civilization spread from the southwest. In the legendary era Japan was all in the zone of evergreen broad-leaved forests. The first Emperor subdued western Japan starting from the southern area of Kyushu to Nara and, until a century ago, the Emperor's seat was in the area around Nara, Osaka and Kyoto. Here the natural vegetation is evergreen broad-leaved forest. Karizumi (1970) identified the plant names appearing in the description of Japan in a Chinese book of the third century. He translated part of the book as follows:

"As trees and shrubs, there are *Machilus thunbergii*, *Quercus serrata*, *Cinnamomum camphora*, *Chaenomeles japonica*, *Quercus acutissima*, *Torreya nucifera*, *Cyclobalanopsis* spp., *Cudrania cochinchinensis*, *Acer* spp.; as bamboo there are *Sasa* spp. and other *Sasa* spp. used for making arrows, and palm, *Trachycarpus fortunei*. There are ginger, *Citrus tachibana*, *Zanthoxylum piperitum*, *Zingiber mioga*, but they do not know that they are good as food."

Karizumi noted that all these plants belong to the temperate broad-leaved evergreen forest zone. His reconstruction of the physiognomy of the forest vegetation from the information in the old Chinese book was as follows:

"The tree layer consists of broad-leaved evergreen trees such as *Machilus thunbergii*, *Cinnamomum camphora* and *Cyclobalanopsis* spp. and *Castanopsis* spp. [though not mentioned in the book they could easily be thought as *Cyclobalanopsis* spp.], and partly of conifers like *Torreya nucifera*. On open land grow deciduous trees such as *Quercus serrata*, *Q. acutissima* and *Acer* spp. As shrubs *Chaenomeles japonica*, *Cudrania cochinchinensis*, *Zanthoxylum piperitum* and *Sasa* spp. grow. In the herb layer grow plants of Zingiberaceae. *Citrus tachibana* and *Trachycarpus fortunei* are found from place to place."

According to Oono (1934), books of Japanese history including the legendary era written in the early eighth century mention trees of 27 families, 40 genera and 53 species. These include broad-leaved evergreen trees like *Castanopsis* spp., *Cinnamomum camphora*, *Cyclobalanopsis gilva*, *C. myrsinaefolia*, *C. sessilifolia*; conifers growing in the broad-leaved evergreen forests, such as *Podocarpus macrophyllus* and *Torreya nucifera*; evergreen lianes, such as *Kadsura japonica*, *Sinomenium acutum*, *Trachelospermum asiaticum*; and such epiphytes as *Neottopteris rigida*.

Wood of *Cinnamomum camphora* is mentioned as a good material for making boats. Indeed, canoes made of wood of this tree about 2000 years ago were found in the vicinity of Osaka (Toyama, 1936). Most farming tools found in ruins of ancient villages were made of wood of species of *Cyclobalanopsis*. These descriptions show that broad-leaved evergreen trees were part of the life of ancient Japanese people, and their wood was used in many ways. The region occupied by broad-leaved evergreen forests has been populated from ancient times and there is a long history of human influence on these forests.

Since the temperate broad-leaved evergreen forests occur naturally at lower altitudes, most are already exploited except on islands and in remote counties. They are absent from flat land except in the small precincts of Buddhist temples and Shinto shrines. On hills and mountains, many of the forests on good soils have been converted to conifer plantations. Most broad-leaved evergreen trees in this zone do not have straight boles and have many thick branches. Their growth rate is small, so that they are not good as timber trees. Conifers like *Torreya nucifera* and *Podocarpus macrophyllus*, especially the former, give very precious wood, but their growth rate is also poor. *Abies firma* and *Tsuga sieboldii* cannot compete with *Cryptomeria japonica* and *Chamaecyparis obtusa* as timber trees. Consequently, most broad-leaved evergreen forests have been replaced by plantations of these two conifers. Especially after the use of firewood and charcoal decreased, the rate of conversion of forests of this type into coniferous plantations accelerated. By 1972, more than half of the forests of Kyushu (54%) and Shikoku (51%), the main areas of forests of this type, were coniferous plantations. These plantations play an important role in timber supply for Japan.

Significance of fire

The fire hazard is high in spring when the soil is still cold but the air is warm and a high saturation deficit occurs in the boundary layer. The season of fire hazard moves from south to north. However, extensive forest fires as occur in other parts of the world scarcely happen in Japan. They do not cause significant destruction of forests over large areas, mainly because of the frequent rainfall and the very irregular topography.

Current changes in distribution and relationships to global forest resource

The area occupied by forests of this type is decreasing due to conversion into coniferous plantation. Their contribution to the global forest resource is unimportant, so far as wood production is concerned. Considering the scarcity of forests of this type in the world, their value as a source of biological information is very great.

GROSS CHARACTERISTICS

Trees and understorey species composition

Warm-temperate broad-leaved evergreen forests consist of many species. Obata (1961), for example, recorded 40 species of trees above 2 cm in diameter in a 0.471 ha sample plot of a forest of this type, from which large trees of *Cyclobalanopsis* spp. were cut more than 100 years ago. The 80 dominant trees were of 18 species. These included 51 trees of 5 species of *Cyclobalanopsis*, 10 trees of 2 species of *Castanopsis*, and 5 trees of 2 species of *Machilus*. The 460 secondary storey trees above 7 cm in trunk diameter were of 26 species. These included 121 trees of 5 species of *Cyclobalanopsis*, 40 trees of 2 species of *Machilus* and 165 trees of *Distylium racemosum*. The 1167 trees between 2 and 6 cm in trunk diameter of the lower strata were of 36 species, including 214 trees of *Distylium racemosum*, 206 trees of *Cleyera japonica* and 341 trees of *Machilus japonica*. Another example of this diversity is a sample plot 20 × 100 m of broad-leaved evergreen forest described by Kitazawa et al. (1959). Here there were 335 trees above 4 cm in diameter representing 26 species. These included 41 trees of *Machilus thunbergii*, 64 trees of *Litsea acuminata*, and 36 trees of *Meliosma rigida*.

The main species of the overstorey of broad-leaved evergreen forests are evergreen members of the Fagaceae, such as *Castanopsis cuspidata*, *Cyclobalanopsis acuta*, *C. gilva*, *C. myrsinaefolia*, *C. salicina*, *C. sessilifolia*, *C. sieboldii*, *Pasania edulis* and *Quercus phylliraeoides*. Trees of Lauraceae, such as *Cinnamomum camphora*, *C. japonica*, *Machilus japonica* and *M. thunbergii*, and many broad-leaved evergreen trees like *Camellia japonica*, *Distylium racemosum* and *Myrica rubra* are also present. Conifers, such as *Podocarpus macrophyllus*, *P. nagi*, and *Torreya nucifera*, occur also in these broad-leaved forests. In the coastal area *Pinus thunbergii* and, on the ridges of the northern and higher altitudes, *Abies firma* and *Tsuga sieboldii* grow in mixture with the broad-leaved evergreen species.

The medium to lower strata contain small trees and shrubs of such broad-leaved evergreen species as *Aucuba japonica*, *Cleyera japonica*, *Illicium religiosum* and *Neolitsea sericea*. The ground vegetation contains small shrubs like *Ardisia crenata*, *A. japonica*, *Damnacanthus indicus* and *Maesa japonica*, the vines *Hedera rhombea* and *Trachelospermum asiaticum*, many kinds of ferns including *Polystichum polyblepharum* and *Arachniodes aristata*, and herbaceous plants like *Alpinia japonica* and *Liriope platyphylla*. Epiphytes, for example, *Bulbophyllum inconspicuum*, *Dendrobium moniliforme* and *Neottopteris rigida* are present, as well as many lianas including *Kadsura japonica* and *Stauntonia hexaphylla*.

When the soil is exposed, for example, by a landslide or fire, many deciduous trees like *Alnus firma*, *Mallotus japonicus* and *Rhus javanica* invade as well as many herbaceous plants. These are followed by *Pinus densiflora*, *Quercus acutissima*, *Q. serrata* and *Q. variabilis*. Under these pioneer species broad-leaved evergreen trees become established and the succession proceeds towards a climax forest. When clear-cut, the stumps of some of the original trees sprout but deciduous trees also invade, and the vegetation moves to climax broad-leaved evergreen forests. However, when stumps of the evergreen trees do not sprout, the time to return to evergreen forests is much longer, especially in the north. This delay is because seeds of most evergreen species are too large to be scattered over a long distance and they lose viability when dried.

Ecophysiology of dominant species

Most dominant species of temperate broad-leaved evergreen forests are, like dominant species of climax forests of other types, tolerant to shade (Honda, 1900a). Kusumoto (1957b) determined the light–photosynthesis curves for the main species of the broad-leaved evergreen forests of southern Kyushu. These included *Castanopsis sieboldii*, *Cinnamomum camphora*, *Cyclobalanopsis glauca* and *Machilus thunbergii*. He concluded that trees of most of these species showed general characteristics of shade-tolerance in that the maximum rate of photosynthesis, respiration and compensation point are relatively small. *Cinnamomum camphora* seemed a little less shade-tolerant. In another experiment, Kusumoto (1963) found that the light-photosynthesis curve of *Quercus phylliraeoides*, which grows on dry sites at the sea coast, showed characteristics of shade-intolerant trees. Kusumoto (1957a) studied also temperature–photosynthesis curves for many broad-leaved evergreen trees in summer as well as in winter. He concluded that trees which are resistant to low temperature, *Camellia japonica* for example, differ little in photosynthetic rate between summer and winter, while trees limited to the southern part of the temperate broad-leaved evergreen zone, such as *Cinnamomum camphora*, showed marked differences. The optimum temperature for photosynthesis in *Quercus phylliraeoides*, which grows mostly in littoral areas, was greater than for *Cyclobalanopsis salicina*, which is found mostly at higher elevations (Kusumoto, 1963).

From these measurements, seasonal changes in the photosynthetic production by leaves for many broad-leaved evergreen trees were synthesized and some are shown in Fig. 9.3 (Kusumoto, 1957a). This production was at a maximum in May, decreased in August and September somewhat, and then increased again in October before decreasing again. From April to September, the difference between tree species was not great but the winter decrease differed according to tree species. The photosynthetic production of leaves in January

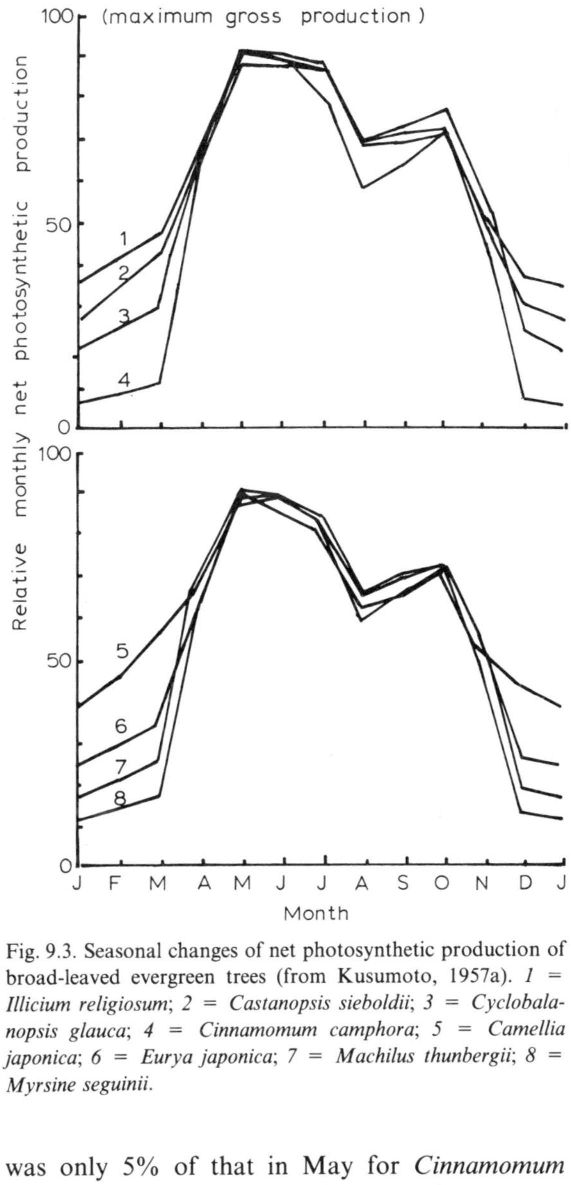

Fig. 9.3. Seasonal changes of net photosynthetic production of broad-leaved evergreen trees (from Kusumoto, 1957a). *1* = *Illicium religiosum*; *2* = *Castanopsis sieboldii*; *3* = *Cyclobalanopsis glauca*; *4* = *Cinnamomum camphora*; *5* = *Camellia japonica*; *6* = *Eurya japonica*; *7* = *Machilus thunbergii*; *8* = *Myrsine seguinii*.

was only 5% of that in May for *Cinnamomum camphora* while it was 38% for *Camellia japonica.*

According to Honda (1901), trees of *Cyclobalanopsis* species grow well on fertile soil with an adequate moisture supply. They have tap-roots and are deep-rooted. In the early stage, growth is slow but increases when the trees are about 3 m tall. They have a single stem but, when grown in the open, the branches from the lower stem are thick. Sprouts grow from the stump after cutting.

There are 300 to 800 seeds weighing 600 to 800 g l^{-1}. *C. gilva* grows mainly in the south and *C. myrsinaefolia*, which tolerates cold winters, grows in the northern part of the zones.

Quercus phylliraeoides grows on rather dry warm, littoral sites, and not in the moist sites of bottomland. Growth is very slow.

Castanopsis spp. grow all over the zone, sometimes as pure stands. There are about 1000 seeds weighing about 750 g l^{-1}.

Cinnamomum camphora grows on fertile, moist, deep, sandy loam near the sea coast. It is very susceptible to cold, drought and exposure to wind. In a liter there are about 4000 seeds weighing 500 g. In the open it attains a large size, the maximum recorded being 22.4 m in girth at 1.5 m and 54.5 m in height. *C. japonicum* is more resistant to coldness than *C. camphora. Machilus thunbergii* is more tolerant of site conditions than *C. camphora.* A liter of seed would contain about 1800 seeds and weigh about 600 g.

Forest structure

The temperate broad-leaved evergreen forests are layered, multi-aged forests having a mixture of many species. The frequency distribution of tree age, diameter at breast height and height in four sample plots reported by Miyoshi (1955), are given in Fig. 9.4. These were believed to be primeval forests but in the fourth, which looked untouched by man, a very old ruin of a charcoal kiln was found. All show a variety of tree size and age. Similar variability was found in tree age and diameter in a sample plot by Obata (1961) and of trunk diameter in sample plots by Kitazawa et al. (1959). Whilst the forests consisted of trees of widely different size and age, the leaves were concentrated in the upper crown. Table 9.2 shows the vertical distribution of leaves in a climax broad-leaved evergreen forest with *Distylium racemosum* dominant, *Castanopsis sieboldii*, *Cyclobala-*

TABLE 9.2

Vertical distribution of leaves in a broad-leaved evergreen forest in southern Kyushu (Kimura, 1960)

Strata	Height (m)	Dry weight ($kg\ m^{-2}$)	Leaf area ($m^2\ m^{-2}$)
upper	> 13	8.52	6.54
middle	4–13	2.38	1.83
lower	< 4	0.50	0.38
total		11.40	8.75

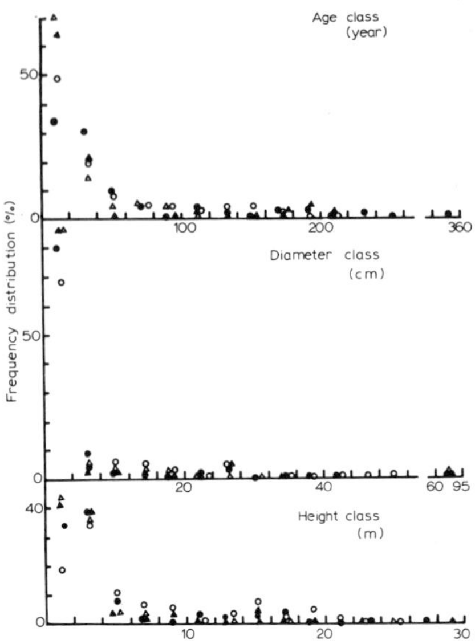

Fig. 9.4. Frequency distribution of age class, diameter class, and height class of four temperate broad-leaved evergreen forests (from tables of Miyoshi, 1955). Closed circles: stems 7–360 years old, density 5400 ha^{-1}, a primeval forest; closed triangles: 6–218 years old, density 4770 ha^{-1}, a primeval forest; open triangles: 6–206 years old, density 6300 ha^{-1}, a primeval forest; open circles: 6–218 years old, density 4000 ha^{-1}, looks primeval but there is an evidence of ancient cutting (very old ruins of a charcoal kiln).

nopsis acuta and *C. salicina* in the upper crown, and the lower crown containing *Camellia japonica*, *C. sasanqua*, *Machilus japonica* and *Cleyera japonica* together with the four upper crown species.

In the warm-temperate broad-leaved evergreen forest zone, the flat lands are used for agriculture and industry. In populated areas, hillside forests on good soils have been replaced by orchards and conifer plantations and on poor soils by deciduous forest and forests of *Pinus densiflora*. Relatively unchanged forests of this type can only be found in remote mountainous areas and some protected areas.

Faunal components

Temperate broad-leaved evergreen forests are rich and variable in flora and fauna. The following large mammals are present: the monkey *Macaca fuscata*, the deer *Cervus nippon*, the wild boar *Sus leucomystax leucomystax*, *Nyctereutes procyoncides*, *Meles anakuma*, the weasel *Mustela sibirica itatsi*, the fox *Vulpes vulpes*, and the hare *Lepus brachyrus brachyrus*. However, because of the large human population and its activities, their territories are reduced and they are few in number. No thorough investigation of the fauna of these forests has been made. Preliminary surveys in the research forest of the Japanese International Biological Programme (IBP) for temperate broad-leaved evergreen forests at Minamata, southern Kyushu, have provided some information. This forest is of 60 to 70 years old growth dominated by *Castanopsis cuspidata*. The small mammals present were the voles *Apodemus speciosus speciosus* and *A. argenteus argenteus* and *Urotrichus talpoides hondonis* (Doi et al., 1970). Kubo et al. (1969) observed 25 species of birds in five days in late May. Surveys in May and October gave 28 orders in the soil fauna (Tanaka and Hamada, 1969). On the shrubs 10 families of spiders and 10 orders of insects were caught (Higashi et al., 1969). Mishima (1970) recorded 16 species of land snails.

In hill forests of the Boso Peninsula in the northern part of the warm-temperate broad-leaved evergreen forest zone, 19 species of fresh-water fish, 10 species of amphibians, 8 species of reptiles, 92 species belonging to 32 bird families, and 12 species belonging to 8 families and 6 orders of Mammalia were listed (Mizuguchi et al., 1973). More than 100 species of spiders belonging to the Araneae (Shinkai and Kurihara, 1975) and 52 species of Myriapoda (Ishii, 1975) were recorded also.

Pests and diseases

No serious pests and diseases have been reported in the climax broad-leaved evergreen forests. However, *Pinus densiflora* and *P. thunbergii*, which appear in the earlier stages of succession, are seriously damaged by bark beetles. Recently it was discovered that the death of many pines previously thought as being caused by bark beetles, is due to a nematode, *Bursaphelenchus lignicolus*. The damage is fatal. This nematode is carried by a long-horned beetle, *Monochamus alteratus*, and multiplies in the wood and causes the tree to wilt. Beetles from infested wood carry the nematodes to other healthy pine trees.

Variability

The zone of temperate broad-leaved evergreen forest extends over thirteen degrees of latitude, and the difference in mean annual temperature between the two extremes is 8°C. Though precipitation is abundant in general, regions surrounding the Seto Naikai inland-sea have a somewhat Mediterranean climate. There are large differences in climate between the Pacific side and the Sea of Japan side (Fig. 9.2). The topography is very complicated, and climatic difference between southern and northern exposures is great. Difference of soil along slopes is also conspicuous.

These differences in physical environment are reflected in the variability of the forests. *Cyclobalanopsis* species are found all over this zone, but *C. gilva* grows mainly in the south and *C. myrsinaefolia* in the north of this zone. *Quercus phylliraeoides* occurs near the sea coast. *Cinnamomum camphora* grows mainly in the south, whereas *Machylus thunbergii* grows in both the northern and southern parts of the zone.

The patterns of species composition also differ altitudinally. Kitazawa et al (1960, 1961) studied the change of dominance of tree species along altitudinal gradients in the mountains of southern Kyushu. On Mount Kirishima the percentages of broad-leaved evergreen trees of the total number of species, total number of trees and basal area decrease with elevation until they are replaced by deciduous forest, mainly consisting of *Fagus crenata* (Fig. 9.5). In the Osumi Peninsula *Machilus thunbergii* is dominant in lowland below 200 m above sea level, and *Cyclobalanopsis acuta* at altitudes over 750 m (Fig. 9.6), whilst *Distylium racemosum* is dominant at intermediate altitudes. *Cyclobalanopsis salicina* is wide-ranging but less abundant. *Castanopsis sieboldii* occurs on lowland, is rare on hillsides but abundant on ridges. On hillsides at higher altitudes, it becomes more abundant as the slope becomes steeper and drier. As seen from Fig. 9.7, the soil meiofauna decreased with elevation (Kitazawa, 1961).

According to Suzuki (1975), temperate broad-leaved evergreen forests belong to the Alliance of Shiion sieboldii and are divided into the following four sub-alliances:

(1) The sub-alliance of Querceto-phylliraeoides is physiognomically sclerophyllous forest, but is floristically a sub-alliance of Shiion sieboldii. In this sub-alliance the Lauraceae are separated from the Fagaceae. Dominant species are *Quercus phylliraeoides*, *Pinus thunbergii* and *Eurya emarginata*. There are three associations distributed on the coastal areas of the Pacific and the Seto Naikai.

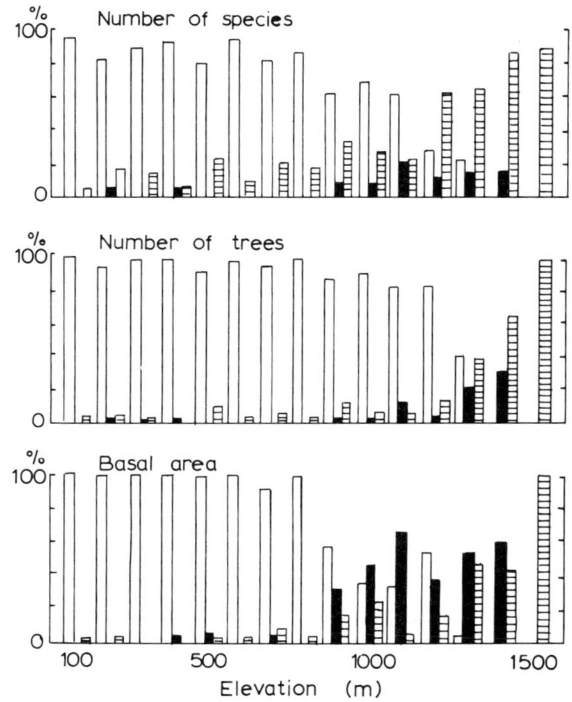

Fig. 9.5. Change of dominance of broad-leaved evergreen trees (white columns), conifers (black columns) and deciduous trees (hatched columns) of primeval forests along the elevational gradient on Mount Kirishima (Kitazawa et al., 1961).

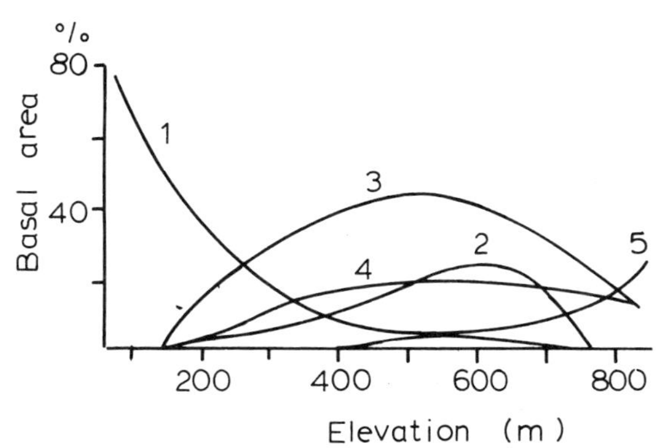

Fig. 9.6. Distribution of the percentages of basal area of dominant species along the elevational gradient of slopes in broad-leaved evergreen forests of Osumi Peninsula (Kitazawa et al., 1960). *1* = *Machilus thunbergii*; *2* = *Castanopsis sieboldii*; *3* = *Distylium racemosum*; *4* = *Cyclobalanopsis salicina*; *5* = *C. acuta*.

(2) The second typical sub-alliance has no distinct characteristic species of its own, but is the main component of the Alliance. *Camellia japonica*, *Castanopsis cuspidata*, *C. sieboldii*, *Cinnamomum japonica*, *Cyclobalanopsis acuta*, *C. gilva* and *Ma-*

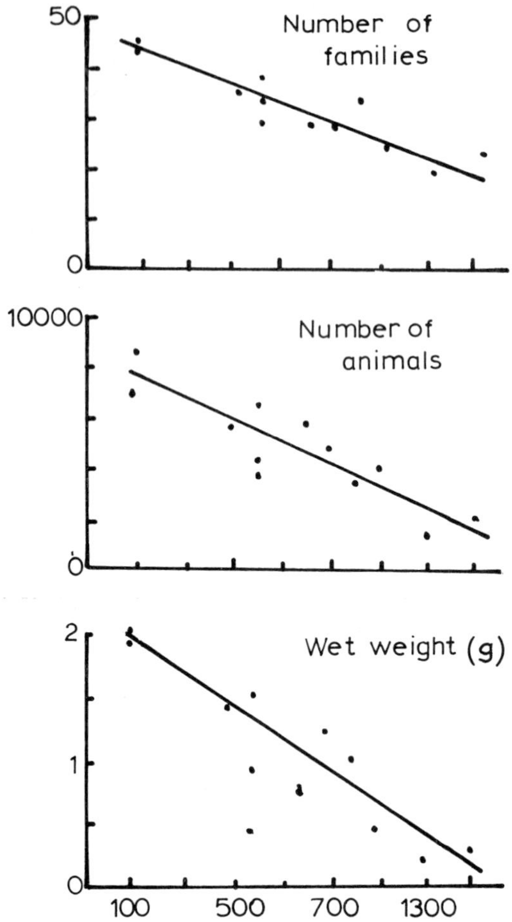

Fig. 9.7. Change of soil meiofauna (average for 1 m^2) along the elevational gradient in southern Kyushu (Kitazawa, 1961).

chilus thunbergii are important members. It is found on lowlands, plateau and foothills but on flat lands of developed areas it is only seen in the precincts of temples and shrines or around large farmhouses. There are five associations.

(3) The Cyclobalanopsion stenophyllae sub-alliance occurs near the upper limit of the Alliance and is relatively more resistant to cold. *Camellia japonica, Castanopsis sieboldii, Cyclobalanopsis acuta, C. salicina, Distylium racemosum* and *Machilus thunbergii* are its important species. There are three associations, of which *Aucubo-Cyclobalanopsietum stenophyllae* is the only one present in the snowy climate of the mountains facing the Sea of Japan.

(4) The *Machilus thunbergii* sub-alliance occurs on lowlands and is hygrophilous and shade-tolerant. *Machilus thunbergii, Camellia japonica, Castanopsis sieboldii, Cinnamomum japonicum* and *Pasania edulis* are its important members. It seems the most thermophilic of the four sub-alliances, for in the south there are subtropical stands of the sub-alliance near the boundary of the *Ficus superba* and *F. microcarpa* forests. Nevertheless, it occurs also further north. There are three associations.

FOREST FUNCTIONAL PROCESSES

In spite of the abundant data available for forest functional processes of coniferous forests (Satoo, 1971), data are very scarce for broad-leaved evergreen forests. This is because they are not of economic importance and good examples of them are found only in limited areas. However, it is expected that abundant information will become available when the results of the investigations on the IBP special area at the Minamata broad-leaved evergreen forest are published.

Organic production and decomposition

Patterns of organic production and decomposition are clearly demonstrated by increases of organic matter and nutrient elements in soil and vegetation. Tezuka (1961) compared organic matter and nutrient contents for "desert" and scrub on the lava flow of the 1778 eruption of O-shima (Izu-shoto) and evergreen–deciduous broad-leaved mixed forests and climax broad-leaved evergreen forests of *Castanopsis* and *Machilus* on lava flows from about the year 684. In Table 9.3 the results of his study are summarized. Organic matter and nutrient elements contained in vegetation and soil increased with the development of vegetation. However, so far as forests are concerned, biomass itself is not an index of productivity. Not only do the life spans of trees differ by species, but also the life spans of tree parts such as trunks, branches and leaves differ. The biomass of a forest is the total sum of accumulation over specified time periods.

Kira and Shidei (1967) postulated that the rate of annual gross production of broad-leaved forests, ranging from deciduous forests to tropical rain forests, is proportional to the product of leaf area index and the length of vegetation season in months. Within the zone of broad-leaved evergreen forests, except for some islands in the south,

TABLE 9.3
Development of vegetation and accumulation of organic matter and nutrient elements in plants and soil on O-shima Volcano (Izu-Shoto) (compiled from tables and figures of Tezuka, 1961)

Site	Desert	Scrub	Evergreen–deciduous broad-leaved mixed forest	Evergreen–deciduous broad-leaved mixed forest	Evergreen broad-leaved climax forest
Degree of cover (%)	10	100	100	100	100
Height (m)	0.5	4	6	9	13
Number of trees (ha^{-1})	0	8000	28 000	13 000	2500
Basal area (m^2 ha^{-1})	0	18	52	50	84
Number of tree species in 100 m^2	0	7	12	13	3
Leaves (t ha^{-1})	0.13	3	11	8	13
Trunks and branches (t ha^{-1})	0	36	125	131	307
Organic matter in soil[1] (t ha^{-1})	15	47	200	325	288
Total nitrogen in foliage (kg ha^{-1})	2.2	62	199	166	144
Total nitrogen in soil[1] (kg ha^{-1})	20	1440	5150	9900	7430
Phosphorus in foliage (kg ha^{-1})	0.2	6.2	15.5	12.9	14.0
Phosphorus in soil[1] (kg ha^{-1})	5.7	113	499	460	340

[1]Organic matter and nutrient contained in the litter layer were omitted.

the frost-free season is about eight months and does not vary greatly (Table 9.1). Consequently, the leaf area index could be a good indicator of organic production. The net production of forests of *Betula maximowicziana*, a deciduous broad-leaved species, has been shown to be proportional to the leaf area index (Satoo, 1970), and similar results have been obtained for woodlands of *Pinus densiflora* (Satoo, 1968a) and plantations of *Larix leptolepis* (Satoo, 1971). Similar trends but on an individual tree basis have been obtained for these species and in a stand of *Cinnamomum camphora* (Satoo, 1968b). Consequently, the annual net production of broad-leaved evergreen forests probably depends on the leaf area index. The available data on the leaf area index of different dominant genera occurring in the overstorey tree layer of temperate broad-leaved evergreen forests in Japan are summarized in Fig. 9.8. Differences depending on the genera of the dominant trees are unclear, and the leaf area index varies widely—from 3 to 13. The dry weights of leaves range from 4.7 to 13.5 t ha^{-1}, with a mean value of 8.9. The amount of leaf in coniferous forests is known to be influenced by conditions such as tree age, soil fertility, fertilization, elevation, relative stand density and air pollution (Satoo, 1971, 1973).

Many studies based on harvesting methods have been done to determine the net production of coniferous plantations in this zone, but there have been relatively few studies for broad-leaved evergreen forests. Annual net production values of 17.8

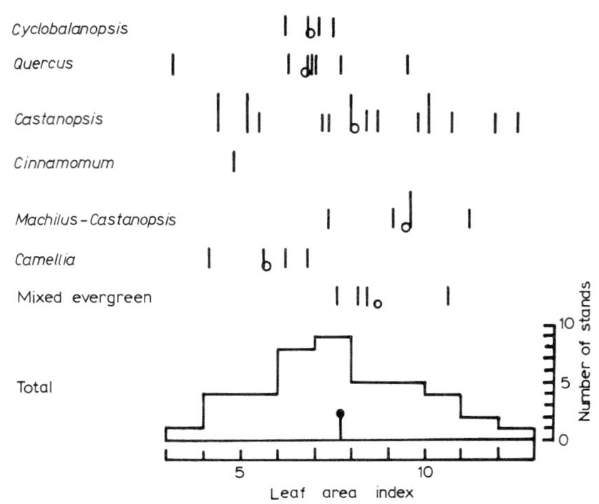

Fig. 9.8. Frequency distribution of the leaf area index of temperate broad-leaved evergreen forests with different dominant genera. Indices for the individual stands studied are indicated by the vertical strokes above the histograms. *Cinnamomum camphora* is a plantation, and four stands of *Castanopsis cuspidata* are of coppice. Open circles: mean values for each genus; closed circles: mean value of whole samples. Sources: Kitazawa et al. (1959), Kimura (1960), Tezuka and Kusumoto (1960), Tezuka (1961), Tadaki et al. (1962, 1964), Kan et al. (1963), Tadaki (1965, 1968), Kan et al. (1967), Kusumoto (1968), Satoo (1968b).

t ha^{-1} have been recorded for a young coppice stand of *Castanopsis cuspidata* in Kyushu (Tadaki, 1965) and 13.6 t ha^{-1} for a mature plantation of *Cinnamomum camphora* with 1.67 t ha^{-1} of undergrowth (Satoo, 1968b). The net production of natural forests of this type has not been determined by harvesting but Kimura (1960) estimated the gross production of a natural forest dominated by *Distylium racemosum* using a mathematical model as 73 t ha^{-1} yr^{-1}.

The available data on the supply of organic matter to the soil by litter fall are given in Fig. 9.9. These values are only of leaves and twigs and do not include larger branches and dead trunks. The values range from 400 to 810 g m^{-2} yr^{-1}, with a mean value of 627. There is no clear evidence of significant differences according to the dominant species or to location in the zone. The organic matter supplied to the soil by litter fall is decomposed by various organisms, the activity of which is measured as "soil respiration". Data of soil respiration in the forests of this type are inadequate. Kirita (1971) found that soil respiration in temperate broad-leaved evergreen forests ranged from 917 to 1556 g m^{-2} yr^{-1} of carbon. Carbon dioxide released from the A_0 soil layer, which contained 17% of the organic carbon in the soil, was 35% of the total soil respiration. The A_0 layer contained 1300 g m^{-2} of organic carbon and released 442 g m^{-2} yr^{-1} of carbon as carbon dioxide; the mineral soil to a depth of 50 cm contained 6400 g m^{-2} of organic carbon and released 830 g m^{-2} yr^{-1} of carbon as carbon dioxide. These values are far larger than the amount of organic carbon supplied annually by litter fall, even if the amount of carbon supplied by dead branches, trunks and roots is taken into account. The reason for this is unclear but similar results have been obtained for other forest types.

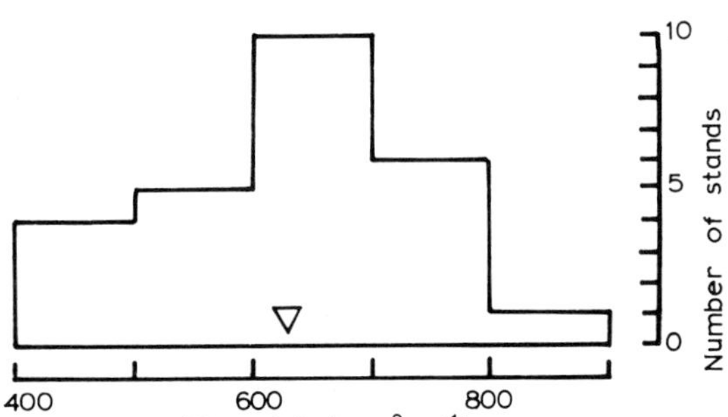

Fig. 9.9. Frequency distribution of the amount of litter fall of broad-leaved evergreen forests. Triangles: mean value of the whole sample. Sources: Tadaki (1965, 1968), Kirita and Saito (1969), Kawahara (1970), Omura and Ando (1970), Kirita (1971), Kabaya et al. (1973).

Energy flow

There are no reliable data on energy flow for Japanese broad-leaved evergreen forest ecosystems. Table 9.4 is based on the mean values of four estimates of gross production (Kimura, 1960; Tadaki, 1965, 1968; Satoo, 1968a, b), two estimates of net production by the harvest method (Tadaki, 1965; Satoo, 1968a, b), and many measurements of litter fall (Fig. 9.9) together with the mean value of soil respiration measured by Kirita (1971). Whilst the data in the Table are based on unsupported assumptions, it provides some idea of energy flow in temperate broad-leaved evergreen forest ecosystems.

Nutrient circulation

The determination of nutrient circulation in a forest ecosystem involves the measurement of the movement of nutrient elements from soil to plant and from plant to soil, of the input into the ecosystem as rain water and as fixation of nitrogen

TABLE 9.4

Approximate calculation of the energy flow in broad-leaved evergreen forests (sources: Kimura, 1960; Satoo, 1968a, b; Tadaki, 1965, 1968)

	Dry matter (t ha^{-1})	Energy (10^9 J ha^{-1})	Ratio energy: total incident energy
Incident solar energy at Kumamoto	—	52 400	1
Gross production	56	938	0.0179
Respiration loss	36	603	0.0115
Net production	20	335	0.0064
Loss by shedding			
leaf and twig	6	101	0.0019
bough	?	?	?
root	?	?	?
Loss by grazing	?	?	?
Accumulation of dry matter	14–?	235–?	0.0045–?
Soil respiration	23?	385?	0.0074?

by micro-organisms, and of the output as discharged water and by the harvesting of timber, if any. In typical broad-leaved evergreen forests, only the movement of nutrient elements from plant to soil by litter fall has been measured. Kawahara (1970) studied the supply of nutrient elements to the soil by litter fall in a broad-leaved evergreen forest dominated by *Castanopsis cuspidata* and recorded an annual return per hectare of 72.5 kg of nitrogen, 2.8 kg of phosphorus, 16.4 kg of potassium, 72.6 kg of calcium and 13.8 kg of magnesium. Iwatsubo and Tsutsumi (1968) and Iwatsubo et al. (1969) studied the nutrient circulation in a woodland mainly consisting of *Ilex pedunculosa* 5 to 7 m tall. Their results are summarized in Table 9.5, but this woodland is not a good example of temperate broad-leaved evergreen forest ecosystems.

Water circulation

Shirai and Takeshita (1968) studied the water economy of a catchment in southern Kyushu covered by a broad-leaved evergreen forest consisting mainly of *Castanopsis* spp., *Cyclobalanopsis salicina* and *Machilus thunbergii*. They gave mean values of observations taken over six years. Of 3073 mm of annual precipitation, 1976 mm was discharged as run-off and of the remaining 1095 mm, 771 mm was calculated as evapotranspiration and the remaining 324 mm was intercepted by the forest canopy. Observations for one year in a small woodland mainly of *Ilex pedunculosa* (Iwatsubo and Tsutsumi, 1967) gave 1462 mm for annual precipitation, 262 mm for interception by the crown canopy and 1200 mm reached the ground as throughfall and stemflow, with 750 mm discharged as run-off into a stream. In another year, the following values were obtained: annual precipitation 1793 mm, throughfall 1015 mm and stemflow 418 mm.

Significance of forest functional processes in relation to global dynamics

Since the potential land area of forests of this type in Japan is not large on a global scale, its potential contribution to global dynamics is negligible. In fact, the present land area occupied by this type of forest is far less than the potential area. If forests of a similar type in the continental area of East Asia are considered together, their effects on global dynamics must be considerable, though information on their status is not available.

PAST, PRESENT AND POTENTIAL USE

Because broad-leaved trees, other than beech and deciduous oaks, are given together in the forest statistics as "other hardwoods", it is impossible to obtain statistical data for broad-leaved evergreen forests alone. Consequently, the figures cited here inevitably also include data from deciduous broad-leaved forests.

TABLE 9.5

Circulation of nutrient elements in a woodland of *Ilex pedunculosa* (compiled from Iwatsubo and Tsutsumi, 1968; and Iwatsubo et al., 1969)

	N	P	K	Ca	Mg
Reservoir (kg ha^{-1})					
soil	3988	542	5798	337	3868
plants	312	19	250	344	121
Circulation (kg ha^{-1} yr^{-1})					
from soil to plants	86.7	4.2	98.6	95.5	26.3
from plants to soil					
by litter fall	40.2	1.8	34.6	41.6	5.9
by leaching	3.0	−0.3	29.9	7.0	4.8
fixed into plants	43.5	2.7	34.1	46.9	15.6
Input by rain (kg ha^{-1} yr^{-1})	5.5	0.5	2.7	8.8	1.3
Output by discharge in run-off (kg ha^{-1} yr^{-1})	0.6	0.1	2.3	5.0	2.3

Wood production

Wood from temperate broad-leaved evergreen forests does not make good timber for use in house-building and other purposes compared with softwoods such as *Cryptomeria japonica* and *Chamaecyparis obtusa*. Many forest stands of this type have already been converted into plantations of softwood species. Especially in Kyushu, genetical improvement of *Cryptomeria japonica* by means of selection and propagation with cuttings has a history longer than a century and many of the cultivars have been used widely.

The wood of *Cyclobalanopsis* species is regarded as the best material for making the paddles and oars of boats, and from Kagoshima alone 30 000 pieces have been exported to other parts of the country (Honda, 1901). The shafts of spears and other hand weapons were also made from this wood. It has been used also for the handles of such tools as hammers and hoes, and for wooden carts. The wood of some other species was used for various small items according to the wood properties. Such uses are now of no importance in an age of mechanization and mass production.

Wood from temperate broad-leaved evergreen forests has been used as fuel in large quantities. Charcoal made of wood of *Cyclobalanopsis* species was regarded as of the highest quality and most expensive. Charcoal made from *Quercus phylliraeoides* wood was used as fuel in preparing food for a favorite Japanese dish. The technique of charcoal-making was so elaborate that for fun charcoal was even made of flowers and leaves, keeping their original shape. However, other energy sources replaced wood as a fuel and the production of firewood and charcoal decreased greatly. Comparing the 1953 and 1974 editions of the "Forestry Statistics Manual", the production — in the part of Japan covered mostly with this type of forest — of firewood and charcoal in 1972 was only 0.58% and 5.9%, respectively, of the production in 1951.

Another important potential use of wood from forests of this type is pulpwood, but, since the terrain is difficult and the cost of logging operations high, it is difficult to compete with imported woodchips. An important use of wood from forests of this type is for the culture of a mushroom, *Cortinellus edodes*. The Japanese word for this mushroom, *siitake*, means mushroom of *sii*, *Castanopsis* spp. This mushroom, which is used in Chinese and Japanese cuisines, is cultured on logs of hardwoods including *Castanopsis* and *Cyclobalanopsis* species. The mushrooms are used fresh or dried. Since separate statistics for evergreen hardwoods are not available, regional statistics for this forest zone have been used. In 1972, about 8600 t of dried mushrooms and 20 300 t of fresh mushrooms were produced, consuming 917 000 m^3 of logs. Possibly some deciduous species like *Quercus acutissima* are included in these figures of wood use.

Other products

Cinnamomum camphora, an important member of broad-leaved evergreen forests, was very valuable as the source of camphor. The history of camphor production is old, and in the late sixteenth century camphor was an important export product from Japan. There are old documents about the exploitation of camphor trees from natural forests and its regulation. The planting of this species was encouraged by the government and later the production of camphor became a government monopoly. Several yield tables were produced for plantations of this species, with volume of trunks, branches and stumps and the amount of camphor produced. In Japan (not including Taiwan, then under Japanese control), 1884 t of camphor and 1592 t of camphor oil were produced in 1933 (Aoki, 1938). In recent years the use of camphor has decreased and it is not produced in Japan any more. Celluloid, a main product from camphor, has been replaced by plastics made from petroleum, and camphor as an insect repellent has been replaced by synthetic materials.

Many kinds of minor products were produced from forests of this type. Some are still produced, but in small quantities, and their money value is so small that no statistics are available. For example, essential oils and high-quality tooth picks are made from *Lindera umbellata*. The forest products include such items as the "bark" of the palm *Trachycarpus fortunei* for fibre (214 t in 1972), fibre of *Broussonetia kazinoki* and *Edgeworthia chrysantha* for extra-quality paper making (540 t), and seeds of *Camellia japonica* to produce good quality oil (91 t). These come mostly from plantations.

Temperate broad-leaved evergreen forests are more significant for the water resources than for the production of wood and other commodities. Of especial importance is the role of the forest in moderating the peak flow, so that water is discharged gradually after the heavy rains which occur frequently. The forests are in an area with heavy precipitation, and in the culture of rice, the most important food crop, a large quantity of water is used.

Aesthetics

Because of their diversity, broad-leaved evergreen forests provide unique scenery, especially in spring with the pale-green fresh leaves contrasting with the dark green of old leaves. Unfortunately they do not display the beauty of autumn color. Forests of this type remaining around shrines and temples provide perspective and also emphasis in the landscape around farming villages.

Recreation

Since these forests are close to heavily populated cities, their recreational use is important. They are used for picnics, hiking and camping. It is said that the number of hunters and fishermen far exceeds the number of game animals in the forest and fish in the brooks. With increasing urbanization, the recreational values of these forests are increasing. Associated with a peculiar landscape of volcanos and hot springs, some of these forests are included in national parks and other nature parks. For example, in Kyushu, where broad-leaved evergreen forests are dominant, there are four major National Parks covering 191 000 ha and seven secondary National Parks covering 152 000 ha. In addition to these parks, many natural forests of this type are also designated as recreational forests and are equipped with the necessary facilities.

MANAGEMENT

Forest legislation and administration

Regulations concerning forestry have been carried out under the "Forest Law" since 1897. This law was completely revised in 1951 and more recently in 1974. Under this law, 15-year plans for the forests of the whole country and 10-year plans for districts are set up and revised every five years. The general policy in forestry is based on the "Basic Law for Forestry" of 1964. There are many other laws for such specific purposes as forest pest control and the regulation of planting stock production. The Forestry Agency of the Ministry of Agriculture and Forestry is responsible for enforcing these laws. Local governments have divisions or sections for forestry administration.

Nature conservation is carried out under the "Law for the Conservation of the Natural Environment" of 1972 and the "Law for Natural Parks" of 1957. For the protection of wildlife there is a "Law for Wildlife Protection and Hunting" of 1918. A Government Environment Agency is responsible for enforcing these laws. There are many bylaws and regulations under these laws.

For forests in nature reserves and nature parks and for protection forests, the restrictions on operations range from strict preservation to limited operations. Other forests are managed under authorized management plans, one for each forest.

Planted and non-planted forests

There are hardly any planted forests of broad-leaved evergreen species, except for *Cinnamomum camphora*, which was planted when camphor was produced, and *Camellia japonica* to produce seeds for oil. These plantations are not large in area, and in forest statistics the planting of broad-leaved evergreen forest, if any, is classified as "others".

Management systems and objectives

When temperate broad-leaved evergreen forests are clear-cut, they are regenerated by coppicing or planted with conifers, mostly the latter in recent years. Where the forests are selectively cut, only some useful species are cut. The objective of coppicing is the production of fuelwood or logs for mushroom culture, and conifers are planted for the production of saw-logs.

Management techniques

Coppice stands are managed either by clear-cutting or selective cutting. In clear-cutting the rotation is 20 to 30 years to get logs of the proper size for firewood, charcoal, or for mushroom culture. In selective cutting only logs of approved sizes are cut and others are left to grow until the next cutting. Since cutting in the growing season, when the reserve food is minimal, results in poor growth of sprouts, this is usually done in winter. When there are too many sprouts per stump, they are thinned to a small number. The improvement of species composition of the stand is done by destroying stumps, taking off sprouts successively, cutting during growing season, or poisoning the stumps to prevent sprouting of unfavorable species. The inplanting of favored species such as *Cyclobalanopsis* spp. is also practised. The techniques used in the conversion of these forests of the plantations to coniferous plantations and the management of the plantations are beyond the scope of this chapter.

Nature of ownership and management controls

Of the 25 million ha of forests in Japan, 8 million are National Forests, 1.1 million belong to prefectures, 1.5 million to municipalities and 14 million to private owners. The percentage of National Forests in the total forest area in this zone is smaller than for other parts of the country. For Kyushu and Shikoku, which belong mostly to the zone of temperate broad-leaved evergreen forests, the National Forests occupy less than 19% of the total forest area. In contrast in northeastern Honshu, in the zone of broad-leaved deciduous forests, the National Forests occupy more than 46%. This difference is not due to the nature of forests but to their history.

National Forests are controlled by the Forestry Agency and administered through fourteen regional forest offices, of which three have responsibilities mostly in this zone and three for some forests in this zone. Under each regional forest office are district forest offices which manage the National Forests directly. In other types of ownership, large owners manage their forests by themselves, and smaller private owners manage their forests by organizing co-operatives, with authorized management plans. Management has to be in accordance with district forest planning. The Government helps forest owners by providing technical aids and advice and by facilitating financing to carry out management in accordance with district forest planning and authorized management plans. Operations are restricted in privately owned forests designated as protection forests, but the owners receive compensation.

The great majority of private ownership is as small holdings. In Kyushu and Shikoku, for example, holdings between 0.1 and 1 ha represent 61.5% of the number of ownerships, 1 to 5 ha 29.7%, 5 to 10 ha 4.7% and 10 to 50 ha 3.8%. In other words, 99.7% of the ownerships are holdings of less than 50 ha. The trend is the same nationally. In addition, even in large holdings most of the forests are split into small pieces. This makes systematic management very difficult.

Afforestation program

There is little, if any, afforestation and reforestation with broad-leaved evergreen trees. According to the statistics for 1972, planting of broad-leaved trees in Kyushu and Shikoku, for example, was 2964 ha. Though tree species are not mentioned, they are most probably fast-growing shade-intolerant deciduous species. It is a general practice that, after clear-cutting broad-leaved evergreen forests, conifers are planted if the soil is good. In 1970, 12 225 ha of natural forests were clear-cut and planted in Kyushu and Shikoku. Most of these natural forests would be broad-leaved evergreen forests and most of the trees planted would be conifers. The best sites on foothills are planted with *Cryptomeria japonica* and the medium sites with *Chamaecyparis obtusa*, in accord with their site requirements. However, in recent years the cutting of forests decreased because of reduced needs of fuelwood and the importation of woodchips. The conversion of native forests to coniferous plantations has also decreased. Future afforestation and reforestation with broad-leaved evergreen species cannot be expected. No broad-leaved evergreen trees are included among the more than 9000 trees selected for seed collection.

Research emphasis

In the past, research on broad-leaved evergreen trees and forests was focused on the improvement of coppice management. Since planting of broad-leaved evergreen trees was rarely done, except for *Cinnamomum camphora*, there has been little research work on planting methods, site requirements and other silvicultural characteristics of tree species other than *C. camphora*.

Because of the importance for water resources of broad-leaved evergreen forests, a forest hydrological station was established in 1956 at Sarukawa, Miyazaki, by the Forest Experiment Station to represent these forests. As important components of Japanese forests, some ecological studies have been made on forests of this type. One of the two IBP special forest research areas, Minamata area in southern Kyushu, was selected to cover such forests. Another consideration in this selection was that very few countries with good research staffs and facilities have forests of this type. Only some results of the research have been published so far; it is expected that this research project will contribute to the understanding of warm-temperate broad-leaved evergreen forest ecosystems. In the Boso Hills in the northern part of this zone, many forest scientists and ecologists are presently working in forests around Tokyo University Forest. Some of the preliminary results of these projects have been given in previous sections.

ECONOMIC STATUS

The fact that wood from broad-leaved evergreen forests is included in the item "others" in the statistics shows clearly the economic status of forests of this type.

Economic basis of forestry operations and contribution to regional economy

The economic value of broad-leaved evergreen forests decreased rapidly when the consumption of wood for fuel diminished. Owing to their diversity in species composition and size structure, efficient operations are very difficult compared to the monocultures of conifers which are replacing forests of this type. Forests of coniferous plantations contribute greatly to the regional economy. The contribution of broad-leaved evergreen forests is not much, except for water supply and recreational use, which do not give money directly. Mushroom culture using logs from forests of this type brings some cash income. Because of the increase in consumption of fresh mushrooms with the development of transport facilities, the production of mushrooms is increasing. However, for the production of logs for mushroom culture, a fast-growing deciduous tree species like *Quercus acutissima* is preferred for planting rather than the slow-growing evergreen *Castanopsis* and *Cyclobalanopsis* species.

Significance in international trade

Camphor produced from *Cinnamomum camphora* in this zone had a significant share in the world production and trade, but camphor is not produced here any more. At present, the only commodity having significance in international trade is dried mushroom, *siitake*. In 1974, 2640 t of *siitake* out of 12 262 t of the national production were exported to Hong Kong, Singapore and the U.S.A. In money value the export amounted to 9727 million yen. More than 90% of the dried mushroom production was from that part of the country having broad-leaved evergreen forests, more than 70% was from Shikoku and Kyushu, 45% was produced from Miyazaki and Oita, two prefectures in Kyushu.

Wood-processing industry

There is no wood-processing industry exclusively using wood of broad-leaved evergreen trees. Chips of these trees are used together with other trees in the pulp and paper industry, though they cannot compete with imported woodchips. Since the wood is hard and there are few large trees, they are not used for veneer.

Development pressure and future utilization

Most of the forest on flat land has been converted into agricultural land except in the precincts of shrines and temples which are not large. Most forests on slopes have been converted into orchards for fruit production, mostly citrus, or tea gardens, and the remainder have been converted into coniferous plantations. With increasing need for recreation, many forests on gentle slopes were turned into golf courses and other recreational facilities. In future, the main use of forests of this type will be recreational, if they are not converted into coniferous plantations.

CONSERVATION STATUS

Exploitation

In historical times the areas covered by natural broad-leaved evergreen forests have decreased. Even in those left as quasi-natural forests, trees of special uses were selectively exploited, *Cinnamomum camphora* and *Cyclobalanopsis* species for example. In forests at remote sites evidence of past felling by axe can usually be found. Among broad-leaved evergreen forests, primeval forest in the strict sense may be very rare, if any, and not much is left even in its looser sense.

Adequacy of protected areas

There are many difficulties in establishing protected areas in an overpopulated country like Japan where there are always severe conflicts in land use. The forest area per capita is only a little more than 0.2 ha including plantations. Only 35% of the wood consumption is from domestic products.

Protected areas are designated by different laws which are independent of each other. The oldest law of this kind is the "Law for Protection of Historical Spots, Scenic Spots, and Natural Monuments", which came into force in 1919 and was revised as the "Law for Protection of Cultural Assets" in 1950. The purpose of this law is to protect objects of high cultural values, including habitats of rare plant or animal species. Some broad-leaved evergreen forests are designated as protection areas under this law, but generally they are rather small areas. The Culture Agency is responsible for enforcement of this law.

Another law is the "Law for National Parks" of 1931, which was succeeded by the "Law for Nature Parks" in 1957. By this law, protected areas including forests are designated, but the law is aimed at the protection of scenic beauty with an emphasis on aesthetic and recreational values. In the zone of broad-leaved evergreen forests there are many Nature Parks including National Parks, but the area strictly protected is only a part of them.

Another law is the "Law for Protection of Natural Environment" of 1972. Under this law, two kinds of protected areas can be designated: "Wilderness Conservation Areas", which have not been affected by any human activities, and "Natural Environment Conservation Areas" to protect areas where natural conditions largely persist; these include natural forests. The area is rather large but not sufficient. As this law is quite new, the number of these protected areas is still small. The only area of broad-leaved evergreen forest protected by this law is Iwaodake in Kagoshima. This forest is a climax forest covering an area of 377 ha, dominated by *Distylium racemosum* and *Cyclobalanopsis salicina*.

In addition, wildlife protection areas are designated by the "Law for Wildlife Protection and Hunting" of 1918. The Environmental Agency is responsible for enforcement of these laws.

Many protection forests for specific purposes, such as the protection of a watershed, have been designated by the "Forest Law" since 1897. All kinds of protection forests according to the purpose have been designated under the law. The Forestry Agency is responsible for enforcement of this law. There are also forests protected for scientific purposes in National Forests and Research Forests belonging to Universities, but they are not protected by any law.

Whilst these protected areas have been established for different purposes, it is a problem that the protection is based on different laws and is carried out by different governmental agencies.

Conservation management

The conservation of protected areas is secured by forbidding various kinds of activities. In article 17 of the "Law for Conservation of Natural Environment", for example, are forbidden such activities as building houses, modification of shape of land, mining and excavation of soil and stones, cutting and planting of trees and bamboos, capture of animals, collection of plants, grazing, use of fire, use of vehicles, horses, and motorboats, etc.

Conservation significance including use as protection forests

The significance of conserving warm-temperate broad-leaved evergreen forests may be, first of all, the fact that forests of this type are not common in the world. Forests of this type are found in the warm-temperate zone with abundant rainfall. The parts of the world having forests of this type are rather heavily populated (e.g. southeastern China, Nepal and Japan) and the impacts of human activities are very strong. Their conservation does not only include the preservation of rare specimens of a type of forest, but also contributes to the knowledge of the functional processes of these forests in comparison with those of other forest types.

The part of Japan having forests of this type is very rainy, especially during the summer typhoon, and fronts bring an extraordinarily large amount of rain in a short while. Under these conditions, forests are very effective in providing protection against flood, washing away of soils, and landslides. Forests along the sea coast are also effective as a barrier against the destructive power of high tides caused by storms and seismic waves and against the movement of coastal sand dunes. The function of forests in securing water resources is especially important, even in this very rainy country, for the main agricultural crop is rice. Rice culture needs a large quantity of water and depends on irrigation systems taking water from small rivers. Water is also important in this highly industrialized country, not to speak of the daily consumption by the large population in the small islands. The significance of forests in water resource management is very high.

Relationships to other forms of land use and their significance as landscape components

As emphasized in previous sections, the land potentially occupied by broad-leaved evergreen forests was replaced by many other land uses. Generally speaking, flat lands are used for agriculture, industries and residential areas; better sites on the bottom of slopes are used for orchards, tea gardens, or coniferous plantations, and the upper slopes with poor soil are left for broad-leaved forests. In the countryside, mosaics of broad-leaved evergreen forests, coniferous plantations, rice fields, and vegetable gardens combined with the topography makes a fine landscape. However, economic growth and urbanization have already resulted in the loss of many such landscapes.

REFERENCES

Aoki, S., 1938. *Planting of Camphor Trees and Camphor Making*. Noda Syobo, Taipei, 256 pp. [+ Appendix 24 pp.] (in Japanese).

Doi, T., Uchida, T. and Yoshida, H., 1970. A preliminary study on the population density and biomass of wild voles in the area. In: *Studies of the Biological Production in Laurel Forests. Interim Report on the Minamata IBP Special Area for 1969*, pp. 88–92 (in Japanese).

Higashi, K., Hamada, R. and Ono, Y., 1969. Estimation of population densities of some insects and spiders in the shrub layer of the area by a new sampling method. In: *Studies of the Biological Production in Laurel Forests. Interim Report on the Minamata IBP Special Area for 1968*, pp. 120–131 (in Japanese).

Honda, S., 1899–1900. Vegetation zones of Japan with special reference to forest zones. *Toyo Gakugei Zassi*, 218: 454–467; 219: 497–514; 220: 28–36 (in Japanese).

Honda, S., 1900a. *Forest Zones of Japan*. Ikeda Syobo, Tokyo, 89 pp. (in Japanese).

Honda, S., 1900b. *Description of Forest Zones of Japan*. Maurice de Brunoff, Paris, 46 pp. (in French).

Honda, S., 1901. *Silviculture. Descriptions on Forest Trees, 2. Broad-leaved Trees*. Miura Syoten, Tokyo, 461 pp. (in Japanese).

Honda, S., 1912. *Forest zones of Japan*. Miura Syoten, Tokyo, 2nd ed., 400 pp. (in Japanese).

Ishii, K., 1975. Myriapoda in the Kiyosumiyama area in the Boso Hills. In: *Nature of the Kiyosumiyama–Takagoyama Areas in the Boso Hills and Impacts of Man on It*, pp. 35–39 (in Japanese).

Iwatsubo, G. and Tsutsumi, T., 1967. On the amount of plant nutrients supplied to the ground by rainwater in adjacent open plots and forests. *Bull. Kyoto Univ. For.*, 39: 110–124 (in Japanese, with English abstract).

Iwatsubo, G. and Tsutsumi, T., 1968. On the amount of plant nutrients contained in run-off water. *Bull. Kyoto Univ. For.*, 40: 140–156 (in Japanese, with English abstract).

Iwatsubo, G., Nishida, Y. and Tsutsumi, T., 1969. Returns to the ground of nutrient substances by litter fall and rain. *Trans. Meet. Jap. For. Soc., 80, Abstr.* No. 325 (in Japanese).

Kabaya, H., Mototani, I., and Staffs of Tokyo University Forest in Tiba, 1973. Litter-fall in evergreen broad-leaved forests, fir–hemlock natural forests, deciduous–evergreen mixed broad-leaved forests, mature plantations of *Cryptomeria japonica* and *Chamaecyparis obtusa*. In: *Nature of the Kiyosumiyama–Takagoyama Areas in the Boso Hills and Impacts of Man on It*, pp. 61–75 (in Japanese).

Kan, M., Shidei, T., Tsutsumi, T, and Suzaki, K., 1963. Productivity of evergreen broad-leaved forests. *Trans. Meet. Jap. For. Soc.*, 74: 171–172 (in Japanese).

Kan, M., Saito, H. and Shidei, T., 1967. Studies of the productivity of evergreen broad-leaved forests. *Bull. Kyoto Univ. For.*, 37: 55–75 (in Japanese, with English abstract).

Karizumi, N., 1970. Notes on the forest vegetation of Japan as appeared in a Chinese book of the third century. *Ringyoo Gizitu*, 334: 15–20 (in Japanese).

Kawahara, T., 1970. Supply of nutrient elements by litter fall in the laurel forest of Minamata. In: *Studies of the Biological Production in Laurel Forests. Interim Report of the Minamata IBP Special Area for 1969*, pp. 65–71 (in Japanese).

Kawakita, J., 1956. *Vegetation. Land and Crops of Nepal Himalaya.* Kyoto, pp. 1–65 (from Yoda, 1967).

Kimura, M., 1960. Primary production of the warm-temperate laurel forest in the southern part of Osumi Peninsula, Kyushu, Japan. *Misc. Rep., Res. Inst. Nat. Resour.*, 52–53: 36–47.

Kira, T., 1949. *Forest Zones of Japan.* Ringyo Gizitu Kyokai, Tokyo, 41 pp. (in Japanese).

Kira, T. and Shidei, T., 1967. Primary production and turnover of organic matter in different forest ecosystems of the western Pacific. *Jap. J. Ecol.*, 17: 70–87.

Kirita, H., 1971. Studies of soil respiration in warm-temperate evergreen broad-leaf forests of southern Japan. *Jap. J. Ecol.*, 21: 230–244 (in Japanese, with English abstract).

Kirita, H. and Saito, H., 1969. Measurement of litter-fall amount in the Minamata IBP Area. In: *Studies of the Biological Production in Laurel Forests. Interim Report of the Minamata IBP Special Area for 1968*, pp. 74–91 (in Japanese).

Kitazawa, Y., 1961. Zonal pattern of ecosystems in the southern part of Kyushu. *Misc. Rep., Res. Inst. Nat. Resour.*, 54–55: 75–85 (in Japanese, with English abstract).

Kitazawa, Y., Kimura, M., Tezuka, Y., Kurasawa, H., Sakamoto, M. and Yoshino, M., 1959. Plant ecology of the southern part of Osumi Peninsula. *Misc. Rep., Res. Inst. Nat. Resour.*, 49: 19–36 (in Japanese, with English abstract).

Kitazawa, Y., Kimura, M., Kusumoto, T. and Yoshino, M., 1960. Distribution of species populations of warm-temperate broad-leaved trees along the elevation gradient of Osumi Peninsula, the southernmost portion of Kyushu, Japan. *Misc. Rep., Res. Inst. Nat. Resour.*, 52–53: 24–36 (in Japanese, with English abstract).

Kitazawa, Y., Kusumoto, T. and Kimura, M., 1961. Distribution of species population of forest trees along an elevation gradient at Mt. Kirishima, southern part of Kyushu, Japan. *Misc. Rep., Res. Inst. Nat. Resour.*, 54–55: 86–92 (in Japanese, with English abstract).

Kubo, H., Uramoto, M., Nishida, T., Iida, K. and Kobayashi, T., 1969. Avifauna in the Minamata IBP Area and population density of some common species. In: *Studies of the Biological Production of Laurel Forests. Interim Report of the Minamata IBP Special Area for 1968*, pp. 95–101 (in Japanese).

Kusumoto, T., 1957a. Physiological and ecological studies on plant production in plant communities. 3. Ecological considerations of the temperature–photosynthesis curves of evergreen broad-leaved trees. *Jap. J. Ecol.*, 7: 126–130 (in Japanese, with English abstract).

Kusumoto, T., 1957b. Physiological and ecological studies on plant production in plant communities. 4. Ecological studies on the apparent photosynthesis curves of evergreen broad-leaved trees. *Bot. Mag., Tokyo*, 70: 299–304.

Kusumoto, T., 1963. Physiological and ecological studies on plant production in plant communities. 8. The activities of photosynthesis and respiration in broad-leaved evergreen trees growing in the littoral region or highland in Kagoshima Prefecture. *Bull. Fac. Educ. Kagoshima Univ.*, 15: 62–73 (in Japanese, with English abstract).

Kusumoto, T., 1968. On the standing crop of forest of *Quercus phylliraeoides* A. Gray in the southern part of Osumi Peninsula. *Bull. Fac. Educ. Kagoshima Univ.*, 19: 25–39 (in Japanese, with English abstract).

Mayr, H., 1891. *The Forests of Japan.* Rieger, München, 59 pp. (in German).

Mishima, Y., 1970. Seasonal fluctuations in population density of some land-snails. In: *Studies of the Biological Production in Laurel Forests. Interim Report of the Minamata IBP Special Area for 1969*, pp. 114–126 (in Japanese).

Miyawaki, A. and Ohba, T., 1963. *Castanopsis sieboldii* forests in Amami Islands. *Sci. Rep. Yokohama Nat. Univ.*, Sect. II, 109: 31–48 (in German).

Miyoshi, S., 1955. Studies on the improvement of management of the natural broad-leaved forests for firewood in Miyazaki district. *J. Jap. For. Soc.*, 37: 537–541 (in Japanese, with English abstract).

Mizuguchi, K., Yamamoto, Y., Sengoku, S., Hamaguchi, T., Uramoto, M., Nganezawa, M. and Nishikata, S., 1973. Vertebrate fauna in Boso Hills. In: *Nature of the Kiyosumiyama–Takagoyama Areas in the Boso Hills and Impacts of Man on It*, pp. 40–46 (in Japanese).

Nakano, H., 1942. Composition of forest vegetation of Japan. *Bot. Mag., Tokyo*, 56: 186–190 (in Japanese).

Obata, S., 1961. *Studies of Growth and Structure of Warm-Temperate Broad-Leaved Forests.* Rinyatyoo, Tokyo, 167 pp.

Omura, S. and Ando, M., 1970. Litter-fall in a forest of *Castanopsis cuspidata*. In: *Studies of the Biological Production of Laurel Forests. Interim Report of the Minimata IBP Special Area for 1969*, pp. 50–55 (in Japanese).

Oono, S., 1934. Trees appearing in Koziki and Nikonsyoki. *J. Jap. For. Soc.*, 16: 302–314 (in Japanese).

Satoo, T., 1968a. Primary production relations in woodland of *Pinus densiflora*. In: H.E. Young (Editor), *Symposium on Primary Productivity and Mineral Cycling in Natural Ecosystems.* University of Maine Press, Orono, Me., pp. 52–80.

Satoo, T., 1968b. Primary production and distribution of produced dry matter in a plantation of *Cinnamomum camphora*: Materials for the studies of growth in stands. *Bull. Tokyo Univ. For.*, 64: 241–275.

Satoo, T., 1970. A synthesis of studies by the harvest method: Primary production relations in temperate deciduous forests in Japan. in: D.E. Reichle (Editor), *Analysis of Temperate Forest Ecosystems. Ecological Studies.* Springer-Verlag, Heidelberg, pp. 55–72.

Satoo, T., 1971. Primary production relations of coniferous forests in Japan. In: P. Duvigneaud (Editor), *Productivity of Forest Ecosystems.* UNESCO, Paris, pp. 191–205.

Satoo, T., 1973. *Production in Terrestrial Plant Communities: Forest.* Kyoritsu Shuppan, Tokyo, 100 pp. (in Japanese).

Shinkai, A. and Kurihara, K., 1975. Spiders of forest border and litter on the forest floor: Araneae of Kiyosumiyama. In: *Nature of the Kiyosumiyama–Takagoyama Areas in the Boso Hills and Impacts of Man on It*, pp. 35–39 (in Japanese).

Shirai, J. and Takeshita, M., 1968. Experiment on forest influences upon stream-flow at Sarukawa. Runoff, water loss and storage of Sarukawa Experimental Forest. *Bull. Gov. For. Exp. Stn.*, 216: 127–166 (in Japanese, with English abstract).

Suzuki, T., 1952. *Forest Vegetation of East Asia.* Kokon Syoin, Tokyo, 137 pp. (in Japanese).

Suzuki, T., 1975. The evergreen broad-leaved forests of Japan. *Phytocoenologia*, 2: 293–300 (in German, with English abstract).

Tadaki, Y., 1965. Studies on production structure of forests. The primary production of a young stand of *Castanopsis cuspidata. Jap. J. Ecol.*, 15: 142–165.

Tadaki, Y., 1968. Studies on the production structure of forest. Third report on the primary production of a young stand of *Castanopsis cuspidata. J. Jap. For. Soc.*, 50: 60–65.

Tadaki, Y., Ogata, N. and Takagi, T., 1962. Studies on production structure of forest. Estimation of standing crop and some analysis on productivity of young stands of *Castanopsis cuspidata. J. Jap. For. Soc.*, 44: 350–359 (in Japanese, with English abstract).

Tadaki, Y., Ogata, N. and Kawazu, A., 1964. Productivity of young stands of *Castanopsis cuspidata. Trans. Meet. Jap. For. Soc.*, 75: 345–347 (in Japanese).

Tanaka, M. and Hamada, R., 1969. Preliminary study on the soil fauna in the area. In: *Studies of the Biological Production in Laurel Forests. Interim Report of the Minamata IBP Special Area for 1968*, pp. 106–117.

Tanaka, Y., 1885. *Report of Investigation of Vegetation Zone of Japan.* Naimusyoo Tirikyoku, Tokyo, 268 pp. [+ Appendix 33 pp.] (in Japanese).

Tezuka, Y., 1961. Development of vegetation in relation to soil formation in the volcanic island of Oshima, Izu, Japan. *Jap. J. Bot.*, 17: 371–402.

Tezuka, Y. and Kusumoto, T., 1960. On the secondary forest of the southern part of Osumi Peninsula Kyusyu. *Misc. Rep., Res. Inst. Nat. Resour.*, 52–53: 48–56 (in Japanese, with English abstract).

Toyama, S., 1936. Microchemical studies of wood of camphor tree from buried canoes (more than 1400 years old). *J. Jap. For. Soc.*, 18: 825–836 (in Japanese, with English abstract).

Wang, C.W., 1961. *The Forests of China with a Survey of Grassland and Desert Vegetation.* Maria Moors Cabot Foundation Publication 5. Harvard University, Cambridge, 313 pp.

Yoda, K., 1967. A preliminary survey of the forest vegetation of eastern Nepal. General description, structure and floristic composition of the sample plots chosen from different vegetation zones. *J. Coll. Arts Sci., Chiba Univ., Nat. Sci. Ser.*, 5: 99–140 [18 plates].

Chapter 10

PEOPLE AND THE TEMPERATE BROAD-LEAVED EVERGREEN FOREST

J.D. OVINGTON

HUMAN INFLUENCE

As the number of people in the world has multiplied and living standards in the developed nations have improved, the total area of forest has diminished. Little, if any, unmodified natural forest remains. Probably about one third of the terrestrial surface of the world can still be regarded as forest land (4126 million hectares) and of this about 3172 million hectares have a relatively close forest cover.

About half of the remaining forest area is outside the tropical zone and this contributes approximately 90% of the wood used commercially in the world. The contribution of temperate broad-leaved evergreen forest to the world supply of wood and other forest products is unknown but probably amounts to less than 20%. However, in Europe and Asia this contribution is of particular importance because of the reduction in transportation costs resulting from the closeness of the forest resource to population centres.

There is considerable potential to increase the production of forest products from the temperate broad-leaved evergreen forest by more intensive management and the re-establishment of forest in cleared areas. Whilst the current level of harvesting gives little indication of potential productivity, large gains in wood production have been obtained where trees of some species of the temperate broad-leaved evergreen forest have been grown as plantations.

If the anticipated increase in the world demand for forest products occurs within the next half century, human impact on the temperate broad-leaved evergreen forest is likely to be greatly enhanced. This emphasizes the urgency for a more detailed understanding of the global importance of this forest in terms of its economic, cultural and conservation significance to human well-being. The impact of people on the temperate broad-leaved evergreen forest will vary immensely.

Areas of the Northern Hemisphere in which temperate broad-leaved evergreen forest is the natural vegetation tend to coincide with ancient centres of human population and have a long history of use and management with increased crowding and industrialization. Consequently, much of the forest has been cleared, and what remains is greatly modified. This deforestation may be seen as part of a wide-scale process leading to progressive soil deterioration, increasing aridity and desertification. Usually the remaining forest is now of little economic importance, its main significance being seen in terms of cultural, aesthetic and grazing values as well as the protection it gives to soil, water, air and wildlife.

Taking the Mediterranean countries as an example, it is evident that deforestation has been caused there by a combination of natural and human factors. Grazing, fire and the removal of wood to provide fuel and structural timber have resulted in extensive forest destruction. The remaining forest is not very productive; whilst there has been some reafforestation, the area of natural forest continues to diminish.

In contrast, human influence has been less pervasive in the Southern Hemisphere, and large areas of relatively undisturbed broad-leaved evergreen forest remain in the temperate zone. As much of this forest has evolved in relative isolation, it is of special scientific interest particularly because of the many endemic and unusual species of plants and animals present.

FUTURE MANAGEMENT NEEDS

Since the broad-leaved evergreen forest of the Northern Hemisphere has been changed so extensively, the main management need is to protect and rehabilitate the remaining forest and especially to improve soil conditions and reduce the adverse effects of fire and grazing.

In some areas, such as around the Mediterranean, there has been a remarkable increase in tourism within the last century with great importance being attached to the maintenance of the existing man-made landscapes. Tourism is placing additional demands on the forest areas and can restrict management options through restraints placed on the distribution of forest and the kinds of forest operations that can be used. New management techniques are required to reconcile tourist and forestry interests.

With an increasing world population and a growing demand for food and wood, attention is being directed more towards the relatively untouched evergreen hardwood forest of the temperate zone of the Southern Hemisphere as a cheap source of timber, for plantation forestry usually of introduced softwoods, and as potential areas for agriculture. The broad-leaved evergreen forest of the Southern Hemisphere constitutes a timber resource of considerable potential economic significance, with large areas of high-quality timber remaining. The wood of *Nothofagus* is commercially valuable, and some eucalypts, the mountain ash and karri which are native only in Australian temperate broad-leaved forest, are the tallest hardwood trees in the world.

If the experience of unplanned destructive exploitation in the Northern Hemisphere is not to be repeated in the Southern Hemisphere, there is a need to implement new management procedures based on conservation principles and more comprehensive research on the structure, functioning and dynamics of the different kinds of native broad-leaved evergreen forest ecosystems, especially with respect to the impact on them of human activities.

Because of the restricted distribution of temperate broad-leaved evergreen forest in the Northern Hemisphere and the limited scientific resources available in the Southern Hemisphere, our knowledge of these forests is remarkably scanty. Even in developed countries such as New Zealand and Australia, where extensive areas of these forest types still remain, silvicultural research has tended to be orientated towards replacement plantations of the exotic *Pinus radiata*, because of the commercial emphasis placed on the production of softwood timber. A much greater research effort with the establishment of major research institutes is required to investigate different management techniques for native forest.

The few data available indicate that some temperate broad-leaved evergreen forest is not very productive either of organic matter or of commercial wood. Paradoxically some tree species of these forests, especially eucalypts, have been widely introduced outside their natural range because of their growth potential. The reasons for the differences in productivities of the various forest types are largely unknown. In some instances poor production seems to have been caused by past selective cutting, which removed the better trees and resulted in slower-growing trees or non-commercial species being left as the growing stock. Silvicultural research could increase forest productivity and the economic return and so encourage forest protection.

Fire is an important factor affecting temperate broad-leaved evergreen forest even in many moist forests. Whilst some tree species such as the oak *Quercus coccifera* and species of eucalypts are pyrophytes, being able to regenerate vegetatively after fire, others fail to regenerate. There is an urgent need to identify the effects of fire and to rationalize the use of fire as a tool of forest management.

Other important issues requiring investigation include the global influence of these forests especially in relation to the hydrological cycle, the revitalization of degraded forest and how best to achieve an appropriate balance between development and the survival of the forest.

The preservation of representative examples of different natural ecosystems of the temperate broad-leaved evergreen forest through the establishment and protection of forest reserves is important in that it provides security for the unique plant and animal genetic diversity it contains. The fauna of temperate broad-leaved evergreen forest ecosystems is often of considerable interest, and protected areas are especially important since many animal species depend for their existence on the maintenance of adequate forest areas. Such areas can also serve as yardsticks against which to monitor the impact of human intervention. A satisfactory network of forest reserves requires international co-ordination to minimize duplication and to ensure reasonable coverage of the main types and some consistency of management.

The continued presence of temperate broad-leaved evergreen forest will depend in the long term on public recognition of the benefits it provides. Consequently, it is important that programmes of public education at local, national and international levels be mounted to make people aware of the significance of this forest and of the implications of its increasing utilization.

SYSTEMATIC LIST OF GENERA[1]

MONERA (PROKARYOTA)

BACTERIA

PLANTS

FUNGI

ASCOMYCOTINA
Coryneliaceae
Coryneliospora
Cyttariaceae
Cyttaria
Englerulaceae
Parenglerula
Schiffnerula
Meliolaceae
Irene [= *Asteridiella*]
Meliola
Microthyriaceae
Asterina
Asterodothis
Lembosia

BASIDIOMYCOTINA
Agaricaceae
Armillaria
Cortinellus [= *Tricholoma*]
Boletaceae
Ganodermataceae
Ganoderma
Polyporaceae
Fomes

DEUTEROMYCOTINA (FUNGI IMPERFECTI)
Agonomycetaceae
Rhizoctonia
Melanconiaceae
Pestalotia
Tuberculariaceae
Fusarium

PHYCOMYCOTINA
Pythiaceae
Phytophthora
Pythium

LICHENES

ASCOLICHENES
Usneaceae
Usnea

BRYOPHYTA

HEPATICAE
Plagiochilaceae
Plagiochila

MUSCI
Dicranaceae
Dicranoloma
Hookeriaceae
Lepidopilum
Leucobryaceae
Leucobryum
Polytrichaceae
Dendroligotrichum
Sphagnaceae
Sphagnum

PTERIDOPHYTA

FILICOPSIDA
Adiantaceae
Adiantum
Aspidiaceae
Arachniodes
Dryopteris
Lastreopsis
Polystichum
Aspleniaceae
Asplenium
Neottopteris
Blechnaceae
Blechnum
Cyatheaceae
Alsophila
Cyathea
Davalliaceae
Rumohra
Dennstaedtiaceae
Histiopteris
Hypolepis
Pteridium
Dicksoniaceae
Dicksonia
Gleicheniaceae
Gleichenia
Hymenophyllaceae
Hymenophyllum
Serpyllopsis
Trichomanes
Lomariopsidaceae
Elaphoglossum
Lophosoriaceae
Lophosoria
Osmundaceae
Todea
Polypodiaceae
Loxogramme
Phymatosorus
Pleopeltis
Polypodium
Schizaeaceae
Lygodium
Thelypteridaceae
Pneumatopteris
Thelypteris
Vittariaceae
Vittaria

LYCOPSIDA
Lycopodiaceae
Lycopodium

PINOPHYTA (GYMNOSPERMAE)

PINOPSIDA (CONIFERAE)
Araucariaceae
Agathis
Araucaria
Cupressaceae
Callitris
Chamaecyparis
Cupressus
Fitzroya
Juniperus
Libocedrus
Pilgerodendron
Tetraclinis
Widdringtonia
Cycadaceae
Cycas
Macrozamia
Pinaceae
Abies
Cedrus
Larix
Picea
Pinus
Pseudotsuga
Tsuga
Podocarpaceae
Dacrycarpus
Dacrydium
Microcachrys
Phyllocladus
Podocarpus
Saxegothaea
Taxaceae
Torreya
Taxodiaceae
Athrotaxis
Cryptomeria

GNETOPSIDA
Ephedraceae
Ephedra

MAGNOLIOPHYTA (ANGIOSPERMAE)

LILIOPSIDA (MONOCOTYLEDONES)
Agavaceae
Cordyline
Dracaena
Amaryllidaceae
Clivia
Haemanthus
Arecaceae
Chamaerops
Cocos
Rhopalostylis
Sabal
Trachycarpus
Bromeliaceae
Fascicularia
Greigia
Tillandsia
Commelinaceae
Commelina
Murdannia
Cyperaceae
Carex
Gahnia
Kobresia
Lepidosperma
Oreobolus
Uncinia
Iridaceae
Aristea
Dietes
Libertia
Juncaceae
Luzula
Liliaceae
Asparagus
Astelia
Collospermum
Dianella
Liriope
Ripogonum
Musaceae
Ensete
Orchidaceae
Angraecum
Bulbophyllum
Dendrobium
Earina
Mystacidium
Spiranthes
Tridactyle
Pandanaceae
Freycinetia
Philesiaceae
Behnia
Lapageria
Luzuriaga
Philesia
Poaceae
Andropogon

---

[1]The taxonomic relationships of all genera of plants and animals mentioned in this Volume on *Temperate Broad-Leaved Evergreen Forests* are shown. Genera are listed alphabetically under families, arranged into phyla and subphyla; orders are included in the animal kingdom.

LILIOPSIDA (*continued*)

Aristida
Arundinaria
Arundinella
Bothriochloa
Brachypodium
Chionochloa
Chrysopogon
Chusquea
Eragrostis
Heteropogon
Isachne
Ischaemum
Microlaena
Oplismenus
Oreobambos
Panicum
Pseudobromus
Sasa
Setaria
Stipa
Themeda
Tripogon
Restionaceae
Leptocarpus
Ruscaceae
Ruscus
Smilacaceae
Smilax
Xanthorrhoeaceae
Kingia
Xanthorrhoea
Zingiberaceae
Alpinia
Zingiber

MAGNOLIOPSIDA (DICOTYLEDONES)
Acanthaceae
Brachystephanus
Dicliptera
Duvernoia
Hypoëstes
Justicia
Mackaya
Mimulopsis
Oreacanthus
Sclerochiton
Strobilanthes
Aceraceae
Acer
Aextoxicaceae
Aextoxicon
Alseuosmiaceae
Alseuosmia
Anacardiaceae
Pistacia
Protorhus
Rhus
Annonaceae
Popowia
Apiaceae
Bupleurum
Heracleum
Osmorrhiza
Pimpinella
Apocynaceae
Carissa
Elytropus
Gonioma
Landolphia
Nerium
Parsonsia
Tabernaemontana
Trachelospermum
Aquifoliaceae
Ilex
Araliaceae
Cussonia
Hedera
Meryta
Polyscias
Pseudopanax
Schefflera
Seemannaralia
Asclepiadaceae
Cynanchum
Riocreuxia
Secamone
Asteraceae
Anaphalis
Artemisia
Baccharis
Bedfordia
Brachylaena
Cassinia
Chiliotrichum
Conyza
Crassocephalum
Dasyphyllum
Eupatorium
Olearia
Senecio
Atherospermataceae
Atherosperma
Aucubaceae
Aucuba
Balsaminaceae
Impatiens
Batidaceae
Batis
Begoniaceae
Begonia
Berberidaceae
Berberis
Mahonia
Betulaceae
Alnus
Betula
Bignoniaceae
Campsidium
Bombacaceae
Bombax
Brassicaceae
Cardamine
Brexiaceae
Ixerba
Buddlejaceae
Buddleja
Nuxia
Buxaceae
Buxus
Sarcococca
Caesalpiniaceae
Cassia
Campanulaceae
Campanula
Canarina
Lobelia
Pratia
Canellaceae
Warburgia
Capparidaceae
Ritchiea
Caprifoliaceae
Lonicera
Viburnum
Caricaceae
Cylicomorpha
Casuarinaceae
Casuarina
Celastraceae
Cassine
Catha
Euonymus
Maytenus
Pleurostylia
Pterocelastrus
Chloranthaceae
Ascarina
Cistaceae
Cistus
Halimium
Helianthemum
Clusiaceae
Garcinia
Hypericum
Mammea
Combretaceae
Combretum
Quisqualis
Terminalia
Connaraceae
Cnestis
Coriariaceae
Coriaria
Cornaceae
Afrocrania
Corynocarpaceae
Corynocarpus
Cucurbitaceae
Melothria
Oreosyce
Cunoniaceae
Anodopetalum
Caldcluvia
Platylophus
Weinmannia
Curtisiaceae
Curtisia
Daphniphyllaceae
Daphniphyllum
Dipsacaceae
Dipsacus
Droseraceae
Drosera
Ebenaceae
Diospyros
Euclea
Elaeocarpaceae
Aristotelia
Elaeocarpus
Eleagnaceae
Eleagnus
Empetraceae
Empetrum
Epacridaceae
Archeria
Ardisia
Cyathodes
Dracophyllum
Lebetanthus
Ericaceae
Agauria
Arbutus
Erica
Gaultheria
Lyonia
Pernettya
Philippia
Rhododendron
Vaccinium
Escalloniaceae
Anopterus
Carpodetus
Escallonia
Quintinia
Eucryphiaceae
Eucryphia
Euphorbiaceae
Beyeria
Bridelia
Croton
Excoecaria
Lachnostylis
Macaranga
Mallotus
Micrococca
Neoboutonia
Suregada
Fabaceae
Adenocarpus
Bossiaea
Calpurnia
Calicotome
Carmichaelia
Ceratonia
Clianthus
Colutea
Coronilla
Craibia
Cytisus
Dalbergia
Desmodium
Genista
Indigofera
Ulex
Fagaceae
Castanopsis
Cyclobalanopsis
Fagus
Lithocarpus
Nothofagus
Pasania
Quercus
Shiia [= *Castanopsis*]
Flacourtiaceae
Aphloia
Azara
Casearia
Dasylepis
Doryalis
Homalium
Kiggelaria
Rawsonia
Scolopia
Trimeria
Gesneriaceae
Asteranthera
Mitraria
Sarmienta
Streptocarpus
Goodeniaceae
Goodenia
Griseliniaceae
Griselinia
Grossulariaceae
Ribes
Gunneraceae
Gunnera
Hamamelidaceae
Altingia
Bucklandia
Distylium
Trichocladus
Hydrangeaceae
Hydrangea
Icacinaceae

MAGNOLIOPSIDA (*continued*)

Apodytes
Cassinopsis
Pennantia
Pyrenacantha
Illiciaceae
Illicium
Iteaceae
Choristylis
Juglandaceae
Engelhardtia
Lamiaceae
Lavandula
Phlomis
Plectranthus
Prostanthera
Rosmarinus
Salvia
Satureja
Thymus
Lardizabalaceae
Boquila
Stauntonia
Lauraceae
Beilschmiedia
Cinnamomum
Cryptocarya
Laurus
Lindera
Litsea
Machilus
Neolitsea
Ocotea
Persea
Loganiaceae
Geniostoma
Loranthaceae
Alepsis
Peraxilla
Malvaceae
Hoheria
Plagianthus
Magnoliaceae
Alcimandra
Magnolia
Manglietia
Michelia
Tasmannia
Meliaceae
Dysoxylum
Ekebergia
Entandrophragma
Lepidotrichilia
Swietenia
Toona
Turraea
Melianthaceae
Bersama
Meliosmaceae
Meliosma
Memecylaceae
Memecylon
Menispermaceae
Sinomenium
Stephania
Mimosaceae
Acacia
Albizia
Sophora
Monimiaceae
Hedycarya
Laurelia
Peumus
Xymalos
Moraceae
Broussonetia
Cudrania
Ficus
Myricaceae
Myrica
Myrsinaceae
Embelia
Maesa
Myrsine
Rapanea
Myrtaceae
Agonis
Amomyrtus
Angophora
Eucalyptus
Eugenia
Leptospermum
Lophomyrtus
Lysicarpus
Melaleuca
Metrosideros
Myrceugenella
Myrceugenia
Myrteola
Myrtus
Neomyrtus
Rhodomyrtus
Syncarpia
Syzygium
Temu
Tepualia
Tristania
Ugni
Myzodendraceae
Myzodendron
Naucleaceae
Mitragyna
Nyctaginaceae
Heimerliodendron
Ochnaceae
Ochna
Oleaceae
Jasminum
Linociera
Nestegis
Olea
Phillyrea
Schrebera
Strombosia
Oliniaceae
Olinia
Onagraceae
Fuchsia
Passifloraceae
Tetrapathaea
Peperomiaceae
Peperomia
Piperaceae
Macropiper
Piper
Pittosporaceae
Pittosporum
Polygalaceae
Polygala
Polygonaceae
Muehlenbeckia
Potaliaceae
Anthocleista
Desfontainia
Proteaceae
Banksia
Cenarrhenes
Dryandra
Embothrium
Faurea
Gevuina
Hakea
Knightia
Lomatia
Persoonia
Ptaeroxylaceae
Ptaeroxylon
Ranunculaceae
Clematis
Ranunculus
Thalictrum
Rhamnaceae
Discaria
Helinus
Pomaderris
Rhamnus
Scutia
Trymalium
Rhizophoraceae
Cassipourea
Rosaceae
Alchemilla
Chaenomeles
Cotoneaster
Crataegus
Fragaria
Hagenia
Leucosidea
Photinia
Potentilla
Poterium
Prunus
Pyrus
Quillaja
Rosa
Rubus
Rubiaceae
Aulacocalyx
Burchellia
Canthium
Coprosma
Damnacanthus
Galiniera
Galium
Gardenia
Lasianthus
Nertera
Oldenlandia
Oxyanthus
Pauridiantha
Pavetta
Psychotria
Randia
Rothmannia
Rubia
Tricalysia
Urophyllum
Vangueria
Rutaceae
Calodendrum
Citrus
Clausena
Fagaropsis
Geijera
Melicope
Skimmia
Teclea
Toddalia
Zanthoxylum
Sambucaceae
Sambucus
Santalaceae
Exocarpus
Sapindaceae
Alectryon
Allophylus
Cupania
Dodonaea
Sapotaceae
Aningeria
Argania
Chrysophyllum
Planchonella
Schisandraceae
Kadsura
Scrophulariaceae
Bowkeria
Halleria
Striga
Solanaceae
Discopodium
Solanum
Staphyleaceae
Turpinia
Sterculiaceae
Brachychiton
Cola
Dombeya
Styracaceae
Symplocaceae
Symplocos
Ternstroemiaceae
Ternstroemia
Theaceae
Balthasaria
Camellia
Cleyera
Eurya
Ficalhoa
Gordonia
Schima
Thymelaeaceae
Dais
Daphne
Edgeworthia
Gnidia
Ovidia
Peddiea
Wikstroemia
Ulmaceae
Celtis
Urticaceae
Australina
Elatostema
Laportea
Myrianthus
Parietaria
Pilea
Urera
Valerianaceae
Valeriana
Verbenaceae
Clerodendrum
Rhaphithamnus
Vitex
Violaceae
Melicytus
Viola
Vitidaceae
Cissus
Rhoicissus
Winteraceae
Drimys
Pseudowintera

ANIMALS

ASCHELMINTHES

NEMATODA
- *Bursaphelenchus*

ARTHROPODA

ARACHNIDA
- Araneae
  - Phalangida
  - Cyphophthalmi

CRUSTACEA
- Branchiopoda
  - Cladocera
- Malacostraca
  - Amphipoda
  - Isopoda

DIPLURA
- Japygidae
  - *Heterojapyx*

INSECTA
- Coleoptera
  - Cerambycidae
    - *Cerambyx*
    - *Monochamus*
    - *Promises*
  - Chrysomelidae
    - *Chrysophthara*
  - Curculionidae
    - *Balaninus*
  - Scarabaeidae
    - *Anoplognathus*
    - *Coroebus*
    - *Costelytra*
  - Scolytidae
    - *Platypus*
- Diptera
  - *Pyrrosia*
- Hemiptera
  - Aphididae
  - Cicadidae
    - *Amorbus*
    - *Kikihia*
  - Coreidae
    - *Gelonus*
    - *Virgilia*
  - Psyllidae
    - *Cardiaspina*
    - *Glycaspis*
    - *Hylinaspis*
- Hymenoptera
  - Pergidae
    - *Perga*
  - Tenthredinidae
- Lepidoptera
  - Anthelidae
  - Geometridae
    - *Epiphryne*
  - Hepialidae
    - *Leto*
    - *Wiseana*
  - Incurvariidae
    - *Perthida*
  - Lasiocampidae
    - *Malacosoma*
  - Limacodidae
    - *Doratifera*
  - Lycaenidae
    - *Helleia*
  - Lymantriidae
    - *Euproctis*
    - *Lymantria*
  - Micropterigidae
  - Nolidae
    - *Uraba*
  - Notodontidae
  - Oecophoridae
    - *Cremnogenes*
    - *Proteodes*
  - Papilionidae
    - *Ctenopseustis*
  - Pyralidae
    - *Eudonia*
    - *Orocrambus*
    - *Scoparia*
  - Saturniidae
    - *Dodonidia*
  - Thaumatopoeidae
    - *Thaumatopoea*
  - Tortricidae
    - *Tortrix*
- Orthoptera
  - Stenopelmatidae
    - *Hemideina*
- Phasmatodea
  - Phasmatidae
    - *Didymuria*
    - *Podacanthus*

CHORDATA

AMPHIBIA
- Salientia
  - Bufonidae
    - *Bufo*
  - Leptodactylidae
    - *Anhydrophryne*
    - *Heleophryne*
    - *Rhabdomanthus*
  - Microhylidae
    - *Breviceps*
    - *Probreviceps*
  - Ranidae
    - *Arthroleptella*
    - *Arthroleptis*
    - *Natalobatrachus*
  - Rhacophoridae
    - *Leptopelis*

REPTILIA
- Rhynchocephalia
  - Sphenodontidae
    - *Sphenodon*
- Squamata
  - Chamaeleonidae
    - *Chamaeleo*
  - Colubridae
    - *Dispholidus*
    - *Philothamnus*
    - *Polemon*
  - Elapidae
    - *Dendroaspis*
    - *Naja*
    - *Natriciteres*
  - Gekkonidae
    - *Lygodactylus*
  - Leptotyphlopidae
    - *Leptotyphlops*
  - Scincidae
    - *Mabuya*
    - *Melanoseps*
  - Typhlopidae
    - *Typhlops*
  - Viperidae
    - *Atheris*
    - *Bitis*

AVES
- Accipitriformes
  - Accipitridae
    - *Accipiter*
    - *Aviceda*
    - *Buteo*
    - *Hieraaetus*
    - *Stephanoaetus*
  - Falconidae
    - *Falco*
- Anseriformes
  - Anatidae
    - *Anas*
- Apterygiformes
  - Apterygidae
    - *Apteryx*
- Columbiformes
  - Columbidae
    - *Aplopelia*
    - *Columba*
    - *Hemiphaga*
- Coraciiformes
  - Alcedinidae
    - *Halcyon*
  - Bucertotidae
    - *Tockus*
- Cuculiformes
  - Cuculidae
    - *Chrysococcyx*
    - *Cuculus*
- Galliformes
  - Phasianidae
    - *Alectoris*
    - *Pternistis*
- Gruiformes
  - Rallidae
    - *Gallirallus*
    - *Notornis*
    - *Sarothrura*
- Musophagiformes
  - Musophagidae
    - *Tauraco*
- Passeriformes
  - Acanthisittidae
  - Callaeatidae
    - *Callaeas*
  - Campephagidae
    - *Coracina*
  - Laniidae
    - *Chloropheneus*
  - Meliphagidae
  - Muscicapidae
    - *Cossypha*
    - *Lioptilus*
    - *Muscicapa*
    - *Rhipidura*
    - *Terpsiphone*
    - *Trochocercus*
    - *Turnagra*
  - Oriolidae
    - *Oriolus*
  - Pycnonotidae
    - *Andropadus*
    - *Phyllastrephus*
  - Rhinocryptidae
    - *Pteroptochos*
    - *Scelorchilus*
  - Sturnidae
    - *Onychognathus*
  - Sylviidae
    - *Prinia*
  - Turdidae
    - *Pogonocichla*
    - *Turdus*
  - Zosteropidae
    - *Zosterops*
- Piciformes
  - Picidae
    - *Campephilus*
    - *Campethera*
    - *Mesopicos*
- Psittaciformes
  - Psittacidae
    - *Cyanoramphus*
    - *Nestor*
    - *Poicephalus*
    - *Strigops*
- Strigiformes
  - Strigidae
    - *Bubo*
    - *Ciccaba*
    - *Ninox*
- Struthioniformes
  - Diornithidae

MAMMALIA
- Artiodactyla
  - Bovidae
    - *Cephalophus*
    - *Rupicapra*
    - *Taurotragus*
    - *Tragelaphus*
  - Cervidae
    - *Alces*
    - *Cervus*
    - *Dama*
    - *Hippocamelus*
    - *Odocoileus*
    - *Pudu*
  - Suidae
    - *Potamochoerus*
    - *Sus*
- Carnivora
  - Canidae
    - *Nyctereutes*
    - *Vulpes*
  - Felidae
    - *Felis*
    - *Panthera*
  - Mustelidae
    - *Lutra*
    - *Meles*
    - *Mellivora*
    - *Mustela*
  - Viverridae
    - *Herpestes*
    - *Nandinia*
- Chiroptera
  - Mystacinidae
    - *Mystacina*
  - Pteropodidae
    - *Eidolon*
    - *Epomophorus*
    - *Epomops*
    - *Rousettus*
- Hyracoidea
  - Procaviidae
    - *Dendrohyrax*
    - *Procavia*
- Insectivora
  - Chrysochloridae
    - *Amblysomus*
    - *Chrysochloris*
  - Macroscelididae
    - *Rhynchocyon*
  - Soricidae
    - *Myosorex*
  - Talpidae

CHORDATA (*continued*)

Urotrichus
Lagomorpha
Leporidae
Lepus
Marsupialia
Caenolestidae
Rhyncholestes
Didelphidae
Dromiciops
Phalangeridae
Gymnobelideus
Phascolarctos
Trichosurus
Primates
Cercopithecidae
Cercopithecus
Macaca
Papio
Galagidae
Galago
Pongidae
Gorilla
Proboscidea
Elephantidae
Loxodonta
Rodentia
Cricetidae
Hystricidae
Hystrix
Muridae
Apodemus
Beamys
Dendromus
Hybomys
Lophuromys
Malacomys
Mus
Otomys
Praomys
Rattus
Thamnomys
Muscardinidae
Graphiurus
Sciuridae
Funisciurus
Heliosciurus

AUTHOR INDEX[1]

[1]Page references to text are in roman type, to bibliographic entries in italics.

SYSTEMATIC INDEX[1]

[1]In this index, no attempt has been made, for larger taxonomic entities, to list all the pages where subordinate taxa are mentioned. These may be found by use of the Systematic List of Genera (pp. 195–199).

Eucalyptus (*continued*)

GENERAL INDEX

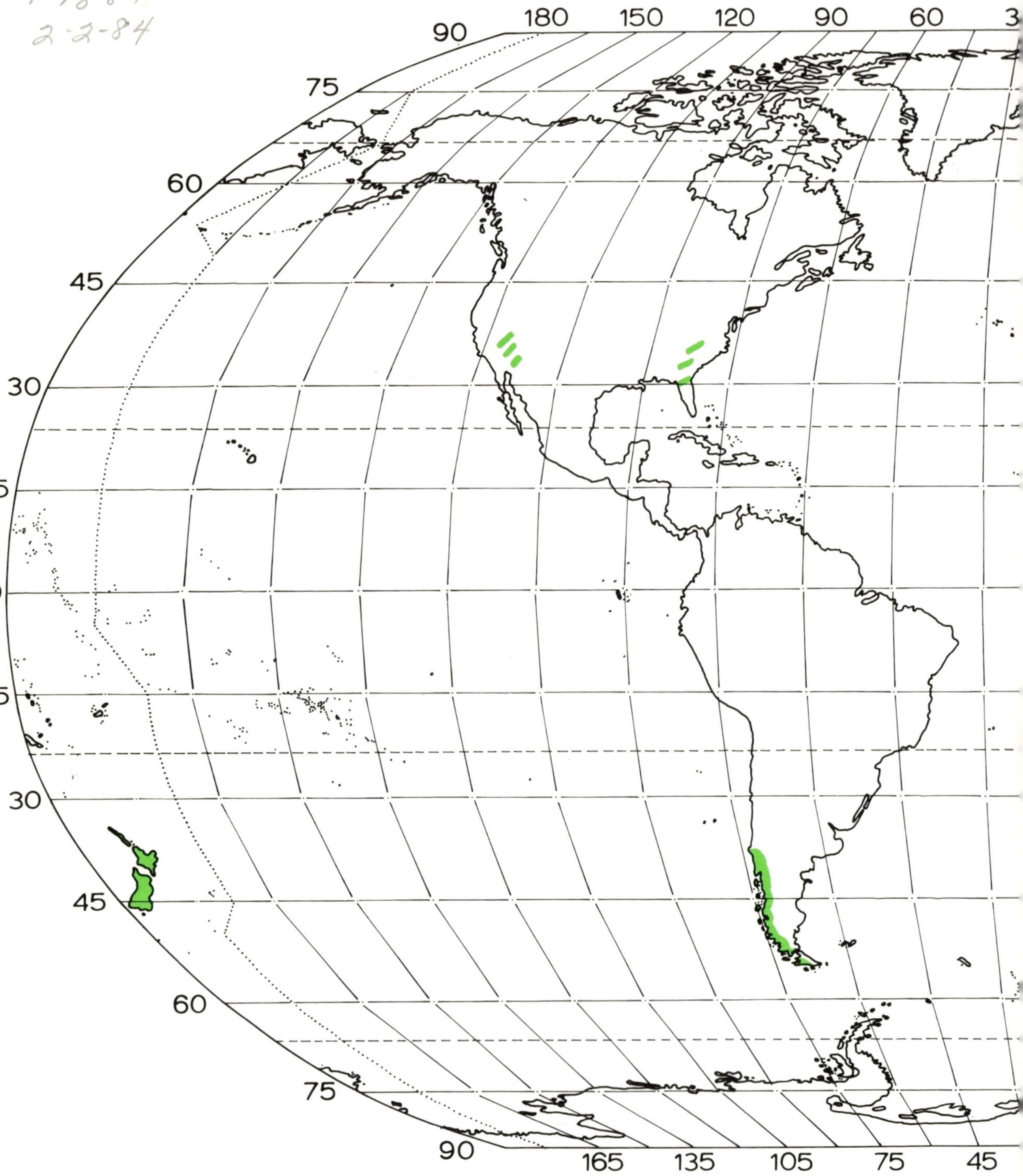

Main areas of temperate br